A General Mathematical Theory

of the

Dynamic Global Political Economy

A General Mathematical Theory of the Dynamic Global Political Economy

David W K Yeung
Saint Petersburg State University, Russia
Hong Kong Shue Yan University, China

Leon A Petrosyan
Saint Petersburg State University, Russia

World Scientific

NEW JERSEY • LONDON • SINGAPORE • BEIJING • SHANGHAI • HONG KONG • TAIPEI • CHENNAI

Published by

World Scientific Publishing Co. Pte. Ltd.
5 Toh Tuck Link, Singapore 596224
USA office: 27 Warren Street, Suite 401-402, Hackensack, NJ 07601
UK office: 57 Shelton Street, Covent Garden, London WC2H 9HE

Library of Congress Control Number: 2024039301

British Library Cataloguing-in-Publication Data
A catalogue record for this book is available from the British Library.

A GENERAL MATHEMATICAL THEORY OF THE DYNAMIC GLOBAL POLITICAL ECONOMY

ISBN 978-981-12-9455-6 (hardcover)
ISBN 978-981-12-9456-3 (ebook for institutions)
ISBN 978-981-12-9457-0 (ebook for individuals)

For any available supplementary material, please visit
https://www.worldscientific.com/worldscibooks/10.1142/13879#t=suppl

Desk Editors: Soundararajan Raghuraman/Nijia Liu

Typeset by Stallion Press
Email: enquiries@stallionpress.com

Preface

William Jevons' "A General Mathematical Theory of Political Economy" (1862) established a ground-breaking work in political economy that earned praise from Irving Fisher as the start of the mathematical method in economics. Milton Friedman praised Jevons' contrition to modern economics as important as Adam Smith, David Ricardo, and Alfred Marshall. Jevons pointed out that the economy being concerned with quantities has always of necessity been mathematical in its subject. A mathematized economics discipline would be a science capable of identifying laws that can explain economic activity. While economics can be regarded as a mathematical discipline, it does not necessarily mean that economic science has to be an exact science. (Science may not be all exact sciences.) Economics is a human science heavily involved with mathematics-oriented analyses.

Unprecedentedly, rapid globalization has appeared in the past three decades. Trade in goods and services as well as capital flows across borders grew to unprecedented levels. This led to the formation of a complex system of the international division of labor, offshoring of productive activities, a combination of advanced technologies with low-cost labor, and global transnational links of the supply chains. The evolution of the global economy, the pattern of production and trade, and the redistribution of economic gains have come to a new era. To deliver globalization to its full potential, all nations have to take serious concerns over international political-economic policies to make their impacts less disruptive and more beneficial for each other.

At the same time, the global economy is under enormous pressure from multiple complex and interconnected crises. These challenges include climate change, military confrontations, cyber-attacks, supply chain disruptions, trade wars, and pandemics. Indeed, almost all political economy issues nowadays can be classified under global political economy. The current global political economy possesses typical features which include (i) a vast number of participating nations, (ii) abundant behavioral choices, (iii) the actions of the nations affect each other's well-being, (iv) the actions of the nations also affect the conditions/states of the economy, (iv) a complex dynamic interactive evolution of the states of the economy, (v) long periods of interactivity, (vi) the effects on the well-being of the nations and the states of the economy can last for some time (or even permanently), (viii) some non-durable strategies that have effects in one single period, and (ix) some durable strategies that have durable effects in multiple periods.

In sum, the global political economy forms a complex dynamic interactive process of change. An empirically meaningful theory must therefore incorporate these features. Appropriate mathematics techniques are required to accommodate the real-life features to formulate a general mathematical theory for the dynamic global political economy nowadays. The durable-strategies dynamic game paradigm developed by us entails these features and forms the foundation of a general mathematical theory for the dynamic global political economy. This would likely open a wide range of potential applications in real-life and theoretical problems.

We would like to thank Dr Cynthia Yingxuan Zhang for her excellent assistance in compiling this book. We thank Nina and Ovanes (LP) and Stella and Patricia (DY) for their love and patience during this book project which on occasion might have diverted our attention from them.

Reference

Jevons, W. S. (1862). *A General Mathematical Theory of Political Economy.*

About the Authors

David W. K. Yeung is the Director of SRS Consortium for Advanced Study in Cooperative Dynamic Games at Shue Yan University and the Co-director of Centre of Game Theory at Saint Petersburg State University. Yeung obtained his PhD in mathematical economics from York University. He studied in the Doctor of Sciences (Doktor Nauk — Continental European higher doctorate above PhD) program in applied mathematics at Saint Petersburg State University and was awarded the University's highest degree of Dr.h.c. for his outstanding contributions in differential games. He has held Faculty positions at Queen's University, University of Hong Kong, and National University of Singapore.

Yeung's main areas of research are game theory, optimization, operations research, and stochastic processes, and he has published extensively in these areas. He serves as Managing Editor of *International Game Theory Review*; Guest Editor of Annals of *Operations Research*, *Journal of the Operations Research Society of China*, and *Journal of Computational and Applied Mathematics*; Associate Editor of *Dynamic Games and Applications*; and Associate Editor of *Operations Research Letters*. He and Leon Petrosyan have coauthored the world's first books on *Cooperative Stochastic Differential Games (Springer), Subgame Consistent Economic Optimization (Birkhauser), Subgame Consistent Cooperation (Springer), and Durable-Strategies Dynamic Games (Springer).*

Leon A. Petrosyan is a Professor and Head of the Department of Mathematical Game Theory and Statistical Decisions, and Dean of the Faculty of Applied Mathematics-Control Processes at St. Petersburg State University. Petrosyan obtained his PhD at Vilnius State University, and Master's and Dr. of Sci. at St. Petersburg State University. In 2008, he was elected as the President of the International Society of Dynamic Games (ISDG) and a foreign member of the Academy of Sciences of the Republic of Armenia. He was also awarded the Issacs Award, the highest prize in dynamic games in 2014.

Petrosyan's main areas of research are theory of games, control theory, operations research, cooperative games, differential games, and mathematical ecology, and he has published extensively in these areas. He serves as the Editor of *International Game Theory Review*, *Game Theory and Applications*, and Chief Editor of *Mathematical Game Theory and Applications* and of the *Viestnik Petrburgskogo Universiteta* (Applied Mathematics, Control, Informatics). He and David Yeung have coauthored the world's first books on *Cooperative Stochastic Differential Games (Springer), Subgame Consistent Economic Optimization (Birkhauser), Subgame Consistent Cooperation (Springer), and Durable-Strategies Dynamic Games (Springer).*

Contents

Chapter 1

Introduction

"Nothing in the world takes place without optimization, and there is no doubt that all aspects of the world that have a rational basis can be explained by optimization methods." — Leonhard Euler

As early as 1744, St. Petersburg University mathematical genius Leonhard Euler stated that "Nothing in the world takes place without optimization, and there is no doubt that all aspects of the world that have a rational basis can be explained by optimization methods". Many seminal contributions of Euler's contribute to the mathematical foundation of optimization theory. Economics is a discipline that has a strong rational basis and every economic activity takes place with optimization. In 1866, a brilliant economist, William Jevons made a ground-breaking work in political economy which Irving Fisher praised as the start of the mathematical method in economics. Milton Friedman praised Jevons' contributions to modern economics as important as Adam Smith, David Ricardo, and Alfred Marshall. In his book: "A General Mathematical Theory of Political Economy (1862)", Jevons described the nature of a theory of economy that will reduce the main problems of economic science to a mathematical form. The economy, being concerned with quantities, has always of necessity been mathematical in its subject. He believed that a mathematized economics discipline would be a science capable of identifying laws that can explain economic activity. The detailed mathematical formulation of Jevons' theory of economics is provided in his 1871 book "Theory of Political Economy".

In this book, we concentrate on one of the most (if not the most) fast-evolving and increasingly important elements in political economy — the dynamic global political economy. Unprecedentedly, rapid globalization has appeared in the past few decades. Trade in goods and services as well as capital flows across borders grew to unprecedented levels. There was also an unprecedented process of economic convergence between the developed and the emerging and developing economies took place. Moreover, production dispersion in cost-effective internationalized supply becomes the only way to remain competitive. This complex form of the international division of labor has become a standard form in the production processes of almost all nations. In addition, environmental degradation, climate change, nuclear threats, and water crisis are issues that are influenced by interactions between different nations. The dynamic and interactive (across nations) nature of these phenomena in political economy is commonly observed. Indeed, almost all studies in the political economy may fall into the study of dynamic global political economy.

While the world has changed significantly, new mathematics developed by the authors of this book at St. Petersburg University is used to formulate a general mathematical theory for the dynamic global political economy nowadays. A distinctive feature of this book is that it combines advanced mathematics, game-theoretic concepts, and economics to develop a general mathematical theory supporting the study of the dynamic global political economy.

This chapter is organized as follows. We first discuss the currently most focused issues in the global political economy in Section 1. The mathematical structure of the global political economy is introduced in Section 2. The organization of this book is given in Section 3.

1. Pressing Issues in Global Political Economy

The global economy is under enormous pressure from multiple complex and interconnected crises. These challenges — which include climate change, the wars in Europe and the Middle East, supply chain disruptions, trade wars, and the COVID-19 pandemic, to name a few — have compounded into what experts have called a "polycrisis" — triggering problems that decades of progress had sought to solve. Some of today's most pressing world problems in the global political economy include the following:

1.1. *Nuclear weapons and threats*

Nuclear arms are the most destructive man-made weapons that can completely wipe out the human race. Despite some progress, we have not reduced the threat of nuclear war enough. After the Cold War, there are still around 10,000 nuclear warheads in the world, with the United States, Russia, and China in possession of more than 90% of the world's nuclear weapons. Nuclear weapons have the potential to kill hundreds of millions of people directly, and billions due to subsequent effects on the environment. The potential for a 'nuclear winter' poses some unknown risk of human extinction, or of a social collapse from which we never recover. There's a non-trivial increase in the number of nations that possess nuclear weapons in the past six decades, and this contributes to a major risk factor for a nuclear confrontation. A nuclear war of whatever scale would probably be the most destructive event in history, shattering our world. It could even threaten us with extinction. We cannot rule out the possibility of such a war or a major nuclear disaster. Technical accidents or diplomatic misunderstandings could spark a conflict that quickly escalates to a state of war. International tension could cause leaders to decide they may be better off fighting than negotiating.

1.2. *Environmental degradation and climate change*

Beyond the suffering it's already causing, worse climate change could increase existential risks and affect the well-being of people living in the future for a long time. Around the world, over half of young people worry that, as a result of climate change, humanity is doomed. Climate change matters so much to so many people around the globe, not just because of the suffering and injustice it is already causing but also because it is a major issue that has obvious potential to affect our world over many future generations. Safeguarding future generations is a key moral priority and should be a crucial consideration in prioritizing global problems on which to work. If climate change could threaten the existence of civilization, then that would mean future generations might have to live in a permanently worse world. Hence, preventing climate change should be one of the most urgent issues to deal with.

1.3. *Space race with militarization of new frontier*

Space programs are widely seen as a sign of national prestige, as they project geopolitical and military power as well as have scientific and commercial significance. Powers such as China, the European Union, France, Germany, India, Japan, NATO, Russia, the United Kingdom, and the United States have publicly announced space forces and continue to build space infrastructure, with plans for at least five new space stations by 2030. In addition, new spacefaring powers will emerge as more nations begin to see opportunities to expand both geopolitical and commercial influence in this arena. Among those that have expressed interest are Argentina, Brazil, and Mexico in Latin America; Egypt, Iran, Turkey, Saudi Arabia, South Africa, and the United Arab Emirates in Africa and the Middle East; and Australia, Indonesia, Malaysia, New Zealand, South Korea, and Vietnam in the Asia-Pacific region.

Satellites and space infrastructures can be used for multiple purposes that include climate and natural resource monitoring, broadband internet, radio and television broadcasting, positioning of navigation, and timing services. Undoubtedly, the critical military importance of space will continue to grow. Armed forces have long relied on space-based or space-supported technologies — including Global Positioning System (GPS) for navigation, dedicated military satellite-based communications for secure digital connectivity, and spy satellites for intelligence — making such systems tempting military targets and spurring the need for enhanced defensive measures. For example, the launch in tandem with SpaceX is the first of five spy satellites South Korea plans to put into orbit by 2025 to form its surveillance system over North Korea. Asia's space race is intensifying with South Korea set to launch its first home-grown military satellite mounted on a SpaceX rocket, while North Korea claims it has carried out successful tests of solid-fuel engines for a new intermediate-range ballistic missile.

1.4. *Development of artificial intelligence*

The development of artificial intelligence is likely to greatly influence the course we take as a society. Undoubtedly, AI has improved scientific advancements and human productivity substantially. The observation that human intelligence may be surpassed by AI within this century is regarded to be highly possible. The questions that remain are the following: "Will they support humanity's attempts to do good or might we lose our control over AI in the future?" The potential development of rival intelligence on Earth should at least be of concern for the possibility of an existential threat to the human race.

1.5. *Cyberattacks*

Cyberattacks targeting computer information systems, computer networks, infrastructures, personal computer devices, or smartphones have become increasingly frequent. The attacker managed to access data, functions, or other restricted areas of the system without authorization, potentially with malicious intent. Cyberattacks can be part of cyber warfare or cyberterrorism. A cyberattack can be employed by sovereign states, individuals, groups, societies, or organizations and it may originate from an anonymous source. Cyberattacks have increased rather rapidly over the last few years. A cyberattack may alter, steal, or destroy a specified target by hacking into a private network or otherwise susceptible system. Cyberattacks can range from installing spyware on a personal computer to attempting to destroy the infrastructure of entire nations. Legal experts are seeking to limit the use of the term to incidents causing physical damage, distinguishing it from the more routine data breaches and broader hacking activities. Cyberattacks have become increasingly sophisticated, hazardous, and expensive to recover from.

1.6. *Global water crisis*

Water is in crisis — as it is both a driver and victim of climate change. Around 1.1 billion people worldwide lack access to water.

Around 31% of schools worldwide lack access to safe water and adequate sanitation. What is more, water cuts across all sectors, from industry and agriculture to housing and energy. For example, 80% of global wastewater is not adequately treated while many regions suffer water scarcity. Concrete water solutions have cross-sectoral implications. The planet's water cycle is a global common good and protecting the global water cycle is for all people and generations. Communities and nations are connected regionally and globally, and water is increasingly intertwined with climate change, the depletion of the planet's natural resources, water is critical to food security, and water cannot be put back on a sustainable trajectory without cooperation in every corner of the globe. Mobilizing resources to build a sustainable water future requires multistakeholder partnerships between government, business, and civil society.

1.7. *Spread of false information on social media*

Social media are interactive technologies that facilitate the creation and sharing of content, ideas, interests, and other forms of expression through virtual communities and networks. Social media refers to new forms of media that involve interactive participation. Social media helps the development of online social networks by connecting a user's profile with those of other individuals or groups. Social media can be viewed as online facilitators or enhancers of human networks — webs of individuals who enhance social connectivity. It helps improve an individual's sense of connectedness with real, virtual, or any online communities and serves as an effective communication (or marketing) tool for corporations, entrepreneurs, non-profit organizations, advocacy groups, political parties, and governments. There has been a rise in social movements using social media as a tool for communicating and organizing in times of political unrest.

The algorithms that social media companies employ to curate content may be contributing to harmful instability and erosion of trust in many societies. It is plausible that political discourse is significantly affected by the way many people now receive their information: through algorithms owned and run by social media companies.

This could have harmful effects which include the making of people's views increasingly far from reality. Social media can also be used to read or share news, whether it is true or false. It serves as an effective means of spreading fake news or undesirable and inaccurate views.

1.8. *Biotechnological developments*

Biotechnological developments pose a potential threat to make much deadlier pandemics possible, due to accidental leaks or malicious use of engineered pathogens. Some of the deadliest events in history have been pandemics. Due to developments in technology, as well as our growing interconnectedness, we are facing the possibility of future biological disasters that are even worse. Global catastrophic biological risks (including risks of severe pandemics) can be serious enough to threaten and change the future of humanity.

1.9. *Trade war*

A trade war is an economic conflict often resulting from extreme protectionism in which nations raise or create tariffs, trade barriers, or sanctions against each other. A trade war between the biggest economies would often cause disruptions in the global economy. Globally, foreign direct investment would slow down. A trade war would hurt the economies that are not directly involved, lower the growth of the directly involved nations, and reduce trade volume directly or indirectly. It would cut off the global supply chain, create uncertainty in the financial market, reduce technology transfer, and lower world production of goods and services. Trade wars are not free and someone or even everyone is going to pay. Tariffs are an application of the coercive power of the state. They are not only imposed on other countries, they are imposed on a nation's own citizenry as taxes on goods coming and going across borders. Nations that impose them are pushing prices higher — and production lower — on all goods, not just the imported ones but the ones produced domestically that rely on outsourcing some aspects of the production process. Trade wars will never be won; they are fought at the expense of the consumers, producers, and investors in all nations.

2. Mathematical Structure of the Global Political Economy

A review of the global political economy issues in Section 1 shows that they have certain common features. These include the following:

(i) a vast number of participating nations,
(ii) abundant behavioral choices,
(iii) the actions of the nations which affect each other's well-being,
(iv) the actions of the nations which also affect the conditions/states of the economy,
(v) a complex dynamic interactive evolution of the states of the economy,
(vi) long periods of interactivity,
(vii) the effects on the well-being of the nations and the states of the economy which can last for some time (or even permanently),
(viii) some non-durable strategies that have effects in one single period,
(ix) some durable strategies that have durable effects in multiple periods.

In sum, the dynamic political economy forms a complex dynamic interactive process of change. An empirically meaningful theory must therefore incorporate these features.

For today's global political economy, game theory is perhaps the most sophisticated and fertile paradigm mathematics can offer to study and analyze decision-making under real-world conditions. One particularly complex and fruitful branch of game theory is dynamic games, which investigates interactive decision-making over time. Since human beings live in a span of time and decisions generally lead to effects over time, it is only a slight exaggeration to claim that "life is a dynamic game". To incorporate all the features stated above, the durable-strategies dynamic games developed by Petrosyan and Yeung, which encompass these features, would be the framework for establishing a general mathematical theory of the dynamic global economy. Durable strategies that have effects lasting over an extended period of time are prevalent in real-life situations in the global political economy. Revenue-generating investments, toxic waste disposal, durable goods, emission of pollutants, regulatory measures, coalition agreements, diffusion of knowledge, contracts, trade barriers, tariffs, and investments to build up physical capital are vivid examples of durable controls. In reality, the existence of durable strategies in the decision-maker's controls is likely to be a rule rather than an exception.

Petrosyan and Yeung (2020) and Yeung and Petrosyan (2019, 2021a, 2021b, 2022) introduced control lags in dynamic games and developed a new paradigm of durable-strategies dynamic games. This game paradigm allows strategies with lagged effects to affect both the player's payoffs and the state dynamics. In the presence of durable strategies, significant modification of the dynamic optimization techniques is required to accommodate the lagged effects of the strategies. A dynamic optimization theorem for solving problems with durable controls affecting both the payoff and state dynamics is provided for the game theoretic analysis. A novel theorem characterizing the feedback Nash equilibrium for durable-strategies dynamic games is derived. This opens up a wide range of potential applications in real-life interactive problems involving durable strategies, especially in the complicated global political economy.

It is well known that non-cooperative behaviors among players would, in general, lead to an outcome that is not Pareto optimal. In a dynamic world, non-cooperative behaviors guided by short-sighted individual rationality could be a source of a series of undesirable or even disastrous consequences, especially when the effects of their strategies are durable. It is a rather common observation that many nations are policy-wise inward-looking and focusing almost exclusively on their own national interests. Consequently, military conflicts, climate disasters, financial crises, resource depletion, ecosystem degradation, and trade wars are increasing around the world. The negative impacts of these events often can be immediate as well as long term. Cooperation suggests the possibility of obtaining socially optimal and group-efficient solutions to decision problems involving durable strategic actions. To guarantee dynamic stability (which does not happen automatically), a cooperative solution has to be subgame consistent so that the solution for a subgame that starts at a later time with the state brought about by prior optimal behaviors would remain optimal (see Yeung and Petrosyan, 2004, 2010, 2016). In the presence of durable strategies, a subgame consistent cooperative solution has to guarantee that the solution policy in a subgame starting at a later time with the state and previously executed strategies brought about by prior optimal behaviors would remain optimal. A Payoff Distribution Procedure (PDP) leading to subgame consistent solutions under durable strategies is presented. Cooperative dynamic games with durable strategies are presented and their subgame-consistent solutions are characterized. This is extremely important for the establishment of sustainable global policy.

The development of effective and efficient and sustainable policy designs to handle world issues is very much in order. Sustainable policies do not only reduce the risk of the future but also provide incentives for nations to agree to participate in the scheme. Hence, sustainable policy plays a crucial role in the success of international policy accords. Subgame-consistent global political economy policy constitutes an important foundation for formulating sustainable policy in the international arena.

The mathematics in this book may look complicated and advanced to some economists. Critics of 'economic sciences' sometimes refer to the development of a 'pseudoscience' of economics, arguing that it uses the trappings of science, like dense mathematics, but only for show. This book is about mathematical theory, and its mathematics is certainly not for show. Economics, as a science that is heavily involved with mathematics, must have correct mathematical reasoning to support its analyses. Nobel Economics laureate Paul Romer (2015) coined the term "mathiness" as follows: "Like mathematical theory, mathiness uses a mixture of words and symbols, but instead of making tight links, it leaves ample room for slippage between statements in natural versus formal language and between statements with theoretical as opposed to empirical content". Romer gave the following warning: "But after readers have been disappointed too often by mathiness that wastes their time, they will stop taking seriously any paper that contains mathematical symbols. In response, authors will stop doing the hard work that it takes to supply real mathematical theory. If no one is putting in the work to distinguish between mathiness and mathematical theory, why not cut a few corners and take advantage of the slippage that mathiness allows? The market for mathematical theory will collapse. Only mathiness will be left. It will be worth little, but cheap to produce and might survive as entertainment".

The insightful view of Romer provides the readers of this book with the needed impetus to work on and take note of the mathematical theory presented which is needed in analyzing the current dynamic global political economy. The Nobel Economics Prize-winning works of Leonid Kantorovich, John Nash, Robert Aumann, and Lloyd Shapley are genuine examples of mathematical theory in economics. Most importantly, the foresighted warning of Romer is crucial to the future development of economics and critical to the survival of mathematical theory in economics.

3. Organization of this Book

This book is the world's first text focusing on a general mathematical theory for the representation and analysis of the dynamic global political economy. It presents a comprehensive description of this mathematical paradigm. Mathematical theorems and solution concepts of major problems in the dynamic global political economy are depicted. The organization of this book is given as follows.

Chapter 2 presents a discourse on Global Political Economy (GPE). Political economy is the study of how a nation — the public's household — is managed or governed, considering both political and economic factors. It is an important and long-established subject of intellectual inquiry and was formally established as an academic discipline in the 16th century. As globalization, international trade, and cross-national political interactions increased rapidly in the last five decades, the rise of the global political economy has become inevitable. Section 1 discusses the political and economic systems and key elements in political economies. A history of the developments in the political economy is given in Section 2. The relationship between political economy and economics is examined in Section 3. The rise of the global political economy and current globalization and GPE analysis are provided in Section 4. The theoretical underpinning of an approach toward subgame-consistent GPE is scrutinized in Section 5. A revisitation of William Jevons' "General Mathematical Theory of Political Economy" is provided in Section 6.

Chapter 3 studies the uncoordinated interaction of nations in a dynamic world. We first provide a description of the dynamic global political economy. Section 1 provides a description of the dynamic GPE which includes the state of the economy and its evolution, target controls/strategies, the evolution of the state variables, the payoffs of the nations, and government policy instruments. Section 2 provides the mathematical representation of a dynamic GPE and presents a general mathematical theory for analysis. Section 3 examines policy analysis and the cost of individually rational self-seeking behaviors without coordination. Section 4 explicates the theoretical analysis and mathematical theory of uncoordinated outcomes in a dynamic GPE through the study of environmental degradation. A computational illustration is given in Section 5 and chapter notes are given in Section 6.

Chapter 4 considers coordination and cooperation in a dynamic GPE. In a global environment, policymakers from many nations develop individual policy designs, statements, and actions that are either not coordinated or not effectively coordinated. Coordination and cooperation are very much needed for resolving global problems. Section 1 presents the theoretical underpinnings for coordination and cooperation in the dynamic global political economy. Section 2 develops the mathematical theories for cooperative optimization for solving Pareto optimal target controls under coordination. Subgame-consistent sharing with individual rationality and group optimality are presented. In Section 3, a Payoff Distribution Procedure (PDP) leading to the realization of subgame-consistent cooperation is provided. Price of Anarchy (PoA) and cost of delay are derived in Section 4. Section 5 gives an explication of cooperation and coordination in global climate agreements. In Section 6, we consider subgame consistent cooperation among nations in tackling the global environmental degradation problem and present different gain-distribution optimality principles. Section 7 proposes an organization of a global environmental pact guided by Pareto optimal cooperative target controls under a grand coalition of all nations to pursue a comprehensive subgame-consistent cooperative scheme of global environmental management. A computational illustration is provided in Section 8 and chapter notes are given in Section 9.

Chapter 5 considers establishing the mathematical theory of coalitions and blocs with an explication of trade networks. In many real-life politics, groups or nations with common interests (or value systems) form coalition blocs by agreement for mutual support and joint actions. Coalition blocs are formed by nations with common interests to enhance their gains through cooperation. Section 1 considers international coalition blocs that share similar aims and interests and act together over some issues. Section 2 provides the mathematical theory of GPE with coalitions. Section 3 examines the notion of grand coalition cooperation. A grand coalition cooperation of the coalitional blocs would improve the payoffs of all the coalitions substantially. In Section 4, we explicate the theoretical analysis and mathematical theory of coalition in trade networks involving China's Belt-Road Initiative (BRI) and G7 countries' Build Back Better World (B3W) Initiative. A grand coalition cooperation that would improve the payoffs of BRI and B3W is provided in Section 5. A computational illustration is given in Section 6 and chapter notes are given in Section 7.

Chapter 6 provides the underlying theory and techniques for dealing with problems in international economics in which there exist random termination and durable strategies. The terminal time of many international issues is not known with certainty. Commonly observed examples of horizontal uncertainties include the terms of offices of elected authorities, contract renewal, continuation of agreements subjected to periodic negotiations, trade wars, international sanctions, and the end of military conflicts. In Section 1, we discuss the issue of termination of international institutions. Game formulation and mathematical solution theory are provided in Section 2. Analysis of dynamic cooperation under random termination is given in Section 3. Group optimality conditions, subgame consistent solutions, and payment mechanisms to guarantee dynamical stability in a dynamic cooperation scheme under random termination are derived. An explication of international joint venture is presented in Section 4. A computational illustration is given in Sections 5 and 6. Chapter notes are provided in Section 7.

Chapter 7 contains mathematical solution mechanisms and the proofs that are used to establish a general mathematical theory of dynamic global political economy. In particular, we present the mathematical solution mechanisms for (i) durable-strategies dynamic optimization, (ii) durable-strategies dynamic games, (iii) dynamically consistent solutions and payment mechanism, (iv) random horizon durable-strategies dynamic optimization, (v) random horizon durable-strategies dynamic games, and (vi) subgame consistent payment mechanism under random termination.

References

Jevons, W. S. (1862). *A General Mathematical Theory of Political Economy.*

Petrosyan, L. A. & Yeung, D. W. K. (2020). Cooperative dynamic games with durable controls: Theory and application, *Dynamic Games and Applications*, 10: 872–896. DOI: 10.1007/s13235-019-00336-w.

Romer, P. M. (2015). Mathiness in the theory of economic growth. *The American Economic Review*, 105(5), 89–93. DOI: 10.1257/aer.p20151066.

Yeung, D. W. K. & Petrosyan, L. A. (2004). Subgame consistent cooperative solutions in stochastic differential games. *Journal of Optimization Theory and Applications*, 120(3), 651–666.

Yeung, D. W. K. & Petrosyan, L. A. (2010). Subgame consistent solutions for cooperative stochastic dynamic games. *Journal of Optimization Theory and Applications*, 145, 579–596.

Yeung, D. W. K. & Petrosyan, L. A. (2016). *Subgame Consistent Cooperation: A Comprehensive Treatise.* Springer, Singapore, 520 p.

Yeung, D. W. K. & Petrosyan, L. A. (2019). Cooperative dynamic games with control lags. *Dynamic Games and Applications*, 9(2), 550–567. https://doi.org/10.1007/s13235-018-0266-6.

Yeung, D. W. K. & Petrosyan, L. A. (2021a). Asynchronous horizons durable-strategies dynamic games and tragedy of cross-generational environmental commons. *International Game Theory Review*, 23(4). DOI: 10.1142/s0219198921500201.

Yeung, D. W. K. & Petrosyan, L. A. (2021b). Generalized dynamic games with durable strategies under uncertain planning horizon. *Journal of Computational and Applied Mathematics*, 395, 113595. DOI: 10.1016/j.cam.2021.113595.

Yeung, D. W. K. & Petrosyan, L. A. (2022). *Durable-strategies Dynamic Games: Theory, Solution Techniques and Applications.* Springer, Switzerland, 303 p.

Chapter 2

A Disquisition on Global Political Economy

"Political economy may be defined as the science which determines what laws men ought to adopt in order that they may, with the least possible exertion, procure the greatest abundance of things useful for the satisfaction of their wants, may distribute them justly and consume them rationally." — Emile de Laveleye

Political economy is a major branch of social science that studies the relationships and interactions between individuals, society, markets, institutions, and governments (local and across nations). A diverse set of instruments, techniques, and methodologies drawn from economics, political science, sociology, statistics, and mathematics is employed in its analyses. Political economy is often understood to be the study of how a nation — the public's household — is managed or governed, considering both political and economic factors. Political economy is an important and long-established subject of intellectual inquiry and was formally established as an academic discipline in the 16th century. This chapter is organized as follows. We first consider the major political economic system and the key elements of a political economy in Section 1. An overview of the development of political economy is presented in Section 2. The rise of the global political economy is discussed in Section 3. Section 4 considers the theoretical underpinning for moving toward a time-consistent global political economy. Section 5 is a revisit of the ground-breaking text in political economy — "A General Mathematical Theory of Political Economy (1866)" by William Stanley.

1. Political Economic Systems

There are three most known structures of political economy throughout recent history: capitalism, communism, and socialism.

(i) Under **Capitalism**, the government tries to minimize its involvement in the operation of the market. The government may use incentives and subsidies to encourage investment or innovation in specific fields. Yet they largely leave the placement of resources, wealth, capital, profits, and goods to the participants in the market. Often, capitalism may contribute to the widening of the gaps of income inequality due to the unequal distribution of wealth.

(ii) Under **Communism**, the government directly controls and oversees the distribution of resources, income, wealth, and services. Economic and political freedoms are very low and this economic system often paves the way for more authoritarian political and economic arrangements.

(iii) Under **Socialism**, the government aims to achieve the ends of communism but uses some of the tools and methodologies of capitalism. The government takes the initiative to redistribute income, wealth, and resources in a more equitable manner than it would in a purely capitalist one. It allows the free market to operate with government interventions to ensure more equitable outcomes.

Five key elements in most political economies are as follows:

(i) **Distribution of resources:** The distribution of income, wealth, and resources within a political economy is one of the top priorities facing the decision-makers. Some nations prefer to have a fairly equitable income or standard of living for all citizens, whereas others prefer to widen the gap between the wealthy and poor for the sake of economic growth.

(ii) **Government intervention:** Certain governments prefer a considerably large degree of control over their economies, whereas others take a more laissez-faire approach and minimal intervention with the market. Political economists aim to study the right menu of controls to avoid financial crises and maintain financial stability under specific political economy structures.

(iii) **International dynamics:** In the era of globalization, interactions among nations are substantial and the entire world has become an international political economy. Macroeconomic factors between multiple nations can create wide ripple effects. The action of one nation on its economy can be a determinant of another nation's economy.

(iv) **Private sector:** On a domestic microeconomic level, the private sector is an essential element of any political economy. Policy concerning the private sector is a vital part of the government's decision-making process.

(v) **Economic development:** Economic development and growth are prime factors of concern for both developing and developed nations. Developing nations may use free trade to relieve poverty and developed nations consider technological advancement and market expansion to raise their GNP.

2. An Overview of the Development of Political Economy

The analysis of political economy (in terms of the nature of state and market relations), both in practical terms and as moral philosophy, can be traced back to Greek philosophers such as **Plato** and **Aristotle** as well as to the Scholastics and those who propounded a philosophy based on natural law.

2.1. *Before the 16th century*

Plato (c. 380–360 BC) described an ideal city-state run by philosopher kings with the specialization of labor for production. He was the first known advocate of a credit theory of money, that is, money as a unit of account for debt. He also argued that collective ownership was necessary to promote the common pursuit of the common interest.

Aristotle (c. 350 BC) analyzed different forms of the state government: monarchy, aristocracy, constitutional government, tyranny, oligarchy, and democracy (instead of philosopher kings). He himself highly disapproved of usury and cast scorn on making money through a monopoly. Though Aristotle agreed with holding some things in common, he argued that "Property should be ... as a general rule, private; for when everyone has a distinct interest, men will not complain of one another and they will make progress, because everyone will be attending to his own business". While it is better that property should be private, Aristotle stated that "but the use of it common; and the special business of the legislator is to create in men this benevolent disposition".

Aristotle believed the appropriate behavior is the "golden mean", that is, the avoidance of "extreme" or unrealistic goals or conduct in the affairs of men. While Aristotle hoped that wise and appropriate policies may improve the conditions and actions of men, he recognized that man possesses a human nature that cannot be molded or bent or transformed to conform to some ideal of a perfect state populated by transformed people in the way that Plato believed was in principle desirable and possible.

Thomas Aquinas (1225–1274), in the treatise "Summa Theologica" (published in 1485), dealt with the concept of a just price, which he considered necessary for the reproduction of the social order. Analogous to the modern concept of long-run equilibrium, a just price was just sufficient to cover the costs of production, including the maintenance of a worker and his family. Aquinas argued it was immoral for sellers to raise their prices simply because buyers had a pressing need for a product. Aquinas was against any form of cheating and recommended paying compensation in lieu of good service. He also held the view that while human laws might not impose sanctions for unfair dealing, divine law did.

Duns Scotus (1265–1308), in his work "Sententiae" (1295), advocated that it was possible to be more precise than Aquinas in calculating a just price, emphasizing the costs of labor and expenses, although the latter might be inflated by exaggeration, because buyer and seller usually have different ideas of a just price. If people did not benefit from a transaction, in Scotus' view, they would not trade. Scotus argued that merchants perform a necessary and useful social role by transporting goods and making them available to the public.

Jean Buridan (c. 1300–after 1358) considered money from two angles: its metal value and its purchasing power, which he acknowledged can vary. He argued that aggregated, not individual, demand and supply determine market prices. Hence, a just price was what the society collectively and not just one individual is willing to pay.

Nicolas d'Oresme (1320–1382) wrote in "De origine, natura, jure et mutationibus monetarum" (1355–1356) about the origin, nature, law, and alterations of money. It is one of the earliest manuscripts on the concept of money. His treatise argued that money or currency belongs to the public and that the government or sovereign of the economy has no right to control the value of the currency just so that they can profit from it.

Saint Antoninus of Florence (1389–1459), in his writings, addressed social and economic development and argued that the state has a duty to intervene in mercantile affairs for the common good, and an obligation to help the poor and needy. In his primary work, "Summa Theological" (1440–1459), he was mainly concerned about price, justice, and capital theory. Like Duns Scotus, he distinguished between the natural value of a good and its practical value. The latter is determined by its suitability to satisfy needs, its rarity, and its subjective value. Due to this subjective component, there cannot only be one just price but a bandwidth of more or less just prices.

2.2. *From 16th to early 20th century*

Thomas More (1478–1535) published "Utopia" (1516) which described an ideal society where land is owned in common and there is universal education and religious tolerance, inspiring the English Poor Laws (1587) and the communism-socialism movement.

Nicolaus Copernicus (1473–1543) published the first known argument for the quantity theory of money in 1517. In 1519, he published the first known form of Gresham's Law: "Bad money drives out good".

Jean Bodin (1530–1596) published "Reply to the Paradoxes of Malestroit" (1586) which contained the first known analysis of inflation, which he claimed was caused by the importation of gold and silver from South America, backing the quantity theory of money.

William Petty (1623–1687) began publishing short works in 1662, applying the rational scientific tradition to economics, requiring that it only use measurable phenomena and seek quantitative precision, coining the term 'political arithmetic,' introducing statistical mathematics, and becoming the first scientific economist.

James Steuart (1713–1780) published "An Inquiry into the Principles of Political Economy" (1767), the first book in English with the term "political economy" in the title and the first complete economics treatise.

The British Enlightenment in the 17th century was accompanied by scientific advancement and spurred the advancement of economic thought.

Richard Cantillon (1680–1734) imitated Isaac Newton's forces of inertia and gravity in the natural world with human reason and market competition in the economic world. In his "Essay on the Nature of Commerce in General" (1730), he argued rational self-interest in a system of freely adjusting markets would lead to order and mutually compatible prices.

John Locke (1632–1704) combined philosophy, politics, and economics into one coherent framework. Locke believed that people contracted into society, which was bound to protect their property rights, which include people's lives and liberties, as well as their wealth. In his "Second Treatise on Civil Government" (1689), he wrote the following: "God hath given the world to men in common... Yet every man has a property in his own person. The labor of his body and the work of his hands we may say are properly his. Whatsoever, then, he removes out of the state that nature hath provided and left it in, he hath mixed his labor with, and joined to it something that is his own, and thereby makes it his property".

He argued that the government should cease interference with people's property (including their "lives, liberties and estates") and should positively work to ensure their protection. His views on price and money were laid out in a letter to a Member of Parliament in 1691 entitled "Some Considerations on the Consequences of the Lowering of Interest and the Raising of the Value of Money" (1691), in which he argued that the "price of any commodity rises or falls, by the proportion of the number of buyers and sellers", a rule which holds universally in all things that are to be bought and sold.

Dudley North (1641–1691) argued that the results of mercantile policy were undesirable. In his "Discourses upon Trade" (1691), he argued against assuming a need for a favorable balance of trade. Trade, he argued, benefits both sides and promotes specialization, division of labor, and wealth for everyone, and regulation of trade interferes with these benefits.

David Hume (1711–1776) agreed with North's philosophy and denounced mercantilist assumptions. He argued that it was not only undesirable to strive for a favorable balance of trade but that it is, in any case, impossible. Hume held that any surplus of exports would be paid for by imports of gold and silver. This would increase the money supply, causing prices to rise. That in turn would lead to a decline in exports until the balance with imports is restored. Hume's "Political Discourses" (1752), which consolidated in his "Essays, Moral, Political, Literary" (1777) were important contributions in political economy at the time.

François Quesnay (1694–1774) believed that trade and industry were not sources of wealth, and instead in his book "Tableau économique" (1758) he argued that agricultural surpluses, by flowing through the economy in the form of rent, wages, and purchases, were the real economic movers. In particular, regulation impedes the flow of income throughout all social classes and therefore hinders economic development. In addition, taxes on the productive classes, such as farmers, should be reduced in favor of rises for unproductive classes, such as landowners, since their luxurious way of life distorts the income flow.

Ferdinando Galiani (1728–1787) published a nearly exhaustive treatise on money entitled "Della Moneta" (On Money 1751). The treatise covered all modern aspects of monetary theory, including the value and origin of money, its regulation, and inflation. The text remained cited by various economists for centuries, as wide-ranging a list as Karl Marx and the scholars in the Austrian School (including Joseph Schumpeter). Della Moneta was published 25 years before Adam Smith's "The Wealth of Nations" and therefore can be regarded possibly as the first modern economic analysis.

Adam Smith (1723–1790) has been regarded as the father of modern political economy by many scholars. In his first book "The Theory of Moral Sentiments" (1759), he argued that people's ethical systems develop through personal relations with other individuals and that right and wrong are sensed through others' reactions to one's behavior. His second book "An Inquiry into the Nature and Causes of the Wealth of Nations" (1776) happened to coincide with the dawn of a new industrial revolution that allowed more wealth to be created on a larger scale than ever before. Adam Smith's famous statement on self-interest said: "It is not from the benevolence of the butcher, the brewer or the baker, that we expect our dinner, but from their regard to their own self-interest. We address ourselves, not to their humanity but to their self-love, and never talk to them of our own necessities but of their advantages".

Smith argued for a "system of natural liberty" where individual effort was the producer of social good. Smith believed even selfishness within society was kept under restraint and worked for the good of all when acting in a competitive market. Prices are often unrepresentative of the true value of goods and services. Following John Locke, Smith thought the true value of things derived from the amount of labor invested in them. He stated the following in his work:

"Every man is rich or poor according to the degree in which he can afford to enjoy the necessaries, conveniences, and amusements of human life. But after the division of labor has once thoroughly taken place, it is but a very small part of these with which a man's own labor can supply him. The far greater part of them he must derive from the labor of other people, and he must be rich or poor according to the quantity of that labor which he can command, or which he can afford to purchase. The value of any commodity, therefore, to the person who possesses it, and who means not to use or consume it himself, but to exchange it for other commodities, is equal to the quantity of labor which it enables him to purchase or command. Labor, therefore, is the real measure of the exchangeable value of all commodities.

The real price of everything, what everything really costs to the man who wants to acquire it, is the toil and trouble of acquiring it. What everything is really worth to the man who has acquired it, and who wants to dispose of it or exchange it for something else, is the toil and trouble that it can save to himself, and which it can impose upon other people. What is bought with money or with goods is purchased by labor as much as what we acquire by the toil of our own bodies. That money or those goods indeed save us this toil. They contain the value of a certain quantity of labor which we exchange for what is supposed at the time to contain the value of an equal quantity. Labor was the first price, the original purchase money that was paid for all things. It was not by gold or by silver, but by labor, that all the wealth of the world was originally purchased; and its value, to those who possess it, and who want to exchange it for some new productions, is precisely equal to the quantity of labor which it can enable them to purchase or command".

Smith's vision of a free market economy, based on secure property, capital accumulation, widening markets, and division of labor, contrasted with mercantilism which attempts to "regulate all evil human actions". Tory Prime Minister, 1783–1801, William Pitt, based his tax proposals on Smith's ideas and advocated free trade as in "The Wealth of Nations". Smith was appointed a commissioner of customs, and within 20 years, Smith had a new generation of follower-writers who were intent on building the science of political economy.

Jean-Baptiste Say (1767–1832) whose book "A Treatise on Political Economy" (1803) contained a brief passage, which later became orthodoxy in political economics until the Great Depression, now known as Say's Law of markets (supply creates its own demand). Say argued that there could never be a general deficiency of demand or a general glut of commodities in the whole economy. People produce things, to fulfill their own wants rather than those of others, therefore production is not a question of supply but an indication of producers demanding goods. Say was in favor of competition, free trade, and lifting restraints on business. He was also one of the first economists to study entrepreneurship and conceptualized entrepreneurs as organizers and leaders of the economy. Say also believed that money was neutral and its sole role was to facilitate exchanges, therefore, the demand for money was only for buying commodities. Hence, "money is a veil".

David Ricardo (1772–1823) was among the most influential classical economists along with Adam Smith and James Mill. In his book "On the Principles of Political Economy and Taxation" (1817), he criticized barriers to international trade and provided a description of how income is distributed in the population. Ricardo made a distinction between workers (who received a wage fixed to the survival level) and landowners (who earn rent) and capitalists (who own capital and receive a profit). As the population grows, it is necessary to cultivate additional land, whose fertility is lower than that of existing cultivated fields. Therefore, the cost of the production of the wheat would increase, as well as the price of the wheat. Then, rents and wages (indexed to inflation) would increase as well. Profits would decrease to the level at which the capitalists would no longer invest. The economy, Ricardo concluded, is bound to approach a steady/stationary state.

John Stuart Mill (1806–1873) found a middle ground in economic growth between Adam Smith's view of ever-expanding opportunities for trade and technological innovation and Thomas Malthus' view of the inherent limits of population. He set out several possible future outcomes rather than predicting one in particular. Mill was a proponent of utilitarianism, an ethical theory developed by Jeremy Bentham, and contributed to the investigation of scientific methodology. Mill was the dominant figure in political and economic thought of his time. His book "Principles of Political Economy" (1848), which contains a comprehensive summary of economic thoughts of the mid-nineteenth century, was the most authoritative economics text of his time. The book has been adopted as the standard economic text or reference source by most universities, even up to now.

Emile de Lavaleye (1822–1892) was particularly connected with bimetallism (a monetary standard in which the value of the monetary unit is defined as equivalent to certain quantities of two metals, typically gold and silver, creating a fixed rate of exchange between them) and primitive property. He had a special interest in the revival and preservation of small nationalities. His activities included the whole realm of political science, political economy, monetary questions, international law, foreign and Belgian politics, questions of education, religion and morality, travel, and literature. He was especially attracted to England, where he thought many of his ideals of social, political, and religious progress were realized. He was a frequent contributor to the English newspapers and leading reviews. He was one of the co-founders of the Institut de Droit International (The Institute of International Law (IIL)) in 1873 to promote arbitration among nations as a peaceful means of settling disputes.

Henry George (1839–1897) provided the intellectual inspiration for the economic philosophy now known as Georgism. It is an economic ideology holding that the economic rent derived from land — including from all natural resources, the commons, and urban locations — should belong equally to all members of society. George's book "Progress and Poverty" (1879) was one of the most widely printed books in English, which sparked a worldwide reform movement and is sometimes marked as the beginning of the Progressive Era. Though Georgism declined in the second half of the 20th century, there are still Georgist organizations and land reform movements around the world. George's ideas have been incorporated into the philosophies of socialism, libertarianism, and ecological economics. Paul Samuelson listed Henry George as one of only six "American saints" in classical economics.

Karl Marx (1818–1883) is regarded as one of the most influential figures in human history, and his work has been both lauded and criticized. His critical theories about society, economics, and politics, collectively understood as Marxism, hold that human societies develop through class conflict. In the capitalist mode of production, this manifests itself in the conflict between the ruling classes (known as the bourgeoisie) that control the means of production and the working classes (known as the proletariat) that enable these means by selling their labor power in return for wages. Marx argued that capitalism would produce internal tensions like previous socioeconomic systems and that these tensions would lead to its self-destruction and replacement by a new system known as the socialist mode of production. He believed that class antagonism under capitalism would eventually lead to the establishment of a classless, communist society constituted by a free association of producers.

Marx's "Das Kapital" (1867), also known as "Capital: A Critique of Political Economy", is a foundational theoretical text in which Marx aimed to reveal the economic patterns underpinning the capitalist mode of production in contrast to classical political economists, such as Adam Smith, Jean-Baptiste Say, David Ricardo, and John Stuart Mill.

Arthur Pigou (1877–1959) published "Wealth and Welfare", which insisted on the possibility of market failures, claiming that markets are inefficient in the case of economic externalities, and the state must interfere to prevent them. However, Pigou retained free market beliefs, in the face of the economic crisis, he explained in "The Theory of Unemployment" (1933) that the excessive intervention of the state in the labor market was the real cause of massive unemployment because the governments had established a minimal wage, which prevented wages from adjusting automatically. This was to be the focus of the attack from Keynes. Pigou published the paper "The Classical Stationary State" (1943) which popularized the Pigou (real balance) effect, the stimulation of output and employment during deflation by increasing consumption due to a rise in wealth.

William Jevons (1835–1882) was a prolific writer and a leader in the UK both as a logician and as an economist. Jevons' work on "A General Mathematical Theory of Political Economy" (1866) and "The Theory of Political Economy" (1871) was praised as the start of the mathematical method in economics and marked the opening of a new period in the history of economic thought. The work is regarded as the watershed between classical and neoclassical economics, with detailed arguments for the mathematization of economic theory. The economy, indeed, being concerned with quantities, has always of necessity been mathematical in its subject. Alfred Marshall said Jevons' work in economics "will probably be found to have more constructive force than any, say that of Ricardo, that has been done during the last hundred years".

Jevons' work, along with similar discoveries made by **Carl Menger** in Vienna (1871) and by **Léon Walras** in Switzerland (1874), marked the opening of a new period in the history of economic thought. Jevons' contribution to the marginal revolution in economics in the late 19th century established his reputation as a leading political economist and logician of the time.

Alfred Marshall (1842–1924) put economics on a more mathematical footing and improved the mathematical rigor of economics and transformed it into a more scientific profession. His book "Principles of Economics" (1890) was the dominant economic textbook in England for many years. It brought the ideas of supply and demand, marginal utility, and costs of production into a coherent whole. In the 1870s, he wrote several articles on international trade and the problems of protectionism, and many of these works were compiled into "The Theory of Foreign Trade: The Pure Theory of Domestic Values" (1879a). His book "Economics of Industry" (1879b) become widely adopted in England as an economic curriculum. He is also known as one of the founders of neoclassical economics. Upon the death of William Jevons in 1882, Marshall became the leading economist of the scientific school of his time.

Léon Walras (1834–1910) is best known for revolutionizing economics with the mathematical formulation of the mechanisms of the price system. He presented equations that tied together theories of production, exchange, money, and capital. Walras provided a general equilibrium theory in economics known as "Walrasian equilibrium" which is an integral of modern economic theory.

Vilfredo Pareto (1848–1923) made important contributions to economics, particularly in the study of income distribution and in the analysis of individuals' choices. He introduced the notion of "Pareto optimality" stating that the system is having the maximum economic satisfaction when no one can be made better off without making someone else worse off. Pareto optimality is one of the most widely used terms in economics. He was also the first to discover that income distribution follows a Pareto principle that 80% of the wealth in Italy belonged to about 20% of the population. In his "Manual of Political Economy" (1906), he focused on equilibrium in terms of solutions to individual problems of "objectives and constraints". His work reflected the concurrence of the ideas of Léon Walras and William Jevons that economics is essentially a mathematical science.

2.3. *Schools of political-economic thought in the 20th century*

In the early 20th century, some major schools of political-economic thought had been formed and provided profound impacts on the development of political economy thereafter.

2.3.1. *The Austrian School*

While economics at the end of the 19th century and the beginning of the 20th was dominated increasingly by mathematical analysis, Carl Menger (1840–1921), Eugen von Böhm-Bawerk (1851–1914), Friedrich von Wieser (1851–1926), Ludwig von Mises (1881–1973), and Friedrich Hayek (1899–1992) followed a different route, advocating the use of deductive logic instead. This group and its followers became known as the Austrian School of Economics, reflecting the Austrian origin of many of the early adherents, and contrasted the marginal theory in the tradition of Alfred Marshall with the philosophies of the Austrian School.

Carl Menger (1840–1921) restated the basic principles of marginal utility in his publication "Grundsätze der Volkswirtschaftslehre" (Principles of Economics) (1871). In particular, he argued that consumers act rationally by seeking to maximize satisfaction of all their preferences; people allocate their spending so that the last unit of a commodity bought creates no more satisfaction than the last unit bought of something else. His subjective theory of value arrives at what he considered one of the most powerful insights in economics that "both sides gain from exchange". Unlike William Jevons, Menger did not believe that goods provide "utils", or units of utility. Rather, he believed goods are valuable because they serve various uses whose importance differs.

Eugen von Böhm-Bawerk (1851–1914) developed a theory of the origin and determination of the rate of interest and the period of turnover of capital occurring with the attainment of the market clearing wage. It provided the basis of the Austrian school's theory of capital. The theory influenced later writers, such as Knut Wicksell and Irving Fisher, and provided the basis for the modern treatment of interest, which is now seen as stemming from the interaction of (a) the preference for present goods (which inhibits savings and investment) and (b) the productivity of longer periods of turnover of capital (which causes investment funds to be demanded). Böhm-Bawerk was the first economist to refute Karl Marx's view that workers are systematically exploited. While Marx attributed productivity to labor, Böhm-Bawerk attributed productivity to an indirect, or "roundabout", process based on an investment in land and labor.

Friedrich von Wieser (1851–1926) is renowned for his work "Natural Value" (1893) which carefully details the alternative-cost doctrine and the theory of imputation. His explanation of marginal utility theory was decisive, at least terminologically. It was his term "Grenznutzen" that developed into the standard term "marginal utility", not William Jevons' "final degree of utility" or Menger's "value". His use of the modifier "natural" indicates that he regarded value as a "natural category" that would pertain to any society, no matter what institutions of property had been established.

The socialist economic calculation debate (a discourse about how a socialist economy would perform economic calculation given the absence of the law of value, money, financial prices for capital goods, and private ownership of the means of production) started with his notion of the paramount importance of accurate calculation to economic efficiency. Wieser argued that prices represented information about market conditions and were thus necessary for any sort of economic activity. Thus, a socialist economy needs a price system to operate. He also stressed the importance of entrepreneurship to economic change. Unlike most other Austrian School economists, Wieser rejected classical liberalism (laissez-faire), arguing that "freedom has to be superseded by a system of order". Another of Wieser's fundamental contributions to economics is the alternative cost theory (now called the opportunity cost theory).

Ludwig von Mises (1881–1973) developed an integrated, deductive science of economics based on the fundamental axiom that individual human beings act purposively to achieve desired goals. Mises advocated the idea that the only viable economic policy for the human race was a policy of unrestricted laissez-faire, of free markets and the unhampered exercise of the right of private property, with government strictly limited to the defense of person and property within its territorial area. He demonstrated (i) that the expansion of free markets, the division of labor, and private capital investment is the only possible path to the prosperity and flourishing of the human race, (ii) that socialism would be disastrous for a modern economy because the absence of private ownership of land and capital goods prevents any sort of rational pricing or estimate of costs, and (iii) that government intervention, in addition to hampering and crippling the market, would prove counter-productive and cumulative, leading inevitably to socialism unless the entire tissue of interventions was repealed.

Friedrich Hayek (1899–1992) argued that the market is a "spontaneous order" and actively disapproved the concept of "social justice". He believed that all forms of collectivism (even those theoretically based on voluntary cooperation) could only be maintained by a central authority. He argued that centralizing economic decision-making would lead not only to infringements of liberty but also to depressed standards of living because centralized experts could not gather and assess the required knowledge to allocate scarce resources efficiently or productively. Hayek claimed in his book, "The Road to Serfdom" (1944) that socialism required central economic planning and that such planning in turn would lead toward totalitarianism. In his book "The Fatal Conceit" (1988), Hayek attributed the birth of civilization to private property. He also maintained that price signals are the only means of enabling each economic decision-maker to communicate tacit knowledge or dispersed knowledge to each other, to solve the economic calculation problem.

2.3.2. *The Keynesian School*

The central tenet of this school of thought is that government intervention can stabilize the economy. During the Great Depression of the 1930s, existing economic theory was unable to either explain the causes of the severe worldwide economic collapse or provide an adequate public policy solution to jump-start production and employment. **John Maynard Keynes** (1883–1946) published his most influential work "The General Theory of Employment, Interest and Money" (1936) during the Great Depression. Orthodox economics thought that a tightening of spending would restore business confidence and profit levels. Keynes by contrast had argued in "A Tract on Monetary Reform" (1932) that a variety of factors determined economic activity and that it was not enough to wait for the long-run market equilibrium to restore itself. As Keynes famously remarked:

"...this long run is a misleading guide to current affairs. In the long run we are all dead. Economists set themselves too easy, too useless a task if in tempestuous seasons they can only tell us that when the storm is long past the ocean is flat again".

Keynes identified the propensity to consume, inducement to invest, marginal efficiency of capital, liquidity preference, and multiplier effect as variables which determine the level of the economy's output, employment, and price levels. Keynes argued that if savings were being withheld from investment in financial markets, total spending would fall, leading to reduced incomes and unemployment, which would reduce savings again. This would continue until the desire to save becomes equal to the desire to invest, which means a new "equilibrium" is reached and the spending decline stops. This new "equilibrium" is a depression, where people are investing less, have less to save, and have less to spend.

Keynes also argued that employment depends on total spending, which is composed of consumer spending and business investment in the private sector. Consumers only spend "passively", or according to their income fluctuations. Businesses, on the other hand, are induced to invest by the expected rate of return on new investments and the cost of interest payments. Then, if business expectations remained the same, and the government reduced interest rates, investment would increase and would have a multiplier effect on total spending. Interest rates are determined by the quantity of money and the desire to hold money in bank accounts. If not, enough money is available to match the amount people want to hold, and interest rates would rise until enough people are put off. So, if the quantity of money were increased, while the desire to hold money remained stable, interest rates would fall, leading to increased investment, output, and employment. For these reasons, Keynes advocated low interest rates and easy credit, to combat unemployment.

Keynes believed in a more egalitarian distribution of income and taxation on unearned income arguing that high rates of savings are not desirable in a developed economy. Therefore, both monetary management and an active fiscal policy are advocated by Keynes.

The Keynes's Circus — a group of Cambridge economists Keynes — with **Richard Kahn** (1905–1989), **James Meade** (1907–1997), **Joan Robinson** (1903–1983), **Austin Robinson** (1897–1993), and **Piero Sraffa** (1898–1983) was formed immediately following the publication of Keynes's "A Treatise on Money" (1935) to discuss the treatise and to provide feedback on Keynes's continuing theoretical work that would lead to his General Theory of Employment, Interest, and Money. **John Hicks** (1904–1989) was also a Keynesian and proposed the Investment Saving–Liquidity Preference Money Supply Model, which treats the intersection of the IS and LM curves as the general equilibrium in both markets.

In 1977, **Edmund Phelps** (1933–present) and **John B. Taylor** (1946–present) published a paper proving that staggered setting of wages and prices gives monetary policy, a role in stabilizing economic fluctuations if the wages/prices are sticky, even when all workers and firms have rational expectations. This work caused Keynesian economics to return to mainstream economics and advance into New Keynesian economics. Its central theme is the provision of a microeconomic foundation for Keynesian macroeconomics, obtained by identifying minimal deviations from the standard microeconomic assumptions which yield Keynesian macroeconomic conclusions, such as the possibility of significant welfare benefits from macroeconomic stabilization. **Paul Samuelson** (1915–2009) was one of the founders of neo-Keynesian economics and a seminal figure in the development of neoclassical economics. He contributed to various areas of economics and was considered to be the "Father of Modern Economics" and the "foremost academic economist of the 20th century".

George Akerlof (1940–present), **Janet Yellen** (1946–present), **Huw Dixon** (1958–present), **Michael Woodford** (1955–present), **Julio Rotemberg** (1953–present), **Alfred Eichner** (1937–1988), **Sidney Weintraub** (1914–1983), **Paul Davidson** (1930–present), and **E. Ray Canterbery** (1935–present) are prominent scholars in the Post Keynesian School.

2.3.3. *The Chicago School*

The main tenets of the Chicago School are that free markets best allocate resources in an economy and that minimal, or even no, government intervention is best for economic prosperity. It promotes the virtues of free-market principles to better society. The school believes that the reduction or elimination of regulations on business is beneficial to an economy. The Chicago School is libertarian and laissez-faire at its core, disputing the Keynesian notions of governments managing aggregate economic demand to promote growth.

The government-interventionist monetary and fiscal policies that the postwar Keynesian economists recommended came under attack by a group of theorists working at the University of Chicago, which came in the 1950s to be known as the Chicago School of Economics. Before World War II, the Old Chicago School of Strong Keynesians was founded by **Frank Knight** (1885–1972), **Jacob Viner** (1892–1970), and **Henry Calvert Simons** (1899–1946). The second generation was known for a more conservative line of thought, reasserting a libertarian view of market activity that people are best left to themselves to be free to choose how to conduct their own affairs.

Ronald Coase (1910–2013) of the Chicago School of Economics was the most prominent economic analyst of law. His article "The Nature of the Firm" (1937) argued that the reason for the existence of business firms is the existence of transaction costs. Human beings with the ability to make rational decisions trade through bilateral contracts on the markets until the costs of transactions make the use of corporations to produce things more cost-effective. His article "The Problem of Social Cost" (1960) argued that if we lived in a world without transaction costs, people would bargain with one another to create the same allocation of resources, regardless of the way a court might rule in property disputes.

The publications of **Richard Posner** (1939–present) "Economic Analysis of Law" (1974) and "The Economics of Justice" (1982) became widely cited textbooks and had been used by judges in interpreting common law as they were trying to maximize economic welfare.

Milton Friedman (1912–2006) of the Chicago School of Economics is one of the most influential economists. In his work "A Monetary History of the United States" (1963), he believed that the Great Depression was caused by the policies of the Federal Reserve. He was a founder of the school of monetarism. Friedman asserted that laissez-faire government policy is more desirable than government intervention in the economy. Governments should adopt a neutral monetary policy oriented toward long-run economic growth with the gradual expansion of the money supply. He advocated the quantity theory of money and maintained that general prices are determined by money. Active monetary policy (through easy credit, etc.) or fiscal policy (through taxation and government spending) could have unintended negative effects. In "Capitalism and Freedom" (1969), Friedman supported non-intervention by governments with the following argument:

"There is likely to be a lag between the need for action and government recognition of the need; a further lag between recognition of the need for action and the taking of action; and a still further lag between the action and its effects".

Friedman also contributed to consumption theory through "The Permanent Income Hypothesis" (1957). This work argued that rational consumers would spend a proportional amount of what they perceived to be their permanent income. Windfall gains would mostly be saved. Tax reductions, likewise, as rational consumers would predict that taxes would have to rise later to balance public finances. Other important contributions include his critique of the Phillips curve and the concept of the natural rate of unemployment.

George Stigler (1911–1991) developed various theories regarding the impact of government regulation on businesses, including "The Theory of Economic Regulation" (1971). He argued that interest groups and political participants will use the regulatory and coercive powers of government to shape laws and regulations in a way that is beneficial to them. His landmark article, "The Economics of Information" (1961) essentially created a new area of study in information economics.

Gary Becker (1930–2014) developed theories that applied economic reasoning to human behavior and business decision-making. His contributions covered a wide spectrum of economics, including rational choice theory, human capital, crime and punishment, economics of the family and household, organ markets, public finance, and political economy.

Robert E. Lucas, Jr. (1937–present) founded the New Classical Macroeconomics based on Milton Friedman's monetarist critique of Keynesian macroeconomics, and the idea of rational expectations (first proposed by Carnegie economist **John F. Muth** (1930–2005) in "Rational Expectations and the Theory of Price Movements" (1961)). The New Classical Macroeconomics opposed the idea that government intervention can or should stabilize the economy. "The Policy-Ineffectiveness Proposition" (1975) of **Thomas J. Sargent** (1943–present) and **Neil Wallace** (1939–present) disagreed with basic assumption of Keynesian economics.

The "Real Business Cycle Theory" (1982) by **Finn Kydland** (1943–present) and **Edward C. Prescott** (1940–present) explained observed fluctuations in output and employment in terms of real variables, such as changes in technology and tastes. Under competitive markets, real business cycle theory implies that cyclical fluctuations are optimal responses to variability in technology and tastes and that macroeconomic stabilization policies would reduce welfare. Kydland and Prescott also developed the theory of Dynamic Stochastic General Equilibrium (DSGE) in their article "Time to Build and Aggregate Fluctuations (1982)".

Eugene Fama (1939–present) published the paper "The Behavior of Stock Market Prices" (1965), which found that stock market prices follow a random walk, proposing the Efficient Market Hypothesis, that randomness is characteristic of a perfectly functioning financial market.

2.4. *Original mathematics in economics*

There were occasions when mathematical sciences merged into economic sciences in which original mathematics was developed for specific economic analyses. For instance, St. Petersburg mathematician **Leonid Kantorovich**, the first mathematician who received the Nobel Economic Prize, founded linear programming, provided preeminent results in functional analysis and operator theory, and made important contributions to the theory of optimal allocation of resources and organization of production. His mathematical contributions are used in inventory management, production planning, optimization, and mathematical programming. His contributions to mathematics and economics were so rich that many contemporaries assumed it to be the work of two separate individuals: Kantorovich the mathematician and Kantorovich the economist.

John Nash, another Nobel Laureate, made fundamental contributions to game theory. He introduced a number of mathematical game solutions, including Nash equilibrium and the Nash bargaining solution, which are now considered central to game theory and its applications in various fields of science. Nash was the most heavily cited scholar in the field of game theory, and the name Nash appears in most of the game theory papers published.

Lloyd Shapley, another Nobel Laureate, contributed to the fields of mathematical economics and especially game theory. Shapley is regarded as one of the most important contributors to the development of game theory since the work of von Neumann and Morgenstern. The Shapley value is one of the most innovative mathematical game solutions and has been extended to the development of many related game solutions. According to *The Economist*, Shapley "may have thought of himself as a mathematician, but he cannot avoid being remembered for his huge contributions to economics".

Nobel laureate **Robert Aumann's** greatest contribution was in the realm of repeated games, which are situations in which players encounter the same situation over and over again. Aumann was the first to define the concept of correlated equilibrium in game theory, which is a type of equilibrium in non-cooperative games that is more flexible than the classical Nash equilibrium. He collaborated with Lloyd Shapley to establish the Aumann–Shapley value.

3. Rise of Global Political Economy

Global political economy (GPE), also known as international political economy (IPE), is the study of how politics shapes the global economy and how the global economy shapes politics. A key focus in GPE is on the distributive consequences of global economic exchange.

A central assumption of GPE theory is that international economic phenomena cannot be separated from the decision-makers who regulate and control them. In addition to formal economic theories of international economics, trade, finance, and monetary systems which are widely utilized within the discipline, GPE also involves the study of political institutions and power relations in the global economy. Key decision-makers in the global economic arena include individual sovereign nations, international organizations, multinational corporations, political unions, and trade blocs. The substantive issue areas of GPE are frequently divided into four broad subject areas:

1. international trade,
2. the international monetary and financial system,
3. multinational corporations, and
4. economic development and inequality.

Global political economy has deep historical roots in the discipline of political economy, the study of the national economy, and its interactions with governance and politics. Alfred Marshall abandoned the term "political economy" in favor of "economics" in his influential book "Principle of Economics" (1890). This started a path of study in political economy that studies economic phenomena with less consideration of political and social concerns. Economics seemed to begin to diverge from political economy studies in the late 19th century, political economy continued to live as an academic tradition within political science departments, as well as within modern economics as a pluralist approach. While the notable works from John Maynard Keynes' "General Theory" (1936) and "The Great Transformation" (1944) by **Karl Polanyi** (1866–1964) were still written in the tradition of political economy, that is, economics in its narrower form dominating economics departments.

The enhancement of the study on global/international political economy appeared to be the need of the world in the late 1960s and early 1970s when economic interdependence intensified because of the appearance of post-WWII economic institutions including the International Monetary Fund, World Bank, and General Agreement on Tariffs and Trade. Increasing attention within the international arena had been drawn toward the study of these institutions, their governance, and their effects on the world economy. The need for a more comprehensive understanding of global economic governance within political science circles became increasingly apparent due to the serious crises of the 1970s: the end of the gold standard, the 1973 oil crisis, the 1973–1975 recession, and calls for greater trade protection. GPE is also a major field of study within history, especially economic history, where scholars study the historical dynamics of the international political economy.

Unprecedentedly, rapid globalization has appeared in the past three decades. Not only that trade in goods and services as well as capital flows across borders grow to unprecedented levels, but an unprecedented process of economic convergence between the developed and the emerging and developing economies took place. This economic convergence occurred because, during the last few decades, the group of rich countries registered slower growth than before. Some developing countries, which had doubled their ratio of trade to GDP and became more open to Foreign Direct Investment (FDI), were able to raise their per capita GDP at a rate more than twice that of rich nations. These developing nations managed to take the crucial step around the 1980s to integrate into the global economy.

As computing and telecommunication capabilities became cheaper and more easily accessible plus the already inexpensive transportation costs and lower impediments to cross-border trade, it became economically attractive and necessary to separate the previously integrated and concentrated production processes. Production dispersion in cost-effective internationalized supply becomes the only way to remain competitive. The traditional production mode has given way to geographical fragmentation of production supported by continual flows of investment, technology, personnel expertise, information, finance, efficient transportation, and logistics services. In addition, the advancement in information technology has made it inexpensive and speedy to coordinate complex activities situated in different locations around the world.

This complex form of the international division of labor leads to the offshoring of productive activities from which the developed nations can combine their more advanced technologies with the low-cost labor of developing nations to enhance their competitiveness. Developing nations, on the other hand, can use offshore links of the supply chain for industrializing more rapidly without building the entire industrial base. This new organization of production, driven by the Internet and the other tools of IT, does not pertain only to large corporations but is fueling transformations covering the entire value chain in practically all sectors and types of companies. In terms of international trade, a nation's comparative advantage is no longer about finished goods or commodities but about the finer tasks making up the manufacturing, commercial, and financial processes necessary to produce the final goods or commodities. The final product at the end of its supply chain is a conjunction of many nations' capital, labor, technology, infrastructure, finance, and business environments. The evolution of the global economy, the pattern of production and trade, and the redistribution of economic gains have come to a new era. To deliver globalization to its full potential, all nations have to take more serious concerns over political-economic policies to make their impact less disruptive and more beneficial for themselves. Indeed, many (if not almost all) political economy issues nowadays can be classified under global political economy. In this era of globalization, the entire world has become a global political economy.

4. Toward Time-consistent GPE: Theoretical Underpinning

In the current era of globalization, interactions among nations are substantial and the entire world can be regarded as a global political economy. The actions of many nations simultaneously could create wide ripple effects leading to very complex outcomes. The action of one nation on its economy can bring about different effects on different nations' economies. Moreover, the interactions between the nations will continue over time, and many actions will have effects lasting for some time. Indeed, the current global political-economic arena is in a dynamic game situation.

4.1. *Time consistency and subgame consistency*

The notion of time consistency is firmly based on theoretical economic grounds. Consider first the commonly known notion of time consistency. It refers to the property that the incentive of keeping a commitment throughout the committed duration is higher than the incentive of not keeping the commitment. If this is not the case, time inconsistency would arise, and the decision-maker has the incentive to default on the commitment. Time consistency is an important concept in formulating economic decisions. Any time-inconsistent commitment would likely not be fully completed. It is a crucial measure of the credibility of government policies. A policy that is time-consistent is credible and can be trusted to be carried out as intended over the committed duration. Often, time inconsistency would arise when a (time-invariant) policy is optimal to the policymaker at the beginning but is not optimal at a later point in time. This happens when individuals and firms anticipate future policy changes and adjust their behavior according to their anticipation, leading to outcomes that are different from what the policymaker had intended.

Since the issue of time consistency is so important in economic analysis, there has been a considerably lengthy list of studies covering various topics, including Alvarez *et al.* (2004), Asheim (1997), Bhattarai *et al.* (2022), Bianchi and Mendoza (2018), Cochrane (1995), Ekeland *et al.* (2012), Epstein and Ji (2020), Calvo (1978), Calvo and Obstfeld (1988), Ireland (1999), Denant-Boemont *et al.* (2017), Klein *et al.* (2008), Klein and Ríos-Rull (2003), Kotlikoff *et al.* (1988), Lindbeck and Weibull (1988), Millner and Heal (2018), Persson *et al.* (1987), Persson *et al.* (2006), Nicolini (1998), Prast (1996), Shepsle (1992), Shi *et al.* (2023), Stokey (1989), and Turnovsky and Brock (1980).

The term time consistency is often applied in cases where there is a sole or prime decision-maker, though it can also be used in the case of multiple decision-makers in interactive game problems. Using the term subgame consistency may sound more pertinent for game theoretical problems, where there exist time-varying elements and consideration of the subgames as a new game given the current state (and previously executed strategies that are still effective, if any). This is especially the case when there are various uncertainty issues, like stochastic dynamics, random game horizon, and intergeneration players (see Yeung and Petrosyan, 2012, 2016, 2022). A thorough study on time-consistent (or subgame-consistent) non-cooperative dynamic games can be found in Başar and Olsder (1999). Game-theoretic studies involving consistency over time can be found in Groot *et al.* (2003), Başar (1975, 1976, 1989), Başar and Ho (1974), Yeung (1992, 1996, 1999), Plourde and Yeung (1989), and Jørgensen and Yeung

(1996, 1999). Subgame-consistent non-cooperative dynamic games with durable strategies can be found in Yeung and Petrosyan (2022).

Besides time consistency in non-cooperative interaction, cooperative time/subgame consistency is needed for stable cooperation schemes. Application studies with cooperative subgame consistent solutions can be found in Jørgensen and Zaccour (2001, 2002), Petrosyan and Zaccour (2003), and Petrosyan and Yeung (2014). Yeung and Petrosyan (2004, 2006b, 2010, 2011, 2016a) developed a generalized method for the derivation of analytically tractable subgame consistent solutions in various types of dynamic games. This has made possible the rigorous study of subgame consistent solutions in discrete-time dynamic cooperation. Subgame-consistent cooperative dynamic games with durable strategies can be found in Yeung and Petrosyan (2022). Cooperative subgame consistency is crucial in handling global political economy policy formulation in global political economy for establishing sustainable international collaboration. In general, the terms time consistency and subgame consistency are used interchangeably in this book.

4.2. *Approaches of GPE Policy formulation*

There are two commonly adopted approaches to policy formulation. The technically oriented approach regards policy formulation as a practice of correct analysis, searching for the optimal (or a preferred viable) solution to a complicated policy problem. The politically oriented approach regards policy formulation as an endeavor to gain support for a policy issue through the legislative process. The former considers policy formulation in terms of rationality and quantitative investigation, while the latter considers policy formulation in terms of compromise and quality analysis. Both are important in practical policymaking. GPE is certainly one of the most complicated policy problems and one of the most crucial policy issues of our time. The proper formulation of policy helps policymakers to understand the direction and consequences of the policy. Initiating a policy that is contextually relevant and practically rational is crucial to achieving the policy objective. The technically oriented approach to policy formulation forms the basis for the politically oriented approach.

Given the reality of rapid and extensive globalization, rational policy formulation under a time-consistent dynamic framework is, indeed, very much needed. Theoretically, time-consistent GPE is a "rational policy goal" that policymakers would seek to attain. It would shed light on policy analysis concerning complicated global problems that are still unresolved.

5. A General Mathematical Theory of Political Economy Revisited

The works by Wiliam Jevons (1866, 1971) on "A General Mathematical Theory of Political Economy" are ground-breaking contributions to political economy. Irving Fisher praised Jevons' work as the start of the mathematical method in economics. Milton Friedman praised Jevons' contrition to modern economics as important as Adam Smith, David Ricardo, and Alfred Marshall. Jevons' work on the utility theory along with similar discoveries made by Carl Menger in 1871 marked the opening of a new period in the history of economic thought. This book is one of the most important ones for the deepest thoughts of economics and mathematics by William Jevons, who is regarded by many as one of the greatest thinkers of modern economics and logic. The reasoning by William Jevons remains as relevant today as it was then.

In 1862, Jevons presented a paper to a meeting of the British Association that envisioned a radical transformation of the discipline of economics. The paper, entitled "A Brief Account of a General Mathematical Theory of Political Economy" advocated the heavy use of mathematics in economics. It argued that the incorporation of mathematical techniques into the discipline would allow economists to explain virtually all economic activity by deductive reasoning from a basic theory of utility (see the following). In 1871, he elaborated his mathematized approach to economics with mathematical formulae in his book "The Theory of Political Economy" (1871). Jevons then published the 1862 paper in the *Journal of the Statistical Society of London* as "A General Mathematical Theory of Political Economy" (1866). It did not gain much attention and did not explicitly use any mathematical formulae. In 1871, he elaborated his mathematized approach to economics with mathematical formulae in his book "The Theory of Political Economy" (1871).

In his 1871 book, Jevons fleshed out the arguments in favor of the mathematization of economics and articulated the theory of utility and an intellectual edifice of economic theory. This book became very influential and was widely reviewed throughout England, with appraisals of theory running in the pages of newspapers in London and Manchester. Two of eminent economists of the age, Alfred Marshall and John Elliot Cairnes, published their own reviews of Jevons' treatise. Jevons' book together with the work of Leon Walras and Carl Menger marked a turning point for economic thought: Jevons, Walras, and Menger would launch the marginalist revolution and transform economics toward a rigorous mathematical discipline. Jevons' 1862 paper (or 1866 book) on a general mathematical theory of political economy is an important essay in the history of economic thought that drafted an early revolutionary step in the discipline.

A review of the central ideas of Jevons' 1866 book is provided in the following.

5.1. *Economics is inherently mathematical*

Economics, Jevons argues, is inherently a mathematical discipline because it deals with quantities. However, the powerful mathematical techniques which have been successfully applied to "most other sciences" have been neglected by economists.

Jevons' 1866 paper briefly describes the nature of a theory of economy which will reduce the main problem of this science to a mathematical form. Economy, indeed, being concerned with quantities, has always of necessity been mathematical in its subject, but the strict and general statement and the easy comprehension of its quantitative laws have been prevented by the neglect of those powerful methods of expression which have been applied to most other sciences with so much success.

Jevons writes of "quantitative laws" of economics which he believes mathematical techniques can illuminate. And he refers to economics as "this science", before comparing it to "most other sciences", which are mathematical.

"... the nature of a Theory of Economy which will reduce the main problem of this science to a mathematical form. Economy, indeed, being concerned with quantities, has always of necessity been mathematical in its subject, but the strict and general statement, and the easy comprehension of its quantitative laws has been prevented by a neglect of those powerful methods of expression which have been applied to most other sciences with so much success. It is not to be supposed, however, that because the economy becomes mathematical in form, it will, therefore, become a matter of rigorous calculation. Its mathematical principles may become formal and certain, while its data remain as inexact as ever...", Jevons Wrote.

Jevons believed a mathematized economics discipline would be a science capable of identifying laws that can explain economic activity. While economics can be regarded as a mathematical discipline, it does not mean that economic science has to be a precise discipline. We can, in Jevons' view, express relations between economic variables in a mathematical form. Under today's technology, many analyses are involved with mathematics, like optimization, operations research, economic cybernetics, inventory control, production scheduling, and financial technology. These endeavors can be viewed as a precise discipline. However, in many economic problems, we are still unable to use mathematical equations to arrive at precise conclusions about their outcomes because we are unable to fill in the values of these variables. Nevertheless, the advancements in big data and artificial intelligence will increase the precision in predicting economic outcomes. One of the modern ways of interpretation is that economics is a science that applies mathematics as an analytical tool.

5.2. *Pain, pleasure, and motivation*

Jevons' approach to economics begins with the idea that all economic activity is driven by a human desire to acquire pleasure and minimize pain. This desire motivates the degrees to which we labor or leisure, consume or save, and so forth. A large part of such feelings arises periodically from the ordinary wants and desires of body or mind, and from the painful exertion we are continually prompted to undergo so that we may satisfy our wants.

He believed that economists should not consider all the motivations that spur human action: considerations such as morality and compassion are to be left to other disciplines. Instead, they should only consider how a desire to maximize pleasure and minimize pain motivates economic activity while there are motives arising from conscience, compassion, or some moral or religious source, which the economy cannot and does not pretend to treat. While these motivations will remain as outstanding and disturbing forces, they must be treated, if at all, by other appropriate branches of knowledge. This is similar to the argument by John Stuart Mill in favor of economists separating non-economic motivations, leaving them for consideration by other disciplines.

Jevons proceeds to argue that a human's pain and pleasure can be expressed as comparable quantities so economists can treat them scientifically. People's estimation of the comparative amounts of feeling is performed in the act of choice. Their choice of one course out of two or more proves that, in our estimation, this course promises the greatest balance of pleasure.

5.3. *Quantity of feeling and utility*

Jevons mentions how mathematics, in particular integral calculus and simple arithmetic, comes into play in his economic theorizing about the "quantity of feeling". There are two dimensions of feelings: intensity (i.e. how strong the feeling is) and duration (i.e. how long the feeling lasts). If the intensity of the feeling is constant for the duration of time over which it is experienced, one may compute an individual's "quantity of feeling" by multiplying the intensity by the duration. If the intensity of the feeling varies over the duration of time during which it is experienced, one would use integration to compute an individual's "quantity of feeling". When making these calculations, pleasure is a positive quantity while pain is a negative quantity.

Anticipated future pleasure and anticipated future pain resulting from taking a particular action also cause us to feel pleasure and pain today. But the intensity of this pleasure or pain experienced today is reduced/discounted owing to its remoteness and the possibility that it may not come to pass.

"A principle of the mind which any true theory must take into account is that of foresight. Every expected future pleasure or pain affects us with similar feelings in the present time but with an intensity diminished in some proportion to its uncertainty and its remoteness in time. But the effects of foresight merely complicate without altering the other parts of the theory", Jevons wrote.

In the seventh paragraph of Jevons' book, he starts to use the term utility.

"... A second part of the theory proceeds from feelings to the useful objects or utilities by which pleasurable feeling is increased or pain removed". Jevons wrote, "An object is useful when it either affects the senses pleasurably in the present moment, or when, by foresight, it is expected that it will do so at some future time. Thus, we must carefully distinguish actual utility in present use from estimated future utility, which yet, by allowing for the imperfect force of anticipation, and for the uncertainty of future events, gives a certain present utility".

He further introduced the idea of "diminishing marginal utility". Marginal utility refers to the change in one's utility from each additional unit of an item that is consumed. Diminishing marginal utility refers to the idea that the utility an individual acquires from every additional unit of an object consumed diminishes. In particular, Jevons wrote:

"Amount of utility corresponds to the amount of pleasure produced. But the continued uniform application of a useful object to the senses or the desires, will not commonly produce uniform amounts of pleasure. Every appetite or sense is more or less rapidly satiated. A certain quantity of an object received, a further quantity is indifferent to us, or may even excite disgust. Every successive application will commonly excite the feelings less intensely than the previous application. The utility of the last supply of an object, then, usually decreases in some proportion, or as some function of the whole quantity received. This variation theoretically exists even in the smallest quantities, we must recede to infinitesimals, and what we shall call the coefficient of utility, is the ratio between the last increment or infinitely small supply of the object, and the increment of pleasure which it occasions, both, of course, estimated in their appropriate units".

He also emphasized the following: "The coefficient of utility is, then, some generally diminishing function of the whole quantity of the object consumed. Here is the most important law of the whole theory".

Jevons devises his theory of economics from the premises that (i) all economic activities are motivated by a desire to maximize pleasure and minimize pain (with other non-economic motives being disregarded), (ii) the pleasure and pain an individual feels from taking any action can be expressed numerically (as utilities), (iii) the future pain and pleasure associated with any action is discounted when taking the decision in the present, and (iv) marginal utility is diminishing. From this starting point, Jevons' reasons deductively to develop theories meant to explain economic activities. He went on to construct a theory of labor, a theory of rent, a theory of exchange, and a theory of capital.

References

Alvarez, F., Kehoe, P. J., & Neumeyer, P. A. (2004). The time consistency of optimal monetary and fiscal policies. *Econometrica*, 72(2), 541–567.

Asheim, G. B. (1997). Individual and collective time-consistency. *The Review of Economic Studies*, 64(3), 427–443. https://doi.org/10.2307/2971722.

Başar, T. & Ho, Y. C. (1974). Informational properties of the Nash solutions of two stochastic nonzero-sum games. *Journal of Economic Theory*, 7, 370–387.

Başar, T. & Olsder, G. J. (1999). *Dynamic Noncooperative Game Theory*, 2nd edn. SIAM, Philadelphia.

Başar, T. (1975). Nash strategies for M-person differential games with mixed information structures. *Automatica*, 11, 547–551.

Başar, T. (1976). On the uniqueness of the Nash solution in linear-quadratic differential games. *International Journal of Game Theory*, 5, 65–90.

Başar, T. (1989). Time consistency and robustness of equilibria in noncooperative dynamic games. In Van der Ploeg, F. & de Zeeuw, A. (eds.), *Dynamic Policy Games in Economics*. North-Holland, Amsterdam, pp. 9–54.

Bhattarai, S., Eggertsson, G. B., & Gafarov, B. (2022). Time consistency and duration of government debt: A model of quantitative easing. *The Review of Economic Studies*; rdac063. https://doi.org/10.1093/restud/rdac063.

Bianchi, J. & Mendoza, E. G. (2018). Optimal time-consistent macroprudential policy. *Journal of Political Economy*, 126(2). https://doi.org/10.1086/696280.

Bodin, J. (1586). *Reply to the Paradoxes of Malestroit*.

Calvo, G. A. (1978). On the time consistency of optimal policy in a monetary economy. *Econometrica* (JSTOR), 46(6), 1411–1428. https://doi.org/10.2307/1913836.

Calvo, G. A. & Obstfeld, M. (1988). Optimal time-consistent fiscal policy with finite lifetimes. *Econometrica*, 56(2), 411–432. https://doi.org/10.2307/1911079.

Cantillon, R. (1730). *Essay on the Nature of Trade in General.* Indianapolis: Liberty Fund, 2015. Print.

Coase, R. (1937). The nature of the firm. *Economica*, 4, 386–405. https://doi.org/10.1111/j.1468-0335.1937.tb00002.x.

Coase, R. (1960). The problem of social cost. *The Journal of Law & Economics*, 3, 1–44.

Cochrane, J. H. (1995). Time-consistent health insurance. *Journal of Political Economy*, 103(3). https://doi.org/10.1086/261991.

Copernicus, N. (1519). *Bad Money Drives Out Good.*

Denant-Boemont, L., Diecidue, E., & l'Haridon, O. (2017). Patience and time consistency in collective decisions. *Experimental Economics*, 20, 181–208. https://doi.org/10.1007/s10683-016-9481-4.

d'Oresme, N. (1355–1356). De origine, natura, jure et mutationibus monetarum.

Ekeland, I., Mbodji, O., & Pirvu, T. A. (2012). Time-consistent portfolio management. *SIAM Journal on Financial Mathematics*, 3(1). DOI: 10.1137/100810034.

Epstein, L. G. & Ji, S. L. (2020). Optimal learning under robustness and time-consistency. *Operations Research*, 70(3), 1317–1329. https://doi.org/10.1287/opre.2019.1899.

Fama, E. F. (1965). The behavior of stock-market prices. *The Journal of Business* (Chicago), 38(1), 34–105. https://doi.org/10.1086/294743.

Friedman, M. (1957). *Theory of the Consumption Function.* Princeton University Press. http://www.jstor.org/stable/j.ctv39x7zh.

Friedman, M. (1963). A monetary history of the United States 1867–1960. In *A Monetary History of the United States, 1867–1960*, Vol. 14. Princeton University Press.

Friedman, M. (1969). *Capitalism and Freedom.* University of Chicago Press.

Galiani, F. (1751). *Della Moneta (On Money).*

George, H. (1879). *Progress and Poverty: An Enquiry into the Cause of Industrial Depressions, and of Increase of Want with Increase of Wealth; The Remedy.* Henry George & Company, New York.

Groot, F., Withagen, C., & de Zeeuw, A. (2003). Strong time-consistency in the cartel-versus-fringe model. *Journal of Economic Dynamics and Control*, 28(2), 287–306.

Hayek, F. (1944). *The Road to Serfdom.* University of Chicago Press.

Hayek, F. (1988). *The Fatal Conceit.* University of Chicago Press.

Hume, D. (1752). *Political Discourses.*

Hume, D. (1777). *Essays, Moral, Political, Literary.*

Ireland, P. N. (1999). Does the time-consistency problem explain the behavior of inflation in the United States? *Journal of Monetary Economics*, 44(2), 279–291. https://doi.org/10.1016/S0304-3932(99)00026-4.

Jevons, W. (1866). *A General Mathematical Theory of Political Economy.*

Jevons, W. S. (1862). A brief account of a general mathematical theory of political economy. *The Journal of the Royal Statistical Society, London*, XXIX (June 1866), pp. 282–287. Section F of the British Association, 1862.

Jevons, W. S. (1871). *The Theory of Political Economy*, 3rd edn. Macmillan, London.

Jørgensen, S. & Yeung, D. W. K. (1996). Stochastic differential game model of a common property fishery. *Journal of Optimization Theory and Applications*, 90(2), 381–403.

Jørgensen, S. & Yeung, D. W. K. (1999). A strategic concession game. *International Game Theory Review*, 1(1), 103–129.

Jørgensen, S. & Zaccour, G. (2001). Time consistent side payments in a dynamic game of downstream pollution. *Journal of Economic Dynamics and Control*, 25(12), 1973–1987.

Jørgensen, S. & Zaccour, G. (2002). Time consistency in cooperative differential games. In Zaccour, G. (ed.), *Decision & Control in Management Science.* Advances in Computational Management Science, Vol. 4. Springer, Boston. https://doi.org/10.1007/978-1-4757-3561-1_19.

Keynes, J. M. (1932). *A Tract on Monetary Reform.* Macmillan.

Keynes, J. M. (1935). *A Treatise on Money.* Macmillan.

Keynes, J. M. (1936). *The General Theory of Employment, Interest and Money.* Palgrave Macmillan, 472 p.

Keynes, J. M. & Hicks, J. R. (1936). La Teoría de la Ocupación de Mr. Keynes: "The general theory of employment, interest and money." *Trimestre económico*, 3(12), 514–534.

Klein, P. & Ríos-Rull, J.-V. (2003). Time-consistent optimal fiscal policy. *International Economic Review*, 64(1). https://doi.org/10.1111/1468-2354.t01-1-00107.

Klein, P., Krusell, P., & Ríos-Rull, J.-V. (2008). Time-consistent public policy. *The Review of Economic Studies*, 75(3), 789–808. https://doi.org/10.1111/j.1467-937X.2008.00491.x.

Kotlikoff, L. J., Persson, T., & Svensson, L. E. O. (1988). Social contracts as assets: A possible solution to the time-consistency problem. *The American Economic Review*, 78(4), 662–677. http://www.jstor.org/stable/1811166.

Kydland, F. E. and Prescott, E. C. (1982). Time to build and aggregate fluctuations. *Econometrica*, 50(6), 1345–1370. https://doi.org/10.2307/1913386.

Lindbeck, A. & Weibull, J. W. (1988). Altruism and time consistency: The economics of fait accompli. *Journal of Political Economy*, 96(6). https://doi.org/10.1086/261582.

Locke, J. (1689). *Second Treatise on Civil Government.*

Locke, J. (1691). *Some Considerations on the Consequences of the Lowering of Interest and the Raising of the Value of Money.*

Marshall, A. (1879a). *Economics of Industry.* Macmillian and Co., London.

Marshall, A. (1879b). *The Pure Theory of Domestic Values.*

Marshall, A. (1890). *Principles of Economics*, Vol. I. Macmillan and Co.

Marx, K. (1867). *Capital: A Critique of Political Economy.* Progress Publishers, Moscow.

Menger, C. (1871). Grundsätze der Volkswirthschaftslehre, Erster, Allgemeiner Theil. Wien: Wilhelm Braumüller — via Internet Archive.; Translated as *Principles of Economics*, First, General Part. Translated by Dingwall, James; Hoselitz, Bert F. Glencoe. The Free Press, Illinois. 1950 — via Internet Archive.

Mill, J. S. (1848). *Principles of Political Economy with Some of Their Applications to Social Philosophy*, Vol. 2. C.C. Little & J. Brown, Massachusetts.

Millner, A. & Heal, G. (2018). Time consistency and time invariance in collective intertemporal choice. *Journal of Economic Theory*, 176, 158–169. DOI: 10.1016/j.jet.2018.03.002.

More, T. (1516). *Utopia.*

Muth, J. F. (1961). Rational expectations and the theory of price movements. *Econometrica*, 29(3), 315–335. https://doi.org/10.2307/1909635.

Nicolini, J. P. (1998). More on the time consistency of monetary policy. *Journal of Monetary Economics*, 41(2), 333–350. https://doi.org/10.1016/S0304-3932(97)00079-2.

North, D. (1691). *Discourses upon Trade.*

Pareto, V. (1906). *Manual of Political Economy.*

Persson, M., Persson, T., & Svensson, L. E. O. (1987). Time consistency of fiscal and monetary policy. *Econometrica*, 1419–1431. https://doi.org/10.2307/1913564.

Persson, M., Persson, T., & Svensson, L. E. O. (2006). Time consistency of fiscal and monetary policy: A solution. *Econometrica*, 74(1), 193–212. https://doi.org/10.1111/j.1468-0262.2006.00653.x.

Petrosyan, L. A. & Yeung, D. W. K. (2014). A time-consistent solution formula for bargaining problem in differential games. *International Game Theory Review*, 16(4), 1450016.01–1450016.11.

Petrosyan, L. A. & Zaccour, G. (2003). Time-consistent Shapley value allocation of pollution cost reduction. *Journal of Economic Dynamics and Control*, 27, 381–398.

Pigou, A. (1933). *The Theory of Unemployment.* Macmillan, London.

Pigou, A. C. (1943). The classical stationary state. *The Economic Journal* (London), 53(212), 343–351.

Plourde, C. & Yeung, D. W. K. (1989). Harvesting of a transboundary replenishable fish stock: A non cooperative game solution. *Marine Resource Economics*, 6, 57–70.

Polanyi, K. (1944). *The Great Transformation.* New York: Farrar & Rinehart (Incorporated), 305 p.

Posner, R. A. (1974). Economic analysis of law. *Stanford Law Review*, 26(3), 711–711. https://doi.org/10.2307/1227682.

Posner, R. A. (1982). The Economics of Justice [Review of The Economics of Justice]. *Michigan Law Review* (University of Michigan Law School), 80(4), 942–946. https://doi.org/10.2307/1288259.

Prast, H. M. (1996). Time consistency, asymmetric information and monetary policy design: An overview. *De Economist*, 144, 445–472. https://doi.org/10.1007/BF01682836.

Quesnay, F. (1758). *Tableau économique.*

Ricardo, D. (1817). *On the Principles of Political Economy and Taxation.* J. Murray, England.

Sargent, T. J. & Wallace, N. (1975). 'Rational' expectations, the optimal monetary instrument, and the optimal money supply rule. *Journal of Political Economy*, 83(2), 241–254. DOI: 10.1086/260321.

Say, J.-B. (1803). *A Treaties on Political Economy.*

Scotus, D. (1295). *Sententiae.*

Shepsle, K. (1992). A bureaucratic drift, coalitional drift, and time consistency: A comment on Macey. *The Journal of Law, Economics, and Organization*, 8(1), 111–118. https://doi.org/10.1093/oxfordjournals.jleo.a037021.

Shi, Y., Li, D., & Cui, X. (2023). Time consistent in efficiency dynamic mean–variance policy. *Journal of the Operational Research Society*, 74, 195–208.

Smith, A. (1759). *The Theory of Moral Sentiments.*

Smith, A. (1776). *An Inquiry into the Nature and Causes of the Wealth of Nations.*

Steuart, J. (1767). *An Inquiry into the Principles of Political Economy.*

Stigler, G. (1961). The economics of information. *Journal of Political Economy*, 69(3), 213–225. http://www.jstor.org/stable/1829263.

Stigler, G. (1971). The theory of economic regulation. *The Bell Journal of Economics and Management Science*, 2(1), 3–21. https://doi.org/10.2307/3003160.

Stokey, N. L. (1989). Reputation and time consistency. *The American Economic Review*, 79(2), 134–139. http://www.jstor.org/stable/1827745.

Turnovsky, S. J. & Brock, W. A. (1980). Time consistency and optimal government policies in perfect foresight equilibrium. *Journal of Public Economics*, 13(2), 183–212. https://doi.org/10.1016/0047-2727(80)90012-2.

von Wieser, F. (1893). *Natural Value*. Edited with a Preface and Analysis by William Smart. Macmillan, London.

Yeung, D. W. K. (1992). A differential game of industrial pollution management. *Annals of Operations Research*, 37, 297–311.

Yeung, D. W. K. (1996). A differential game model of a market of substitutable products. *European Journal of Operational Research*, 90, 599–608.

Yeung, D. W. K. (1999). A stochastic differential game of institutional investor speculation. *Journal of Optimization Theory and Applications*, 103(1), 463–477.

Yeung, D. W. K. & Petrosyan, L. A. (2004). Subgame consistent cooperative solutions in stochastic differential games. *Journal of Optimization Theory and Applications*, 120(3), 651–666.

Yeung, D. W. K. & Petrosyan, L. A. (2006b). *Cooperative Stochastic Differential Games*. Springer-Verlag, New York, 242 p.

Yeung, D. W. K. & Petrosyan, L. A. (2010). Subgame consistent solutions for cooperative stochastic dynamic games. *Journal of Optimization Theory and Applications*, 145, 579–596.

Yeung, D. W. K. & Petrosyan, L. A. (2011). Subgame consistent cooperative solution of dynamic games with random horizon. *Journal of Optimization Theory and Applications*, 150, 78–97.

Yeung, D. W. K. & Petrosyan, L. A. (2012). Subgame consistent solution for cooperative stochastic dynamic games with random horizon. *International Game Theory Review*, 14(2), 1250012.

Yeung, D. W. K. & Petrosyan, L. A. (2016a). *Subgame Consistent Cooperation: A Comprehensive Treatise*. Singapore: Springer, 520 p.

Yeung, D. W. K. & Petrosyan, L. A. (2022). *Durable-Strategies Dynamic Games: Theory, Solution Techniques and Applications*. Springer, Switzerland, 303 p.

Chapter 3

Uncoordinated Interaction in a Dynamic World: Mathematical Theory and Explication in Environmental Degradation

"The spiders have equivalents of the Prisoners' Dilemma, but they think in terms of intricate interconnectivity, of a world not just of sight but of constant vibration and scent. The idea of two prisoners incapable of communication would not be an acceptable status quo for them, but a problem to overcome: the Prisoners' Dilemma as a Gordian knot, to be cut through rather than be bound by." — Adrian Tchaikovsky

We are living in a dynamic world where policymakers from many nations develop their own policy designs and actions without coordination despite the interactive effects of these actions among the nations. On top of the dynamic nature of many global issues, many of the actions have durable effects. Durable strategies that have effects lasting over an extended period of time are prevalent in real-life situations. Revenue-generating investments, trade barriers, physical infrastructures, toxic waste disposal, enforceable agreements, legal frameworks, durable goods, business contracts, regulations, emission of pollutants, regulatory measures, coalition agreements, diffusion of knowledge, advertisement, and investments to build up physical capital are vivid examples of durable controls. In reality, the existence of durable strategies in the decision-maker's controls is likely to be a rule rather than an exception. In many practical situations, both the decision-maker's payoff and the evolution of the state dynamics can be subjected to effects arising from durable strategies.

Strategic behavior in the human and social world has been increasingly recognized in theory and practice. From a decision-maker's perspective, it becomes important to consider and accommodate the interdependencies and interactions of human decisions. As a result, game theory has emerged as a fundamental instrument in pure and applied research. In addition, since human beings live over time, most of the strategic interactions are dynamic rather than static. One particularly complex and fruitful branch of game theory is dynamic games, which investigates interactive decision-making over time. To establish a practical theory for analysis of the dynamic global political economy, the essential characteristics of (i) a vast number of participating nations, (ii) abundant behavioral choices, (iii) the actions of the nations affect each other's well-being, (iv) the actions of the nations also affect the conditions/states of the economy, (v) a complex dynamic interactive evolution of the states of the economy, (vi) long periods of interactivity, (vii) the effects on the well-being of the nations and the states of the economy can last for some time (or even permanently), (viii) some non-durable strategies that have effects in one single period, and (ix) some durable strategies that have durable effects in multiple periods, we adopt the paradigm of durable-strategies dynamic games established by Petrosyan and Yeung (2020) and Yeung and Petrosyan (2021a, 2021b, 2022).

The organization of this chapter is as follows. Section 1 provides a description of the dynamic GPE. Section 2 presents the mathematical representation of the constituent elements of dynamic GPE and the mathematical solution theory. Section 3 examines policy analysis and the cost of individually rational self-seeking behaviors. Section 4 explicates the theoretical analysis and mathematical theory of uncoordinated outcomes in a dynamic GPE through the study of environmental degradation. A computational illustration is given in Section 5 and chapter notes are given in Section 6.

1. Description of the Dynamic GPE

Our world is a dynamic entity. Most real-life cross-border economic and political activities have certain common features. These include a vast number of nations involved, abundant behavioral choices, a long period of interactivity, actions producing durable effects, and behaviors of one side influencing the outcome of all and vice versa. Furthermore, some behaviors may have effects projecting into the future, thus forming a complex dynamic process of change. Characterizing the nations' behaviors in the dynamic GPE with a realistic and tractable framework is an important part of the formulation and analysis of effective policy. In this section, we describe the major traits of the dynamic GPE.

1.1. *The state of the economy and its evolution*

The world is dynamic. It keeps "evolving", in technical terms, on a continual and often nonlinear trajectory over time. Therefore, if we were to describe what the economy of a specific country is like at a specific point in time, or similarly how the world economy is like at a specific point in time, we may refer to these descriptions as the "state of the national economy" or the "state of the global economy". Thus, the state of a national or global economy can be represented as an array of variables within the radar of policymakers. These are the key independent and dependent variables they include in their working models. Some of the common state variables include the following:

Social variables	Public health conditions Medical infrastructures Schools, universities, and education institutes Recreational facilities Transportation networks
Economic measures	GDP growth Inflation Employment level Aggregate demand Income distribution Labor supply

	Industrialization Innovation Level of technology Level of public and private capitals Foreign direct investment Trade deficit/surplus Degree of currency convertibility Exchange rate Export/import Government trade policies Trade judicial system Tariff barrier
Political measures	Citizen participation Power distribution in society Freedom and liberty Social and economic justice Level of national defense Level of regional security Security from terrorism Cyber-security Crime rate
Environmental measures	Air and water pollution Green technology Clean water supply Sustainability Biodiversity Energy security Environmental and resources security Ecosystems

1.2. *Target controls/strategies*

The strategies or target controls are instruments that a nation can manipulate directly or indirectly. Strategies can be non-durable (which have effects in one period) or durable (which have effects lasting over some periods of time). Durable strategies that have effects lasting over an extended period of time are prevalent in real-life situations. Revenue-generating investments, toxic waste disposal, durable goods, emission of pollutants, regulatory measures, coalition agreements, diffusion of knowledge, advertisement, and investments to build up physical capital are vivid examples of durable controls.

In reality, the existence of durable strategies in the decision-maker's controls is likely to be a rule rather than an exception. In many practical situations, both the decision-makers' payoff and the evolution of the state dynamics can be subjected to effects arising from durable strategies. For instance, durable goods, revenue-generating investments, toxic waste disposal, and long-term effects of business contracts and regulations affect the payoff of the decision-maker. Tariffs have delayed impacts on the economy's growth dynamics, knowledge diffusion takes time in building up knowledge-based capital, conversion of investments into physical capital stock involves a certain number of periods, and advertisement enhances the firm's reputation gradually over time.

There may exist hundreds of target controls. Some of the common target controls include the following:

Non-durable target controls	Factor inputs in industrial production Factor inputs in public goods Factor inputs in government services Food production Production of non-durable goods and services
Durable target controls	Production of durable goods Investments in physical capital Investments in knowledge-based capital Investments in scientific research Investments in education Investments in information technology Investments in ecosystem restoration Financial investments Spending on national defense Outputs of polluting products Ecosystem damaging activities Deforestation Waste disposal Toxic waste management Food production Genetic engineering endeavors Destruction of marine habitats Destruction of animal habitats Irrigations Constructions Electricity generation Nuclear power generation Resource extraction Fishery Emission of greenhouse gas

1.3. *Evolution of the state*

The change in the state variables from one period to the succeeding period depends on the durable and non-durable target controls executed in that period, the durable target controls executed in periods before which are still effective, and the current state variables. For example, the change in the pollution from one period to the succeeding period depends on the polluting outputs produced in that period, the polluting outputs (which have durable effects) produced before, and the level of current pollution (which affects the amount of pollution decayed due to the natural rate of decay).

1.4. *The payoffs of the nations*

The payoffs of nations depend on the durable and non-durable target controls and the states. In each period, the nations' single-period gains are related to the durable and non-durable target controls executed in that period, the durable target controls executed in periods before which are still effective, and the current state variables. Moreover, the nations' payoffs consist of multiple objectives.

1.5. *Government policy instruments*

The government would use some policy instruments to reach the preferred levels of target controls. The commonly adopted ones include imposing taxes, incentive subsidies, direct government production, legal restrictions, public goods provision, and investments.

Policy instruments may be broadly classified into the following categories:

Economic incentive driven	Taxes Levies Subsidies Fines and penalties
Regulation-based	Limits Rationings Activity bans Legislative and regulation measures Constitutional amendments

Direct government provisions	Infrastructure Public goods Government investments Public spending Public medical care/education Pollution abatement Ecosystem rehabilitation
Others	Moral suasion NGO formation Public–private partnership Welfare payments Green technology R&D

2. Game Formulation and Mathematical Solution Theory

The self-interest-seeking behavior of the nations in a dynamic GPE is indeed a durable-strategies dynamic game currently going on. Jevons (1862) argued that economics is a science concerned with quantities and is necessarily mathematical. We can analyze the problem under the forms of mathematical laws in their theoretic perfection and complication, although in practice we have to be contented with approximate and empirical laws. In this section, we first present the constituent elements in the global economic arena in mathematical notation. Then, we formulate the dynamic GPE as a durable-strategies dynamic game mathematically and provide a mathematical solution theorem to characterize the uncoordinated outcome.

2.1. *Mathematical representation of constituent elements*

We first characterize mathematically the constituent elements of a global economic arena involving n nations and a planning horizon of T periods. In particular, T is no less than two and can be very large:

(i) The set of non-durable target controls

The set of non-durable target controls of all the n national governments is denoted by $\underline{u}_k$. It includes the non-durable target controls of nation 1, nation 2, to nation n. For notational convenience, we express $\underline{u}_k = (u_k^1, u_k^2, \ldots, u_k^n)$ for $k \in \{1, 2, \ldots, T\}$, where u_k^i is the vector of non-durable target controls of nation i in period k.

(ii) The set of durable target controls

The set of durable target controls of all the n national governments is denoted by $\underline{\bar{u}}_k$. It includes the durable target control sets of nation 1, nation 2, to nation n. For notational convenience, we express $\underline{\bar{u}}_k = (\bar{u}_k^1, \bar{u}_k^2, \ldots, \bar{u}_k^n)$ for $k \in \{1, 2, \ldots, T\}$, where $\bar{u}_k^i$ is the vector of durable targets controls of nation i in period k.

(iii) The set of effective durable target controls previously executed

The set of durable target controls of all nations executed before period k but still have effects is denoted by $\underline{\bar{u}}_{k-}$. We express $\underline{\bar{u}}_{k-} = (\bar{u}_{k-}^1, \bar{u}_{k-}^2, \ldots, \bar{u}_{k-}^n)$, for $k \in \{1, 2, \ldots, T\}$, where $\bar{u}_{k-}^i$ is the vector of durable targets controls of nation i executed before period k but still have effects.

(iv) State variables and evolution dynamics

The vector of state variables of the economy at period k is denoted by $x_k \subset X$. The evolution dynamics of the vector of state variables are affected by $\underline{u}_k$, $\underline{\bar{u}}_k$, $\underline{\bar{u}}_{k-}$, and x_k. We denote the evolution dynamics by the vector difference equation:

$$x_{k+1} = f(x_k, \underline{u}_k, \underline{\bar{u}}_k; \underline{\bar{u}}_{k-}), \quad x_1 = x_1^0.$$

(v) Single-period gain of nation i in period k

The gain of nation i in period k depends on $(\underline{u}_k, \underline{\bar{u}}_k, \underline{\bar{u}}_{k-}, x_k)$ and is expressed as $\sum_{\varsigma=1}^{\Theta_i} w_k^{i(\varsigma)} g_k^{i(\varsigma)}(x_k, \underline{u}_k, \underline{\bar{u}}_k; \underline{\bar{u}}_{k-})$ if nation i has Θ_i objectives, where $g_k^{i(\varsigma)}(x_k, \underline{u}_k, \underline{\bar{u}}_k; \underline{\bar{u}}_{k-})$ is the ςth objective and $w_k^{i(\varsigma)}$ is the corresponding monetary weight. If $\Theta_i = 1$, the gain of nation i in period k reduces to $w_k^{i(1)} g_k^{i(1)}(x_k, \underline{u}_k, \underline{\bar{u}}_k; \underline{\bar{u}}_{k-})$. The terminal payoff of nation i at period $T+1$ depends on $\underline{\bar{u}}_{(T+1)-}$ and x_{T+1}, and is expressed as $q_{T+1}^i(x_{T+1}, \underline{\bar{u}}_{(T+1)-})$.

(vi) The payoffs of the nations

The payoff nation $i \in \{1, 2, \ldots, n\}$ over the period 1 to period T can be expressed as

$$\sum_{k=1}^{T} \sum_{\varsigma=1}^{\Theta_i} w_k^{i(\varsigma)} g_k^{i(\varsigma)}(x_k, \underline{u}_k, \bar{\underline{u}}_k; \bar{\underline{u}}_{k-})\delta_1^k + q_{T+1}^i(x_{T+1}; \bar{\underline{u}}_{(T+1)-})\delta_1^{T+1}.$$

Note that the influences of durable and non-durable strategies, previously executed durable strategies, and the current state of the nation's objectives and terminal payoffs are known realities. The influence on the evolution of the state by durable and non-durable strategies and previously executed durable strategies, and the current state are commonly observed. The relational contents exhibited in the system indeed reflect real-life phenomena in the global political economy.

2.2. *Mathematical solution theory*

The above scenario of a global political and economic arena involving n nations/participants can be modeled as a durable-strategies dynamic game (see Yeung and Petrosyan, 2022) as follows:

$$\max_{\substack{u_k^i, \bar{u}_k^i \\ k\in\{1,2,\ldots,T\}}} \left\{ \sum_{k=1}^{T} \sum_{\varsigma=1}^{\Theta_i} w_k^{i(\varsigma)} g_k^{i(\varsigma)}(x_k, \underline{u}_k, \bar{\underline{u}}_k; \bar{\underline{u}}_{k-})\delta_1^k \right.$$
$$\left. + \, q_{T+1}^i(x_{T+1}; \bar{\underline{u}}_{(T+1)-})\delta_1^{T+1} \right\}, \tag{2.1}$$

subject to

$$x_{k+1} = f_k(x_k, \underline{u}_k, \bar{\underline{u}}_k; \bar{\underline{u}}_{k-}), \quad x_1 = x_1^0, \text{ for } i \in \{1, 2, \ldots, n\} \equiv N. \tag{2.2}$$

In addition, we assume that the functions $g_k^{i(\varsigma)}(x_k, \underline{u}_k, \bar{\underline{u}}_k; \bar{\underline{u}}_{k-})$ and $f(x_k, \underline{u}_k, \bar{\underline{u}}_k; \bar{\underline{u}}_{k-})$ are differentiable functions in $(x_k, \underline{u}_k, \bar{\underline{u}}_k; \bar{\underline{u}}_{k-})$, and $q_{T+1}^i(x_{T+1}; \bar{\underline{u}}_{(T+1)-})$ is differentiable in $(x_{T+1}; \bar{\underline{u}}_{(T+1)-})$.

To derive a mathematical theory for solving the durable-strategies dynamic game (2.1)–(2.2), we first provide a mathematical theorem that solves the single-player dynamic optimization problem with durable controls:

$$\max_{\substack{u_k^i, \bar{u}_k^i \\ k \in \{1,2,\ldots,T\}}} \left\{ \sum_{k=1}^{T} \sum_{\varsigma=1}^{\Theta} w_k^{(\varsigma)} g_k^{(\varsigma)}(x_k, u_k, \bar{u}_k; \bar{u}_{k-}) \delta_1^k \right.$$
$$\left. + \, q_{T+1}(x_{T+1}; \bar{u}_{(T+1)-}) \delta_1^{T+1} \right\}, \tag{2.3}$$

subject to

$$x_{k+1} = f_k(x_k, u_k, \bar{u}_k; \bar{u}_{k-}), \quad x_1 = x_1^0, \text{ for } i \in \{1, 2, \ldots, n\} \equiv N. \tag{2.4}$$

The solution of the durable-strategies dynamic optimization problem (2.3)–(2.4) can be characterized by a theorem from Yeung and Petrosyan (2022) as follows:

Theorem 2.1. *Let $V(k, x; \bar{u}_{k-})$ be the maximal value of the payoff*

$$\sum_{k=1}^{T} \sum_{\varsigma=1}^{\Theta} w_k^{(\varsigma)} g_k^{(\varsigma)}(x_k, u_k, \bar{u}_k; \bar{u}_{k-}) \delta_1^k + q_{T+1}(x_{T+1}; \bar{u}_{(T+1)-}) \delta_1^{T+1}$$

of for dynamic optimization problem (2.3)–(2.4) starting at period k with state $x_k = x$ and previously executed controls $\bar{u}_{k-}$, then the function $V(k, x; \bar{u}_{k-})$ satisfies the following system of recursive equations:

$$V(T+1, x; \bar{u}_{(T+1)-}) = q_{T+1}(x; \bar{u}_{(T+1)-}) \delta_1^{T+1}, \tag{2.5}$$

$$V(k, x; \bar{u}_{k-}) = \max_{u_k, \bar{u}_k} \left\{ \sum_{\varsigma=1}^{\Theta} w_k^{(\varsigma)} g_k^{(\varsigma)}(x, u_k, \bar{u}_k; \bar{u}_{k-}) \delta_1^k \right.$$
$$\left. + V\left[k+1, f_k(x, u_k, \bar{u}_k; \bar{u}_{k-}); \bar{u}_k, \bar{u}_{(k+1)-} \cap \bar{u}_{k-}\right] \right\}, \tag{2.6}$$

for $k \in \{1, 2, \ldots, T\}$ and $i \in N$.

Proof. See the proof of Theorem I of Mathematical Solution Mechanisms in Chapter 7. □

The solution of the durable-strategies dynamic game (2.1)–(2.2) can be characterized by a theorem from Yeung and Petrosyan (2022) as follows:

Theorem 2.2 (Petrosyan and Yeung). *Let* $(\underline{u}_k^{**}, \bar{\underline{u}}_k^{**})$ *be the set of feedback Nash equilibrium target controls, and* $V^i(k, x; \bar{\underline{u}}_{k-}^{**})$ *be the feedback Nash equilibrium payoff of player* i *at period* k *in the non-cooperative dynamic game* (2.1)–(2.2), *then the function* $V^i(k, x; \bar{\underline{u}}_{k-}^{**})$ *satisfies the following recursive equations:*

$$V^i(T+1, x; \bar{\underline{u}}_{(T+1)-}^{**}) = q_{T+1}^i(x; \bar{\underline{u}}_{(T+1)-}^{**})\delta_1^{T+1}, \tag{2.7}$$

$$\begin{aligned} V^i(k, x; \bar{\underline{u}}_{k-}^{**}) = \max_{u_k^i, \bar{u}_k^i} \Bigg\{ & \sum_{\varsigma=1}^{\Theta_i} w_k^{i(\varsigma)} g_k^{i(\varsigma)} \left(x, u_k^i, \bar{u}_k^i, \underline{u}_k^{**(\neq i)}, \bar{\underline{u}}_k^{**(\neq i)}; \bar{\underline{u}}_{k-}^{**}\right) \delta_1^k \\ & + V^i \Big[k+1, f_k\left(x, u_k^i, \bar{u}_k^i, \underline{u}_k^{**(\neq i)}, \bar{\underline{u}}_k^{**(\neq i)}; \bar{\underline{u}}_{k-}^{**}\right); \\ & \bar{u}_k^i, \bar{\underline{u}}_k^{**(\neq i)}, \bar{\underline{u}}_{(k+1)-}^{**} \cap \bar{\underline{u}}_{k-}^{**} \Big] \Bigg\}, \end{aligned} \tag{2.8}$$

for $k \in \{1, 2, \ldots, T\}$ *and* $i \in N$,

$$\textit{where } \underline{u}_k^{(\neq i)**} = (u_k^{1**}, u_k^{2**}, \ldots, u_k^{i-1**}, u_k^{i+1**}, \ldots, u_k^{n**}) \textit{ and}$$
$$\bar{\underline{u}}_k^{(\neq i)**} = (\bar{u}_k^{1**}, \bar{u}_k^{2**}, \ldots, \bar{u}_k^{i-1**}, \bar{u}_k^{i+1**}, \ldots, \bar{u}_k^{n**}).$$

Proof. Conditions (2.7)–(2.8) show that $V^i(k, x; \bar{\underline{u}}_{k-}^{**})$ is the maximized payoff of player $i \in N$ according to Theorem 2.1. given the game equilibrium strategies of the other $n-1$ players. Hence, a Nash equilibrium results. □

The characterization of the game equilibrium of the game in Theorem 2.2 allows one to derive the game equilibrium target controls $(\underline{u}_k^{**}, \bar{\underline{u}}_k^{**})$.

System (2.7)–(2.8) can be regarded as an advanced form of the Hamilton–Jacobi–Bellman equations for durable-strategies dynamic games. Note about the novelty that the previously executed target controls $\underline{\bar{u}}_{k-}^{**}$ appear on both sides of equations in (2.3)–(2.4). For the sake of clarity, one may call it Hamilton–Jacobi–Bellman–Petrosyan–Yeung equations.

Worth noting is that this class of games cannot be handled by the standard approach of dynamic programming. The definitive contributions of this game theory paradigm include the following:

First, durable-strategies dynamic games provide a novel and realistic dynamic game paradigm game framework that allows the presence of durable strategies. Given that durable strategies would be involved in most of real-life situations, the paradigm yields a better framework in characterizing interactive decision-making over time.

Second, the game paradigm establishes novel mathematical results. A dynamic optimization theorem involving durable strategies is supplied as the essential foundation of corresponding game theory analysis. The derived dynamic optimization technique predominates the classic dynamic programming by providing a more comprehensive solution mechanism. In addition, a novel system of Hamilton–Jacobi–Bellman–Petrosyan–Yeung equations is presented to characterize the feedback Nash equilibrium solution.

Finally, novel results, which cannot be generated by any of the existing dynamic games, can be obtained. For an example application given in Petrosyan and Yeung (2020), the emission of environmentally damaging industrial waste (like toxic and chemical wastes and deforestation) could continue for many years, hence it is possible that pollution would persist or exacerbate even after the operation stops completely. A series of new results are given in Chapters 3–9.

Note that in case where there are no durable strategies, Theorem 2.1 will be reduced to the standard result for dynamic games without durable strategies:

$$\max_{\substack{u_k^i, \bar{u}_k^i \\ k \in \{1,2,\ldots,T\}}} \left\{ \sum_{k=1}^{T} \sum_{\varsigma=1}^{\Theta_i} w_k^{i(\varsigma)} g_k^{i(\varsigma)}(x_k, \underline{u}_k)\delta_1^k + q_{T+1}^i(x_{T+1})\delta_1^{T+1} \right\}, \quad (2.9)$$

subject to

$$x_{k+1} = f_k(x_k, \underline{u}_k), \quad x_1 = x_1^0, \text{ for } i \in \{1, 2, \ldots, n\} \equiv N. \quad (2.10)$$

Theorem 2.3. *Let* $(\underline{u}_k^{**})$ *be the set of feedback Nash equilibrium target controls, and* $V^i(k, x)$ *be the feedback Nash equilibrium payoff of player* i *at period* k *in the non-cooperative dynamic game*

(2.9)–(2.10), *then the function* $V^i(k, x)$ *satisfies the following recursive equations:*

$$V^i(T+1, x) = q^i_{T+1}(x)\delta_1^{T+1}, \tag{2.11}$$

$$V^i(k, x; \underline{u}^{**}_{k-}) = \max_{\substack{u^i_k, \bar{u}^i_k \\ k\in\{1,2,\ldots,T\}}} \left\{ \sum_{\varsigma=1}^{\Theta_i} \bar{w}_k^{i(\varsigma)} g_k^{i(\varsigma)}\left(x, u^i_k, \underline{u}_k^{**(\neq i)}\right)\delta_1^k + V^i\Big[k+1, f_k\Big(x, u^i_k, \underline{u}_k^{**(\neq i)}\Big)\Big]\right\}, \tag{2.12}$$

for $k \in \{1, 2, \ldots, T\}$ *and* $i \in N$, *where* $\underline{u}_k^{(\neq i)**} = (u_k^{1**}, u_k^{2**}, \ldots, u_k^{i-1**}, u_k^{i+1**}, \ldots, u_k^{n**})$.

Remark 2.1. In the case where the effects of the political economy are confined to one nation only, Theorem 2.2 will be replaced by Theorem 2.1 in which the number of nations is $n = 1$.

In the case where the planning horizon is one period, $T = 1$ will appear in Theorems 2.1, 2.2, and 2.3.

3. Policy Analysis and Cost of Individually Rational Self-Seeking

The idea of policy analysis and formulation contains several approaches. The literature typically features two most commonly adopted approaches. The technically oriented approach conducts rigorous technical analysis, finding the optimal solution to complicated political problems. The politically oriented approach analyzes the problem as gaining support for a policy through the cumbersome legislative process. The former casts policy formulation in terms of rationality; the latter in terms of compromise and majority-building. In our analysis, we focus primarily on the technically oriented approach. However, the effects of the politically oriented approach may appear in the accepted valuation of the monetary weights attached to the political objectives. Policy Formulation is the process of arriving at the policy targets by organizing policy resources and supporting information to convert ideas into solutions that are practically implementable to deal with certain activities or address challenges evolved therewith and which are in the direction of the policy objectives. The equilibrium target controls are guiding benchmarks for policy analysis. In addition, policy instruments have to be implemented by nation i to achieve the target controls $(\underline{u}_k^{i**}, \bar{u}_k^{i**})$. The policy instruments listed in Section 2.5 are used to attain the target controls.

In the uncoordinated dynamic setting, each nation aims to maximize its national payoff in (2.1) subject to the state evolution dynamics (2.2) under the influences of the actions of all nations. The game equilibrium target controls of nation i are $(\underline{u}_k^{i**}, \bar{u}_k^{i**})$, which indicate what it can do to maximize its payoffs, given the maximizing behavior of the other nations. The negative externalities of a nation's controls were not taken into consideration. The negative externalities could have long-lasting effects from durable controls. The damaging costs of self-interest-seeking behaviors of participants in a game situation are well known. It has been shown in the cases like "Prisoner's Dilemma" and "Price of Anarchy". Consider the case of global environmental degradation. There are over a hundred polluting nations with harmful durable actions/strategies that would continue to worsen the environment even if such actions stop completely. Worth noting is that the resultant evolution of environment-damaging states, like pollution, toxic wastes, and degradation of the ecosystem are subject not only to current durable and non-durable polluting target controls but also to previously executed durable target controls, which is shown in their dynamic equations. Finally, the game equilibrium objective of nation i, $g_k^{i(\varsigma)}(x_k^{**}, \underline{u}_k^{**}, \bar{u}_k^{**}; \underline{\bar{u}}_{k-}^{**})$, is not completely under the control of the nation i.

4. An Explication in Environmental Degradation

In this section, we explicate the theoretical analysis and mathematical theory of uncoordinated outcomes in a dynamic GPE through the study of environmental degradation.

4.1. *Environmental degradation*

Damages to ecosystems and the global environment, in general, have increasingly become major concerns. Anthropogenic impacts on the environment are the prime contributors to our catastrophe-bound environmental problems. The term anthropogenic designates an effect or object resulting from human activity. Anthropogenic environmental impacts are much more than just pollution accumulation, global warming, or climate change. They also include changes to biophysical environments and ecosystems, depletion and destruction of natural resources, environmental degradation, acidification, massive species extinction, biodiversity loss, toxic soil and water, ecological crises, and even ecological collapse. Altering the environment in the process of satisfying the needs of society can have detrimental effects on the survival of human beings. Recent rapid technological advancement and economic growth have led to an unprecedented level of increment in environmental damages brought about by human deeds through industrial pollution and ecosystem degradation. Due to the geographical diffusion of harmful anthropogenic environmental impacts, a unilateral response on the part of one nation or region is often ineffective. Major international environmental management initiatives, like the Paris Agreement, which focus mainly on emissions reduction and permit trading can hardly be expected to provide an effective means to solve the problem.

Natural ecosystems are made up of abiotic factors and biotic factors. They contain abiotic resources such as air, water, soil, rocks, and energy and interacting life forms of individual organisms such as plants, animals, and micro-organisms that aggregate into populations which aggregate into ecological communities. Ecosystems are dynamic and changing over time. For an ecosystem to be stable and sustainable, it needs to be in equilibrium. Disruptions that cause a shift from one equilibrium to another can be reversible or nonreversible. In fact, our ability to exist continuously requires the coexistence of the biosphere and human civilization. We need a balanced environment in which the exploitation of resources, the direction of investments, the orientation of technological development, and institutional change are all in harmony ecologically so that the current and future potentials to meet human needs and aspirations can be maintained.

As the human population and economic activities are growing rapidly with human beings increasingly dominating available resources, ecosystems begin to show increasing signs of stress. Loss of biodiversity, environmental degradation, and conflict over resources among the dominant species are typical signs that a biological system is nearing a state of change, which could range from the collapse of the dominant species, and the development of alternative biological communities, to the collapse of the entire system. Some changes can be reversed given adequate time for recovery. Some changes are irreversible and their impacts on our ecosystem can be permanent and potentially devastating.

Pollution is one of the most damaging elements to ecosystems and biodiversity. The two most widely spread forms of pollution are air pollution and water pollution. Air pollution is the release of pollutants like chemicals and particulates into the atmosphere. Common gaseous pollutants include carbon monoxide, sulfur dioxide, chlorofluorocarbons (CFCs), and nitrogen oxides produced by industry and motor vehicles. Photochemical ozone and smog are created as nitrogen oxides and hydrocarbons react to sunlight. Water pollution, by the discharge of commercial and industrial wastes (intentionally or unintentionally through spills in accidents) into surface waters. Discharges of untreated domestic sewage, chemical contaminants, chemical fertilizers, pesticides, and human feces into surface runoff flowing to surface waters constitute serious cases of water pollution. The dumping of toxic materials, radioactive water, and plastics into oceans has become a recent source of water pollution. Pollution is not just the emission of harmful pollutants; it also damages the entire ecosystem by reducing its capacity to regenerate. Ultimately, pollution affects the ability of ecosystems to provide ecosystem services, such as nutrient cycling, carbon cycling, and clean water provision, on which the earth and human life are dependent. Millions of deaths are due to pollution and tens of millions of people's health conditions are affected every year.

4.2. *Damaging anthropogenic impacts*

The degradation of ecosystems is also affected by many anthropogenic factors. Damaging anthropogenic impacts on the environment come from six major categories of sources:

The first category of anthropogenic impacts is related to agriculture, fishery, and irrigation, which include the following:

(i) The environmental impacts on agriculture involve a variety of factors ranging from the soil to water, the air, animal and soil diversity, plants, and the food supply itself. Deforestation, irrigation problems, pollutants, soil degradation, genetic engineering, and waste are also damages brought about by agriculture.

(ii) The environmental impacts on fishery include the decrease or demise of biomass, especially through overfishing, unsustainable fisheries, and fisheries mismanagement. The disruption of the marine food chain and negative impact on other elements of the environment, for instance, the destruction of marine habitats such as coral reefs (see Oppenlander, 2013), are also often observed.

(iii) Irrigation projects can have negative side effects that are often overlooked (see Thakkar, 1999). The massive depletion of freshwater severely damages surrounding ecosystems and contributes to the wipe-out of aquatic species (see Pearce, 2006).

The second category of anthropogenic impacts is related to the energy industry, which includes the following:

(i) The environmental impact of coal combustion and coal mining is serious and diverse. Coal contains sulfur and other dangerous metals that are emitted into the air when coal is burned.

(ii) The environmental impact of electricity generation is significant because of the heavy use of electricity in modern society. Although electricity is a relatively safe form of energy, the generation and transmission of electricity affect the environment. Almost all mechanisms for generating electricity release carbon dioxide and other greenhouse gases.

(iii) The environmental impact of nuclear power stems from nuclear fuel cycle processes like mining, processing, transporting, and storing fuel and radioactive fuel waste. The release of radioisotopes poses a health danger to human populations, animals, and plants. The environmental impacts of nuclear plant accidents persist for an extremely long time or even indefinitely.

(iv) The environmental impact of petroleum is highly undesirable because its use and extraction are toxic to almost all forms of life.

The third category of anthropogenic impacts is related to manufactured products, which include the following:

(i) Substantial damaging environmental impacts from manufactured goods like cleaning agents, nano-technological products, paints, paper, plastics, pesticides, electronic products, cars, vehicles, and buildings.
(ii) Significant damages to the ecosystems and deletion of natural resources brought about by high levels of consumption, resources utilized, pollutants emitted, and waste generated cause significant damages to the ecosystems.

The fourth category of anthropogenic impacts is related to mining and transportation, which include the following:

(i) The environmental impact of mining can be devastating because some mining methods have significant negative environmental and public health effects. Soil erosion, formation of sinkholes, loss of biodiversity, and contamination of soil and water by chemicals from mining processes create serious environmental damages.
(ii) The environmental impact of transport is significant because it is a major user of energy and consumes most of the world's petroleum. With faster growth in the number of automobiles around the world, the emission of carbon dioxide in the transportation sector has become one of the largest contributors to global warming.
(iii) The environmental impact of aviation comes from noise, particulates, and gases emitted by aircraft engines. The rapid growth of air travel in recent years contributed to an increase in total pollution attributable to aviation.
(iv) The environmental impact of shipping comes mainly from greenhouse gas emissions and oil spills. Invasive species were brought into new areas through shipping, usually by attaching themselves to the ship's hull.

The fifth category of anthropogenic impacts is overconsumption and over-extraction, which include the following:

(i) A situation where resource use has outpaced the sustainable capacity of the ecosystem.
(ii) Prolonged over-extraction leads to environmental degradation and the eventual loss of resource bases and even irreversible damages to the ecosystem.

The sixth category of anthropogenic impacts is waste disposal, war, and military activities, which include the following:

(i) Waste disposal contributes significantly to the degradation and contamination. The disposal of solid waste, electronic appliances, chemical waste, pharmaceuticals, radioactive materials, plastics, oil products, and toxic waste poses a danger to the environment.
(ii) There are significant environmental impacts from war and military activities like nuclear bomb testing. Scorched earth tactics during or after war are not uncommon in traditional warfare throughout history. And modern technological warfare, with the use of missiles and chemical weapons, can cause even far greater devastation effect.

Worthwhile to point out is that anthropogenic activities affecting pollution and/or the ecosystems cause durable effects. Deforestation by lumber firms would cause increases in carbon dioxide for a certain period of time. Toxic and chemical wastes from industrial production will continue to produce damaging elements that add to the pollution stock for a lengthy period. In addition, anthropogenic activities are carried out by many parties and externalities are prevalent.

4.3. *Modeling the uncoordinated environmental problem*

In this section, we explicate the mathematical theory in Section 2 using the framework of the scenario of global environmental degradation.

4.3.1. *The set of target controls*

The set of non-durable target controls $\underline{u}_k$ includes food production, manufactured products that cause one-shot environmental damages or no environmental damages, resources' extraction, clean transportation, clean productions, government spending, taxes, subsidies, pollution abatement activities, environmental improvement activities, and policy instruments.

The set of durable target controls $\bar{\underline{u}}_k$ includes deforestation, irrigation, genetic engineering, waste disposal, fishery, coal mining and use, electricity generation, nuclear power production, petroleum extraction and use, resources extraction, transportation by land-sea-air, energy consumption, greenhouse gas emissions, oil spills, toxic waste, loss of bio-diversification, over-farming, disposal of solid waste, chemical waste, radioactive materials, plastics waste, and investment in green technology. The set of durable target controls executed before period k but still have effects $\underline{\bar{u}}_{k-}$ but still have effects in period k.

4.3.2. *State variables and evolution dynamics*

Major state variables include air and water pollution, the condition of the ecosystems, the level of green technology, the level of biodiversity, government debt, and public capital. In particular, we can express the vector of states as $x_k = (x_k^1, x_k^2, \ldots, x_k^m)$.

The evolution dynamics of the state can be expressed as

$$\begin{pmatrix} x_{k+1}^1 = f_k^1(x_k, \underline{u}_k, \bar{u}_k; \underline{\bar{u}}_{k-}) \\ x_{k+1}^2 = f_k^2(x_k, \underline{u}_k, \bar{u}_k; \underline{\bar{u}}_{k-}) \\ \vdots \\ x_{k+1}^m = f_k^m(x_k, \underline{u}_k, \bar{u}_k; \underline{\bar{u}}_{k-}) \end{pmatrix}.$$

(i) Evolution of pollutions

The major pollution states include air pollution and water pollution. The evolution of these systems depends on non-durable and durable target controls that generate air pollution and water pollution, previously executed pollution-generating durable target controls that still have effects, the current level of pollution, and pollution abatement activities. In particular, the level of pollution will be increased by non-durable and durable target controls that generate air pollution and water pollution. Previously executed generate pollution durable target controls that still have effects generate pollution. The current level of pollution stock would also affect the increment of the pollution stocks through their natural decay. Pollution abatement activities would reduce pollution.

(ii) Evolution of the condition of the ecosystems

The evolution of the condition of the ecosystem depends on non-durable and durable target controls that damage the ecosystem, previously executed ecosystem damaging durable target controls that still have effects, the current condition of the ecosystem and ecosystem improving activities, and the self-regenerating rate of the system. In particular, the condition of the ecosystem will be worsened by non-durable and durable target controls that damage the ecosystem, previously executed ecosystem damaging durable target controls that still have effects. Ecosystem-improving activities and the self-regenerating rate of the system would enhance the condition of the ecosystem.

(iii) Evolution of the level of green technology

The evolution of the level of green technology depends on non-durable and durable target investments that advance the technology, previous technology investments that still have effects, and the current technology level, in particular, non-durable and durable target investments that advance the technology, and previous technology investments that still have effects enhance the level of green technology. The deterioration of green technology depends on the rate of obsolescence and the current technology level.

(iv) Evolution of the level of biodiversity

The evolution of the level of biodiversity depends on non-durable and durable target improvement activities on biodiversity and previous improvement activities that still have effects, environment-damaging production activities (with durable and non-durable effects), the current levels of biodiversity, and the current level of pollution. In particular, non-durable and durable target improvement activities on biodiversity and previous improvement activities that still have effects, and the current level of biodiversity contribute to protecting biodiversity. The level of pollution and environment-damaging production activities damage biodiversity.

(v) Accumulation of government debt

Government debts would increase due to spending on environmental improvements which include pollution abatement activities, green technology development, biodiversity, and ecosystem improvements. These activities may have durable and non-durable effects.

(vi) Building up of public capital

The building up of public capital like parks, re-built forests, and water purifying facilities is affected by investments (both durable and non-durable) in these public facilities.

4.3.3. *Government objectives and payoffs*

The list of objectives of the government in managing the environment includes (i) air quality improvement, (ii) improving waste management, (iii) minimizing toxic exposure, (iv) good water quality, (v) reducing noise pollution, (vi) resource conservation, (vii) efficiency in energy use, (viii) sustainable ecosystem, (ix) biodiversity, (x) maintaining employment and income, (xi) government revenue, and (xii) economic growth. In particular, achieving these objectives would involve using the set of target controls in Section 4.3.1 and being influenced by the state variables in Section 4.3.2. Attaching the corresponding monetary weight $w_k^{i(\varsigma)}$, the gain of nation i in period k can be expressed as $\sum_{\varsigma=1}^{\Theta_i} w_k^{i(\varsigma)} g_k^{i(\varsigma)}(x_k, \underline{u}_k, \bar{u}_k; \bar{u}_{k-})$.

The monetary weights can be positive or negative. For instance, air quality, waste management, water quality, resource conservation, efficiency in energy use, sustainable ecosystem, biodiversity, employment, and income entail a positive monetary weight. Toxic exposure, different kinds of pollution, a decrease in government revenue, and an increase in related costs would entail a negative monetary weight.

4.3.4. *Characterization of uncoordinated environmental management*

The uncoordinated environmental management by the n nations can be characterized as a durable-strategies dynamic game in which each nation intends to maximize its payoff

$$\sum_{k=1}^{T}\sum_{\varsigma=1}^{\Theta_i} w_k^{i(\varsigma)} g_k^{i(\varsigma)}(x_k, \underline{u}_k, \bar{u}_k; \bar{u}_{k-})\delta_1^k + q_{T+1}^i(x_{T+1}; \bar{u}_{(T+1)-})\delta_1^{T+1}, \tag{4.1}$$

subject to

$$x_{k+1} = f_k(x_k, \underline{u}_k, \bar{u}_k; \bar{u}_{k-}), \quad x_1 = x_1^0, \text{ for } i \in \{1, 2, \ldots, n\} \equiv N, \tag{4.2}$$

with their target controls.

The uncoordinated outcome of the above environmental management situation is represented by the equilibrium of the game (4.1)–(4.2). Theorem 2.1 characterizes the feedback Nash equilibrium target controls $(\underline{u}_k^{**}, \bar{\underline{u}}_k^{**})$ and the feedback Nash equilibrium payoff of nation i as $V^i(k, x; \bar{\underline{u}}_{k-}^{**})$.

The damaging costs of self-interest-seeking behaviors of participants in a game situation are well known. It has been shown in cases like, "Prisoner's Dilemma" and "Price of Anarchy". Consider the case of global environmental degradation. There are over a hundred polluting nations with harmful durable actions/strategies that would continue to worsen the environment even if such actions stop completely.

Worth noting is that the resultant evolution of environment-damaging states, like pollution, toxic wastes, and degradation of the ecosystem, is subject not only to current durable and non-durable target controls of pollution but also to previously executed durable target controls, which is shown in their dynamic equations:

$$x_{k+1} = f_k(x_k, \underline{u}_k^{**}, \bar{\underline{u}}_k^{**}; \bar{\underline{u}}_{k-}^{**}). \tag{4.3}$$

Moreover, the pollution states and ecosystem states are interrelated. The level of pollution damages the condition of the ecosystems and a damaged ecosystem weakens the natural rate of decay of the pollutants. The durable interactions of these state variables could bring about the non-reversibility of the environment back to its original condition. Some of the environmental objectives under uncoordinated actions could reach a highly undesirable status and bring about an immense cost to all the involved nations. The damaging environmental states like pollution and degradation of the ecosystem would continue to worsen even if we stop all the polluting and environmental activities. Supporting observations consistent with the theoretical outcome above are emerging. We quote the "Best of the worst-case scenarios" in Bond's (2017) study on "If We Stopped Emitting Greenhouse Gases Right Now, Would We Stop Climate Change?" in the following:

> *"In any event, it's not possible to stop emitting carbon dioxide right now. Despite significant advances in renewable energy sources, total demand for energy accelerates and carbon dioxide emissions increase. .....*
>
> *.... It's hard to say we're on a new path until we see a peak and then a downturn in carbon emissions. With the approximately*

1°C of warming we've already seen, the observed changes are already disturbing.

There are many reasons we need to eliminate our carbon dioxide emissions. The climate is changing rapidly; if that pace is slowed, the affairs of nature and human beings can adapt more readily.

The total amount of change, including sea-level rise, can be limited. The further we get away from the climate that we've known, the more unreliable the guidance from our models and the less likely we will be able to prepare.

It's possible that even as emissions decrease, the carbon dioxide in the atmosphere will continue to increase. The warmer the planet gets, the less carbon dioxide the ocean can absorb.

Rising temperatures in the polar regions make it more likely that carbon dioxide and methane, another greenhouse gas that warms the planet, will be released from storage in the frozen land and ocean reservoirs, adding to the problem.

If we stop our emissions today, we won't go back to the past. The Earth will warm.

And since the response to warming is more warming through feedbacks associated with melting ice and increased atmospheric water vapour, our job becomes one of limiting the warming.

If greenhouse gas emissions are eliminated quickly enough, within a small number of decades, it will keep the warming manageable. It will slow the change – and allow us to adapt.

Rather than trying to recover the past, we need to be thinking about best possible futures."

In addition, the nations' gains, $g_k^i(x_k, \underline{u}_k^{**}, \underline{\bar{u}}_k^{**}; \underline{\bar{u}}_{k-}^{**})$, also receive negative impacts from the current and previously executed environmentally damaging durable strategies from all nations.

5. A Computational Illustration

To illustrate the computational details of the analysis in this chapter, we present an example of an international political encounter involving n nations and four sectors with explicit functional forms.

5.1. *Problem formulation*

Consider T periods and n nations international political encounter and each nation has three types of industrial sectors and a government sector:

(i) a pollution-generating sector in which an output $\bar{q}_k^i$ is produced in period k that would continue to generate greenhouse gas for three periods,
(ii) a green sector in which an output q_k^i is produced in period k without any pollution,
(iii) a green technology R&D sector in which a research output $\bar{w}_k^i$ is produced in period k that would continue to enhance the technology level for two periods,
(iv) a government sector in which a pollution abatement output u_k^i is produced in period k that would reduce the pollution stock.

The pollution stock in period k is x_k, and the dynamics of pollution accumulation is governed by the difference equation:

$$\begin{aligned} x_{k+1} &= x_k + \sum_{i=1}^{n} a_k^i \bar{q}_k^i + \sum_{i=1}^{n} a_k^{(1)i} \bar{q}_{k-1}^i \\ &\quad + \sum_{i=1}^{n} a_k^{(2)i} \bar{q}_{k-2}^i - \sum_{i=1}^{n} b_k^i u_k^i (x_k)^{\frac{1}{2}} - \gamma_k x_k, \\ x_1 &= x_1^0, \end{aligned} \tag{5.1}$$

where $a_k^i \bar{q}_k^i$ is the amount of pollution created by a unit of nation i's output $\bar{q}_k^i$, $a_k^{(1)i} \bar{q}_{k-1}^i$ is the amount of pollution created by a unit of nation i's output $\bar{q}_{k-1}^i$, $a_k^{(2)i} \bar{q}_{k-2}^i$ is the amount of pollution created by a unit of nation i's output $\bar{q}_{k-2}^i$, $b_k^i u_k^i (x_k)^{1/2}$ is the amount of pollution removed by u_k^i units of abatement effort from nation i, and γ_k is the natural rate of decay of the pollutant stock.

The level of green technology in period k is ϑ_k, and the evolution of the level of clean technology is governed by the difference equation:

$$\begin{aligned} \vartheta_{k+1} &= \vartheta_k + \sum_{i=1}^{n} z_k^i \bar{w}_k^i + \sum_{i=1}^{n} z_k^{(1)i} \bar{w}_k^i - \lambda_k \vartheta_k, \\ \vartheta_1 &= \vartheta_1^0, \end{aligned} \tag{5.2}$$

where $z_k^i \bar{w}_k^i$ is the advancement in clean technology brought about by $\bar{w}_k^i$ in period k, $z_k^{(1)i} \bar{w}_{k-1}^i$ is the advancement in clean technology brought about by $\bar{w}_{k-1}^i$ in period $k-1$, and λ_k is the rate of obsolescence of the level of green technology.

Each nation seeks to maximize the sum of the discounted payoffs over the T periods plus the terminal appraisal. In particular, nation $i \in N$ would seek to maximize the objective

$$\sum_{k=1}^{T} [\alpha_k^i \bar{q}_k^i - \beta_k^i (\bar{q}_k^i)^2 - c_k^i \bar{q}_k^i + 2\varpi_k^i (\vartheta_k q_k^i)^{\frac{1}{2}} - \varepsilon_k^i q_k^i + \chi_k^i \bar{w}_k^i - (\bar{w}_k^i)^2$$
$$- c_k^{(a)i} (u_k^i)^2 - h_k^i x_k] \delta^{k-1} + (h_{T+1}^i x_{T+1} + y_{T+1}^i \vartheta_{T+1} + Q_{T+1}^i) \delta^T, \tag{5.3}$$

where $\alpha_k^i \bar{q}_k^i - \beta_k^i (\bar{q}_k^i)^2 - c_k^i \bar{q}_k^i$ is the net economic profit in the polluting industrial sector that generates greenhouse gases, $2\varpi_k^i (\vartheta_k q_k^i)^{1/2} - \varepsilon_k^i q_k^i$ is the net economic profit in the green industrial sector, $\chi_k^i \bar{w}_k^i - (\bar{w}_k^i)^2$ is the net economic profit in the green technology development sector, $c_k^{(a)i} (u_k^i)^2$ is the cost of the government's abatement effort u_k^i, $h_k^i x_k$ is the damage cost of x_k amount of pollution, $(h_{T+1}^i x_{T+1} + y_{T+1}^i \vartheta_{T+1} + Q_{T+1}^i)$ is the terminal payoff of nation i in period $T+1$, and $\delta = 1/(1+r)$ is the discount factor.

The problem of maximizing the nation's objective (5.3) with target controls $\{q_k^i, \bar{q}_k^i, \bar{w}_k^i, u_k^i\}$ subject to pollution dynamics (5.1) and clean technology development dynamics (5.2) is a dynamic game between these n nations.

5.2. *Equilibrium outcome*

We first obtain the game equilibrium payoffs and target controls of the nations in the following proposition.

Proposition 5.1. (1) *The game equilibrium payoffs of the nations are*

$$V^i(k, x, \vartheta; \underline{q}_{k-}, \underline{w}_{k-}) = (A_k^i x + B_k^i \vartheta + C_k^i) \delta^{k-1},$$

where

$$A_k^i = -c_k^{(a)i}\left(\frac{b_k^i}{2c_k^{(a)i}}A_{k+1}^i\delta\right)^2 - h_k^i$$
$$+ A_{k+1}^i\left((1-\gamma_k) + \sum_{j=1}^{n}\frac{(b_k^j)^2}{2c_k^{(a)j}}A_{k+1}^j\delta\right)\delta,$$
$$B_k^i = \frac{(\varpi_k^i)^2}{\varepsilon_k^i} + B_{k+1}^i(1-\lambda_k)\delta,$$

and C_k^i *is a function involving game equilibrium target controls executed before period* k*, namely,* $\bar{q}_{k-1}^j$*,* $\bar{q}_{k-2}^j$*, and* $\bar{w}_{k-1}^j$*, for* $j \in N$ *and* $k \in \{1, 2, \ldots, T-2\}$; $V^i(T-1, x\vartheta; \bar{q}_{(T-1)-}, \bar{w}_{(T-1)-}) = (A_{T-1}^i x_{T-1} + B_{T-1}^i \vartheta_{T-1} + C_{T-1}^i)\delta^{T-2}$*, where*

$$A_{T-1}^i = -c_{T-1}^{(a)i}\left(\frac{b_{T-1}^i}{2c_{T-1}^{(a)i}}A_T^i\delta\right)^2 - h_{T-1}^i$$
$$+ A_T^i\left((1-\gamma_{T-1}) + \sum_{j=1}^{n}\frac{(b_{T-1}^j)^2}{2c_{T-1}^{(a)j}}A_T^j\delta\right)\delta,$$
$$B_{T-1}^i = \frac{(\varpi_{T-1}^i)^2}{\varepsilon_{T-1}^i} + B_T^i(1-\lambda_{T-1})\delta, \tag{5.4}$$

and C_{T-1}^i *is a function involving target controls executed before period* $T-1$*, namely,* $\bar{q}_{T-2}^j$*,* $\bar{q}_{T-3}^j$*, and* $\bar{w}_{T-2}^j$*, for* $j \in N$; $V^i(T, x, \vartheta; \underline{\bar{q}}_{T-}, \underline{\bar{w}}_{T-}) = (A_T^i x_T + B_T^i \vartheta_T + C_T^i)\delta^{T-1}$*, where*

$$A_T^i = -c_T^{(a)i}\left(\frac{b_T^i}{2c_T^{(a)i}}A_{T+1}^i\delta\right)^2 - h_T^i$$
$$+ A_{T+1}^i\left((1-\gamma_T) + \sum_{j=1}^{n}\frac{(b_T^j)^2}{2c_T^{(a)j}}A_{T+1}^j\delta\right)\delta, \tag{5.5}$$

$B_T^i = \frac{(\varpi_T^i)^2}{\varepsilon_T^i} + B_{T+1}^i(1-\lambda_T)\delta$, *and* C_{T-1}^i *is a function involving target controls executed before period* $T-1$*, namely,* $\bar{q}_{T-2}^j$, $\bar{q}_{T-3}^j$, *and* $\bar{w}_{T-2}^j$, *for* $j \in N$.

(2) *The game equilibrium target controls are*

$$\bar{q}_T^{i**} = \frac{\alpha_T^i - c_T^i + A_{T+1}^i \delta a_T^i}{2\beta_T^i}, \quad q_T^{i**} = \frac{(\varpi_T^i)^2 \vartheta_T}{(\varepsilon_T^i)^2},$$

$$u_T^{i**} = -\frac{b_T^i}{2c_T^{(a)i}} A_{T+1}^i \delta (x_T)^{1/2}, \quad \bar{w}_T^{i**} = \frac{\chi_T^i + B_{T+1}^i \delta z_T^i}{2}, \text{ for } i \in N;$$

$$\bar{q}_{T-1}^{i**} = \frac{\alpha_{T-1}^i - c_{T-1}^i + A_T^i \delta a_{T-1}^i + A_{T+1}^i \delta^2 a_{T-1}^{(1)i}}{2\beta_{T-1}^i},$$

$$q_{T-1}^{i**} = \frac{(\varpi_{T-1}^i)^2 \vartheta_{T-1}}{(\varepsilon_{T-1}^i)^2}, \tag{5.6}$$

$$u_{T-1}^{i**} = -\frac{b_{T-1}^i}{2c_{T-1}^{(a)i}} A_T^i \delta (x_{T-1})^{1/2},$$

$$\bar{w}_{T-1}^{i**} = \frac{\chi_{T-1}^i + B_T^i \delta z_{T-1}^i + B_{T+1}^i \delta^2 z_{T-1}^{(1)i}}{2}, \text{ for } i \in N,$$

$$\bar{q}_k^{i**} = \frac{\alpha_k^i - c_k^i + A_{k+1}^i \delta a_k^i + A_{k+2}^i \delta^2 a_k^{(1)i} + A_{k+3}^i \delta^3 a_k^{(2)i}}{2\beta_k^i},$$

$$q_k^{i**} = \frac{(\varpi_k^i)^2 \vartheta_k}{(\varepsilon_k^i)^2},$$

$$u_k^{i**} = -\frac{b_k^i}{2c_k^{(a)i}} A_{k+1}^i \delta (x_k)^{1/2}, \quad \bar{w}_k^{i**} = \frac{\chi_k^i + B_k^i \delta z_k^i + B_{k+2}^i \delta^2 z_k^{(1)i}}{2}, \tag{5.7}$$

for $i \in N$, *and* $k \in \{T-2, T-3, \ldots, 1\}$.

Proof. See Appendix A in the Mathematical Appendices. □

5.3. Policy instruments and implementation

In this section, we consider the nations' policy instruments in attaining the game equilibrium target controls:

(i) The Polluting Industrial Sector

In the pollution-generating sector, the net economic profit is $\alpha_k^i \bar{q}_k^i - \beta_k^i (\bar{q}_k^i)^2 - c_k^i \bar{q}_k^i$. Maximizing economic profit with respect to $\bar{q}_k^i$ yields the first-order condition

$$\alpha_k^i - 2\beta_k^i \bar{q}_k^i - c_k^i = 0,$$

leading to the industry's profit-maximizing output $\bar{q}_k^i = \frac{\alpha_k^i - c_k^i}{2\beta_k^i}$.

However, the game equilibrium target control in the polluting industrial sector is $\bar{q}_k^{i**} = \frac{\alpha_k^i - c_k^i + A_{k+1}^i \delta a_k^i + A_{k+2}^i \delta^2 a_k^{(1)i} + A_{k+3}^i \delta^3 a_k^{(2)i}}{2\beta_k^i}$, for $k \in \{1, 2, \ldots, T-2\}$.

A Pigouvian tax equaling $\frac{A_{k+1}^i \delta a_k^i + A_{k+2}^i \delta^2 a_k^{(1)i} + A_{k+3}^i \delta^3 a_k^{(2)i}}{2\beta_k^i}$ per unit of output has to be levied on the polluting output to achieve the target control in period $k \in \{1, 2, \ldots, T-2\}$.

Similarly, a Pigouvian tax $\frac{A_T^i \delta a_{T-1}^i + A_{T+1}^i \delta^2 a_{T-1}^{(1)i}}{2\beta_{T-1}^i}$ in period $T-1$ and a Pigouvian tax $\frac{A_{T+1}^i \delta a_T^i}{2\beta_T^i}$ in period T have to be imposed.

(ii) The Green Sector

The net economic profit in the green sector is $2\varpi_k (\vartheta_k q_k^i)^{1/2} - \varepsilon_k^i q_k^i$. Maximizing economic profit with respect to q_k^i yields the first-order condition

$$\varpi_k (\vartheta_k)^{1/2} (q_k^i)^{-1/2} - \varepsilon_k^i = 0,$$

which leads to an industry's profit-maximizing output $q_k^i = \frac{(\varpi_k^i)^2 \vartheta_k}{(\varepsilon_k^i)^2}$.

The game equilibrium target control in the green industry is

$$q_k^{i**} = \frac{(\varpi_k^i)^2 \vartheta_k}{(\varepsilon_k^i)^2},$$

which corresponds to that industry's profit-maximizing output. Hence there is no need for government intervention.

(iii) Green Technology R&D Sector

The net economic profit in the green technology R&D sector is $\chi_k \bar{w}_k^i - (\bar{w}_k^i)^2$. Maximizing economic profit with respect to $\bar{w}_k^i$ yields the first-order condition

$$\chi_k - 2\bar{w}_k^i = 0,$$

which leads to an industry's profit-maximizing R&D $\bar{w}_k^i = \frac{\chi_k^i}{2}$.

However, the game equilibrium target control of R&D is

$$\bar{w}_k^{i**} = \frac{\chi_k^i + B_k^i \delta z_k^i + B_{k+2}^i \delta^2 z_k^{(1)i}}{2}, \text{ for } k \in \{1, 2, \ldots, T-1\}.$$

A Pigouvian subsidy equaling $\frac{B_k^i \delta z_k^i + B_{k+2}^i \delta^2 z_k^{(1)i}}{2}$ per unit of output has to be provided to the green technology sector to achieve the target control in period $k \in \{1, 2, \ldots, T-1\}$.

Similarly, a Pigouvian subsidy $\frac{B_{T+1}^i \delta z_T^i}{2}$ has to be provided in period T.

(iv) The Government Sector for Pollution Abatement

The pollution abatement sector is a government sector in which the government would spend abatement up to the level when the marginal abatement cost equals the marginal benefit from reducing pollution, that is,

$$2c_k^{(a)i} u_k^i = -A_{k+1}^i b_k^i u_k^i (x_k)^{1/2} \delta,$$

which leads to the target control $u_k^{i**} = -\frac{b_k^i}{2c_k^{(a)i}} A_{k+1}^i \delta (x_k)^{1/2}$.

Finally, if the tax revenue collected from the polluting industrial sector is higher than the total subsidy offered to the green technology R&D sector, that is,

$$\frac{A_{k+1}^i \delta a_k^i + A_{k+2}^i \delta^2 a_k^{(1)i} + A_{k+3}^i \delta^3 a_k^{(2)i}}{2\beta_k^i} \bar{q}_k^{i**} \geq \frac{B_k^i \delta z_k^i + B_{k+2}^i \delta^2 z_k^{(1)i}}{2} \bar{w}_k^{i**},$$

the government can keep the balance as reserves or distribute it to the private sector.

On the other hand, if the tax revenue collected from the polluting industrial sector is less than the total subsidy offered to the green technology R&D sector, that is,

$$\frac{A_{k+1}^i \delta a_k^i + A_{k+2}^i \delta^2 a_k^{(1)i} + A_{k+3}^i \delta^3 a_k^{(2)i}}{2\beta_k^i} \bar{q}_k^{i**}$$

$$< \frac{B_k^i \delta z_k^i + B_{k+2}^i \delta^2 z_k^{(1)i}}{2} \bar{w}_k^{i**},$$

the government can bear the loss or impose a general tax.

6. Notes

In this chapter, we present the mathematical theory that characterizes the uncoordinated behavior of the nations in a dynamic GPE. The durable-strategies dynamic game paradigm adopted in this chapter from Petrosyan and Yeung (see Yeung and Petrosyan, 2019, 2020, 2021a, 2021b, 2022) incorporated a crucial element in GPE — strategies with durable effects. To explicate the theory in a real-life scenario, we consider a major dynamic GPE issue that needs urgent attention — global environmental degradation.

Studies of dynamic transboundary games can be found in Plourde and Yeung (1989), Kaitala *et al.* (1991, 1992a, 1992b), Long (1992), Yeung (1992), Hoel (1992, 1993), Dockner and Long (1993), Martin *et al.* (1993), and Yeung and Cheung (1994). Refinement of dynamic environmental games includes those of Haurie (1995, 2005), Dockner and Nishimura (1999), Jørgensen and Yeung (1999), Feenstra *et al.* (2001, 2002), Fredj *et al.* (2004, 2006), Haurie (2005), Krawczyk (2005), Tidball and Zaccour (2005, 2009), Kossioris *et al.* (2008), Jørgensen (2010), Li and Guo (2019), and Tilman *et al.* (2020). Collections of articles on dynamic environmental games can be found in the surveys of Long (2010) and Jørgensen *et al.* (2010). Studies on using green technology and environmental damages have been presented in Billatos and Basaly (1997), Ane (2000), Eis *et al.* (2016), Yeung and Petrosyan (2016a), and Yeung *et al.* (2017).

The durable-strategies dynamic game paradigm adopted in this chapter incorporated a crucial element in GPE — strategies with durable effects. Dynamic environmental game analyses involving durable strategies can be found in Petrosyan and Yeung (2020) and Yeung and Petrosyan (2021a, 2021b, 2022).

Finally, a generalization remark of the model structure is given as follows.

Remark 6.1. The function $w_k^{i(\varsigma)} g_k^{i(\varsigma)}(x_k, \underline{u}_k, \bar{u}_k; \bar{\underline{u}}_{k-})$ can be expressed as a function $w_k^{i(\varsigma)}[g_k^{i(\varsigma)}(x_k, \underline{u}_k, \bar{u}_k; \bar{\underline{u}}_{k-})]$ which reflects a nonlinear monetary weight.

References

Ane, R. P. (2000). Stimulating innovation in green technology. Policy alternatives and opportunities. *American Behavioral Scientist*, 44(2), 188–212.

Billatos, S. & Basaly, N. (1997). *Green Technology and Design for the Environment.* Washington, DC: CRC Press, 312 p.

Dockner, E. & Long, N. V. (1993). International pollution control: Cooperative versus noncooperative strategies. *Journal of Environmental Economics and Management*, 24, 13–29.

Dockner, E. & Nishimura, K. (1999). Transboundary pollution in a dynamic game model. *The Japanese Economic Review*, 50(4), 443–456.

Eis, J., Bishop, R., & Gradwell, P. (2016). Galvanising low-carbon innovation. *A New Climate Economy Working Paper for Seizing the Global Opportunity: Partnerships for Better Growth and a Better Climate.* New Climate Economy, London, 1–28 pp.

Feenstra, T., Kort, P., & de Zeeuw, A. (2001). Environmental policy instruments in an international duopoly with feedback investment strategies. *Journal of Economic Dynamics & Control*, 25, 1665–1687.

Feenstra, T., Kort, P., & de Zeeuw, A. (2002). International competition and investment in abatement: Taxes versus standards. In Marsiliani, L., Rauscher, M., & Withagen, C. (eds.), *Environmental Economics and the International Economy.* Kluwer, Dordrecht, pp. 89–98.

Fredj, K., Martín-Herrán, G., & Zaccour, G. (2004). Incentive mechanisms to enforce sustainable forest exploitation. *Environmental Modelling & Assessment*, 11, 145–156.

Fredj, K., Martín-Herrán, G., & Zaccour, G. (2006). Slowing deforestation pace through subsidies: A differential game. *Automatica*, 40, 301–309. DOI: 10.1016/j.automatica.2003.10.020.

Haurie, A. (1995). Environmental coordination in dynamic oligopolistic markets. *Group Decision and Negotiation*, 4(1), 39–57.

Haurie, A. (2005). A two-timescale stochastic game framework for climate change policy assessment. In Haurie, A. & Zaccour, G. (eds.), *Dynamic Games: Theory and Applications*. GERAD Twenty-Fifth Anniversary Series. Springer, New York, pp. 193–211.

Hoel, M. (1992). Emission taxes in a dynamic international game of CO_2 emissions. In Pethig, R. (ed.), *Conflict and Cooperation in Managing Environmental Resources*. Springer, Berlin, pp. 39–70.

Hoel, M. (1993). Intertemporal properties of an international carbon tax. *Resource and Energy Economics*, 15(1), 51–70.

Jevons, W. S. (1862). A brief account of a general mathematical theory of political economy. *The Journal of the Royal Statistical Society*, London, XXIX (June 1866), pp. 282–287. Section F of the British Association, 1862.

Jørgensen, S. (2010). A dynamic game of waste management. *Journal of Economic Dynamics & Control*, 34, 258–265. DOI: 10.1016/j.jedc.2009.09.005.

Jørgensen, S. & Yeung, D. W. K. (1999). Inter- and intragenerational renewable resource extraction. *Annals of Operations Research*, 88, 275–289.

Jørgensen, S., Martín-Herrán, G., & Zaccour, G. (2010). Dynamic games in the economics and management of pollution. *Environmental Modeling and Assessment*, 15, 433–467. DOI: 10.1007/s10666-010-9221-7.

Kaitala, V., Pohjola, M., & Tahvonen, O. (1991). Transboundary air pollution between Finland and the U.S.S.R.: A dynamic acid rain game. In Hamalainen, R. & Ehtamo, H. (eds.), *Dynamic Games in Economic Analysis*. Springer, Berlin, pp. 183–192.

Kaitala, V., Pohjola, M., & Tahvonen, O. (1992a). Transboundary air pollution and soil acidification: A dynamic analysis of an acid rain game between Finland and the U.S.S.R. *Environmental and Resource Economics*, 2(2), 161–181.

Kaitala, V., Pohjola, M., & Tahvonen, O. (1992b). An economic analysis of transboundary air pollution between Finland and the Soviet Union. *Scandinavian Journal of Economics*, 94, 409–424.

Kossioris, G., Plexousakis, M., Xepapadeas, A., de Zeeuw, A., & Mäler, K.-G. (2008). Feedback Nash equilibria for non-linear differential games in pollution control. *Journal of Economic Dynamics & Control*, 32(4), 1312–1331.

Krawczyk, J. B. (2005). Coupled constraint Nash equilibria in environmental games. *Resource and Energy Economics*, 27(2), 157–181.

Li, H. & Guo, G. (2019). Dynamic decision of transboundary basin pollution under emission permits and pollution abatement. *Physica A: Statistical Mechanics and Its Applications*, 532(15), 121869.

Long, N. V. (1992). Pollution control: A differential game approach. *Annals of Operational Research*, 37, 283–296.

Long, N. V. (2010). *A Survey of Dynamic Games in Economics*. Singapore: World Scientific.

Martin, W. E., Patrick, R. H., & Tolwinski, B. (1993). A dynamic game of a transboundary pollutant with asymmetric players. *Journal of Environmental Economics and Management*, 24, 1–12.

Oppenlander, R. (2013). *Food Choice and Sustainability*. Langdon Street Press, Minneapolis, pp. 120–123.

Pearce, R. (2006). *When the Rivers Run Dry: Water – The Defining Crisis of the Twenty-First Century*. Boston: Beacon Press.

Petrosyan, L. A. & Yeung, D. W. K. (2020) Cooperative dynamic games with durable controls: Theory and application. *Dynamic Games and Applications*, 10, 872–896. DOI: 10.1007/s13235-019-00336-w.

Plourde, C. & Yeung, D. W. K. (1989). Harvesting of a transboundary replenishable fish stock: A non cooperative game solution. *Marine Resource Economics*, 6, 57–70.

Thakkar, H. (1999). *Assessment of Irrigation in India*. World Commission on Dams.

Tidball, M. & Zaccour, G. (2005). An environmental game with coupling constraints. *Environmental Modeling and Assessment*, 10, 153–158.

Tidball, M. & Zaccour, G. (2009). A differential environmental game with coupling constraints. *Optimal Control Applications and Methods*, 30, 197–207.

Tilman, A. R., Plotkin, J. B., & Akçay, E. (2020). Evolutionary games with environmental feedbacks. *Nature Communications*, 11, 915. https://doi.org/10.1038/s41467-020-14531-6.

Yeung, D. W. K. (1992). A differential game of industrial pollution management. *Annals of Operations Research*, 37, 297–311.

Yeung, D. W. K. & Cheung, M. T. (1994). Capital accumulation subject to pollution control: A differential game with a feedback Nash equilibrium. In Basar, T. & Haurie, A. (eds.), *Advances in Dynamic Games and Applications*. Annals of the International Society of Dynamic Games, Vol. 1. Birkhäuser, Boston, pp. 289–300.

Yeung, D. W. K. & Petrosyan, L. A. (2016). *Subgame Consistent Cooperation: A Comprehensive Treatise*. Singapore: Springer, 520 p.

Yeung, D. W. K., Petrosyan, L. A., Zhang, Y. X., & Cheung, F. (2017). *BEST SCORES Solution to the Catastrophe Bound Environment*. Nova Science, New York, 184 p.

Yeung, D. W. K. & Petrosyan, L. A. (2021a). Asynchronous horizons durable-strategies dynamic games and tragedy of cross-generational environmental commons. *International Game Theory Review*, 23(4). DOI: 10.1142/s0219198921500201.

Yeung, D. W. K. & Petrosyan, L. A. (2021b). Generalized dynamic games with durable strategies under uncertain planning horizon. *Journal of Computational and Applied Mathematics*, 395, 113595. DOI: 10.1016/j.cam.2021.113595.

Yeung, D. W. K. & Petrosyan, L. A. (2022). *Durable-Strategies Dynamic Games: Theory, Solution Techniques and Applications.* Springer, Switzerland, 303 p.

Chapter 4

Coordination and Cooperation: Mathematical Theory and Explication in Global Climate Agreement

"Cooperation, collaboration and coordination are together more powerful than competition." and "In a permaculture economy, it's less about competition and more about collaboration, cooperation and coordination." — Hendrith Vanlon Smith Jr.

"Competition has been shown to be useful up to a certain point and no further, but cooperation, which is the thing we must strive for today, begins where competition leaves off."
— Franklin D. Roosevelt

We live in a global environment where policymakers from many nations develop individual policy designs, statements, and actions that are either not coordinated or not effectively coordinated. Coordination and cooperation are very much needed for resolving global problems. In this chapter, we present the mathematical theory for analyzing dynamic coordinated cooperative policy with the incorporation of durable strategies. We explicate the mathematical exposition in a global climate agreement setting to facilitate the exposition of the materials and avoid unnecessary repetition. Anthropogenic eco-degradation has become one of the world's most complex and urgent problems to be dealt with by scientists, policymakers, and economists alike. Self-seeking maximization by nations portrays the situation of an industrial civilization on the verge of suicide, destroying its environmental conditions of existence with people being held as prisoners on a runaway catastrophe-bound train. Unilateral response of a single nation is deemed to be ineffective; only through cooperation with an effective solution-driven coordination mechanism can a viable solution be made. The organization of this chapter is as follows. Section 1 presents the theoretical underpinnings for coordination and cooperation in the dynamic global political economy. Section 2 develops the mathematical theories for cooperative optimization for solving Pareto optimal target controls under coordination. Subgame-consistent sharing with individual rationality and group optimality are presented. In Section 3, a Payoff Distribution Procedure (PDP) leading to the realization of subgame-consistent cooperation is provided. Price of Anarchy (PoA) and cost of delay are derived. Section 5 gives an explication of cooperation and coordination in global climate agreements. Section 7 proposes an Organization of a Global Environmental Pact. A computational illustration is provided in Section 8 and chapter notes are given in Section 9.

1. Theoretical Underpinnings for Coordination and Cooperation

The global economy is under enormous pressure from multiple complex and interconnected crises. Some of today's most pressing world problems in the global political economy have compounded into what experts have called a "polycrisis" — triggering problems that decades of progress had sought to solve. These pressing problems (as discussed in Chapter 1) include (i) nuclear weapons and threats, (ii) environmental degradation and climate change, (iii) space race with militarization of new frontier, (iv) development of artificial intelligence, (v) cyberattacks, (vi) global water crisis, (vii) spread of false information on social media, (viii) biotechnological developments, and (ix) trade war.

Moreover, the seriousness of some current problems in the global political economy and the urgent call for collaborated efforts to develop solutions may be manifested by United Nations Secretary-General António Guterres' address on 20 September 2022, which said, "Our world is in big trouble. We have a duty to act, and yet we are gridlocked in colossal global dysfunction. The international community is not ready or willing to tackle the big dramatic challenges of our age. These crises threaten the very future of humanity and the fate of our planet. Crises like the war in Eastern Europe and the multiplication of conflicts around the globe. Crises like the climate emergency and biodiversity loss. Crises like the dire financial situation of developing countries and the fate of the Sustainable Development Goals. And crises like the lack of guardrails around promising new technologies to heal disease, connect people and expand opportunity. ... Our world is in peril – and paralyzed. ... No cooperation. No dialogue. No collective problem-solving ... But the reality is that we live in a world where the logic of cooperation and dialogue is the only path forward. ... So let's develop common solutions to common problems — grounded in goodwill, trust, and the rights shared by every human being. Let's work as one, a coalition of the world".

We live in a global scenario where policymakers from many nations develop individual policy designs, statements, and actions that are either not coordinated or not effectively coordinated. It is well known (as shown in Chapter 3) that non-cooperative behaviors among participants would, in general, lead to an outcome that is not Pareto optimal. Worse still, highly undesirable outcomes (like the prisoner's dilemma) and even devastating results (like the tragedy of the commons) could appear when the involved parties only care about their individual self-interests in a non-cooperative situation. In a dynamic world, non-cooperative behaviors guided by short-sighted individual rationality could be a source of a series of disastrous consequences in the future. The phenomenon of the "inter-temporal tragedy of temporal individual rationality" is emerging in our era of rapid globalization. Cooperation suggests the possibility of obtaining socially optimal and group-efficient solutions to decision problems involving strategic actions. The calls for cooperation had not been scarce:

"You look at the large problems that we face — that would be overpopulation, water shortages, global warming and AIDS, I suppose - all of that needs international cooperation to be solved". — Molly Ivins.

"Competition has been shown to be useful up to a certain point and no further, but cooperation, which is the thing we must strive for today, begins where competition leaves off". — Franklin D. Roosevelt.

"When times are tough and people are frustrated and angry and hurting and uncertain, the politics of constant conflict may be good, but what is good politics does not necessarily work in the real world. What works in the real world is cooperation". — William J. Clinton.

"The only thing that will redeem mankind is cooperation". — Bertrand Russell.

"Nature is based on harmony. So it says if we want to survive and become more like nature, then we actually have to understand that it's cooperation versus competition". — Bruce Lipton.

"When it comes to the fundamental issues that humanity faces, I think that solutions involve shifting consciousness towards cooperation". — Jeremy Gilley.

Since the global problems mentioned above are dynamic, therefore coordination and cooperation needed for resolving dynamic global problems have to be dynamic. Worth noting is that dynamic cooperation cannot be sustainable if there is no guarantee that the participants will always be better off within the entire duration of the cooperation. More than anything else, it is due to the lack of this kind of guarantee that cooperative schemes fail to last till their end. Dynamic cooperation represents one of the most intriguing forms of optimization analysis. The complexity of the problem leads to great difficulties in the derivation of satisfactory solutions. In dynamic cooperation, two fundamental factors — individual rationality and group optimal — have to be maintained throughout the cooperation duration. Group optimality ensures that all potential gains from cooperation are captured. Individual rationality is required to hold so that the payoff allocated to any participant under cooperation will be no less than his non-cooperative payoff. Failure to fulfill group optimality leads to the condition where the participants prefer to deviate from the agreed-upon solution plan to extract the unexploited gains. Failure to guarantee individual rationality leads to the condition where the concerned participants would reject the agreed-upon solution plan and act non-cooperatively. Yet for dynamic cooperation, group optimality and individual rationality have to be satisfied at all time instants during the entire cooperation duration.

On top of individual rationality and group optimality, a sustainable cooperative scheme requires the satisfaction of the property of subgame consistency. A cooperative solution is a subgame consistent if an extension of the solution policy to a subgame with a later starting time and a state brought about by prior optimal behaviors would remain optimal. Petroysan (1997) introduced the notion of agreeable solutions in cooperative differential games in which subgame-consistent (then referred to as time consistent) solution was developed. Crucial to the derivation of a subgame-consistent cooperative solution is the formulation of a payment distribution mechanism that would lead to the realization of the solution. Yeung and Petrosyan (2004) developed a generalized method for the derivation of analytically tractable subgame-consistent solutions in stochastic differential games. This has made possible the rigorous study of subgame-consistent solutions in continuous-time dynamic cooperation. Applications and exegeses of subgame-consistent solutions in differential games and stochastic differential games can be found in Yeung (2005, 2006, 2010) and Yeung and Petrosyan (2006a, 2007a, 2007b, 2007c, 2008, 2013, 2014). Subgame-consistent solution of cooperative durable strategy dynamic games can be found in Petrosyan and Yeung (2020) and Yeung and Petrosyan (2021, 2022).

Global cooperative policy instruments are formulated to achieve the target controls. The commonly adopted ones include imposing international taxes, incentives for investment, global subsidies, joint technology development, and collaborative government provisions. Some examples of global cooperative policy instruments in an international context are provided in the following:

Economic incentive	International tax scheme Cross-national levies Fines and penalties Cross-national transfer of subsidies Adaptation funds
Regulation-based	International emission limits Rationings across border Activity bans Cross-national legislation and regulation

Collaborative government provisions	Infrastructure building Trans-boundary public goods Government investments Public spending Public medical care/education Pollution abatement Ecosystem rehabilitation
Others	Moral suasion Multinational NGO formation Public–private partnership Cross-national welfare payments Joint Green technology R&D

2. Mathematical Theory for Cooperative Optimization

In this section, we provide the mathematical theory for subgame-consistent cooperative optimization. We first present the optimization techniques for deriving Pareto optimal target controls under coordination with durable strategies. Then we present the mathematics for deriving a payoff distribution scheme that yields subgame-consistent cooperation.

2.1. *Pareto optimal target controls under coordination*

Consider the layout of the durable-strategies dynamic game presented in Section 2 of Chapter 3. The Pareto optimal target controls, which lead to group optimality, can be obtained by solving the joint maximization problem:

$$\max_{\substack{u_k^j, \bar{u}_k^j, j \in N \\ k \in \{1,2,\ldots T\}}} \left\{ \sum_{j=1}^{n} \left[\sum_{k=1}^{T} \sum_{\varsigma=1}^{\Theta_j} w_k^{j(\varsigma)} g_k^{j(\varsigma)}(x_k, \underline{u}_k, \underline{\bar{u}}_k; \underline{\bar{u}}_{k-}) \delta_1^k \right.\right.$$
$$\left.\left. + \, q_{T+1}^j(x_{T+1}; \underline{\bar{u}}_{(T+1)-}) \delta_1^{T+1} \right] \right\} \tag{2.1}$$

subject to

$$x_{k+1} = f(x_k, \underline{u}_k, \underline{\bar{u}}_k; \underline{\bar{u}}_{k-}), x_1 = x_1^0. \tag{2.2}$$

Note that $\sum_{\varsigma=1}^{\Theta_j} w_k^{j(\varsigma)} g_k^{j(\varsigma)}(x_k, \underline{u}_k, \bar{u}_k; \bar{\underline{u}}_{k-})\delta_1^k$ and $f(x_k, \underline{u}_k, \bar{u}_k; \bar{\underline{u}}_{k-})$ may not be the same as those in the non-cooperative scenario if cooperative actions, like joint R&D, lead to changes in these functions.

Invoking Theorem 2.1 in Chapter 3, the Pareto optimal solution can be characterized by the following theorem.

Theorem 2.1. *Let $W(k, x; \bar{\underline{u}}_{k-})$ be the maximal value of the joint payoff*

$$\sum_{j=1}^{n}\sum_{t=k}^{T}\sum_{\varsigma=1}^{\Theta_j} w_k^{j(\varsigma)} g_t^{j(\varsigma)}(x_t, \underline{u}_t, \bar{u}_t; \bar{\underline{u}}_{t-})\delta_1^t + \sum_{j=1}^{n} q_{T+1}^j(x_{T+1}; \bar{\underline{u}}_{(T+1)-})\delta_1^{T+1},$$

for the joint payoff maximization problem (2.2)–(2.3) starting in period k with state $x_k = x$ and previously executed controls $\bar{\underline{u}}_{k-}$, then the function $W(k, x; \bar{\underline{u}}_{k-})$ satisfies the following system of recursive equations:

$$W(T+1, x; \bar{\underline{u}}_{(T+1)-}) = \sum_{j=1}^{n} q_{T+1}^j(x; \bar{\underline{u}}_{(T+1)-})\delta_1^{T+1}, \tag{2.3}$$

$$\begin{aligned} &W(k, x; \bar{\underline{u}}_{k-}) \\ &= \max_{\underline{u}_k, \bar{u}_k}\left\{\sum_{j=1}^{n}\sum_{\varsigma=1}^{\Theta_j} w_k^{j(\varsigma)} g_k^{j(\varsigma)}(x, \underline{u}_k, \bar{u}_k; \bar{\underline{u}}_{k-})\delta_1^k \right. \\ &\qquad \left. + W[k+1, f_k(x, \underline{u}_k, \bar{u}_k; \bar{\underline{u}}_{k-}); \bar{\underline{u}}_{(k+1)-}]\right\} \\ &= \max_{\underline{u}_k, \bar{u}_k}\left\{\sum_{j=1}^{n}\sum_{\varsigma=1}^{\Theta_j} g_k^{j(\varsigma)}(x, \underline{u}_k, \bar{u}_k; \bar{\underline{u}}_{k-})\delta_1^k \right. \\ &\qquad \left. + W[k+1, f_k(x, \underline{u}_k, \bar{u}_k; \bar{\underline{u}}_{k-}); \bar{\underline{u}}_k, \bar{\underline{u}}_{(k+1)-} \cap \bar{\underline{u}}_{k-}]\right\}, \\ &\text{for } k \in \{1, 2, \ldots, T\}. \end{aligned} \tag{2.4}$$

Proof. See the proof of Theorem I of Mathematical Solution Mechanisms given in Chapter 7. □

The characterization of the group optimal outcome in Theorem 2.1 allows one to derive the optimal target controls. We denote the Pareto optimal target controls derived from Theorem 2.1 by $\{(\underline{u}_k^{i*}, \bar{u}_k^{i*});\ for\, i \in N\}$. Policy instruments have to be implemented by all national governments cooperatively to achieve the optimal target controls $(\underline{u}_k^*, \bar{u}_k^*)$. Note that the evolution of the environmental states under cooperation will be affected by previously executed activities:

$$x_{k+1} = f(x_k, \underline{u}_k^*, \bar{u}_k^*; \bar{u}_{k-}^*). \tag{2.5}$$

We use $\{x_k^*\}_{k=1}^{T+1}$ to denote the Pareto optimal trajectory of the state path. The maximal value of the joint payoff along the cooperative trajectory $\{x_k^*\}_{k=1}^{T+1}$ can be expressed as follows:

$$W(k, x_k^*; \bar{u}_{k-}^*), \text{ for } k \in \{1, 2, \ldots, T.\}. \tag{2.6}$$

Global cooperative policy instruments like international taxes, incentives for investment, global subsidies, joint technology development, and collaborative government provisions are formulated to achieve the target controls $(\underline{u}_k^*, \bar{u}_k^*)$.

2.2. *Subgame-consistent sharing*

Now consider the case when the nations agree to distribute the cooperative payoff among themselves according to an optimality principle. We use $\xi^i(k, x_k; \bar{u}_{k-})$ to denote the agreed-upon payoff that nation $i \in N$ conditional on the state x_k and the previously executed durable strategies $\bar{u}_{k-}$. Individual rationality will be satisfied if $\xi^i(k, x_k; \bar{u}_{k-}) \geq V^i(k, x_k; \bar{u}_{k-})$. Group rationality will be satisfied if $\sum_{i=1}^n \xi^i(k, x_k; \bar{u}_{k-}) = W(k, x_k; \bar{u}_{k-})$.

Given that the nations are asymmetrical and their payoffs under non-cooperation can vary greatly, a practicable gain-sharing optimality principle is to share the total cooperative payoff proportional to the nation's non-cooperative payoffs. According to this optimality principle, the payoff to nation i at the start of the cooperation scheme is

$$\xi^i(1, x_1^0; \bar{u}_{1-}^0) = \frac{V^i(1, x_1^0; \bar{u}_{1-}^0)}{\sum_{j=1}^n V^j(1, x_1^0; \bar{u}_{1-}^0)} W(1, x_1^0; \bar{u}_{1-}^0), \tag{2.7}$$

for $i \in N$.

Sharing the cooperative gain over the cooperation duration plays an important role in achieving a sustainable cooperation scheme. On top of the requirements of individual rationality and group optimality, the sustainability of dynamic cooperation requires the satisfaction of a stringent condition — subgame consistency — being satisfied throughout the cooperation duration. A cooperative solution is subgame-consistent if an extension of the solution policy to a subgame with a later starting time and a state brought about by prior optimal behaviors would remain optimal. In particular, it implies that the specific optimality principle agreed upon at the outset must remain effective at any instant of time throughout the game along the optimal state trajectory. To satisfy subgame consistency, the optimality condition in (2.7) has to be maintained throughout the cooperation duration along the cooperative trajectory $\{x_k^*\}_{k=1}^{T+1}$, that is,

$$\xi^i(k, x_k^*; \underline{\bar{u}}_{k-}^*) = \frac{V^i(k, x_k^*; \underline{\bar{u}}_{k-}^*)}{\sum_{j=1}^n V^j(k, x_k^*; \underline{\bar{u}}_{k-}^*)} W(k, x_k^*; \underline{\bar{u}}_{k-}^*), \tag{2.8}$$

for $k \in \{1, 2, \ldots, T\}$ and $i \in N$.

Note that $\xi^i(k, x_k^*; \underline{\bar{u}}_{k-}^*)$ in (2.8) satisfies individual rationality throughout the cooperation duration because

$$\xi^i(k, x_k^*; \underline{\bar{u}}_{k-}^*) = \frac{V^i(k, x_k^*; \underline{\bar{u}}_{k-}^*)}{\sum_{j=1}^n V^j(k, x_k^*; \underline{\bar{u}}_{k-}^*)} W(k, x_k^*; \underline{\bar{u}}_{k-}^*) \geq V^i(k, x_k^*; \underline{\bar{u}}_{k-}^*), \tag{2.9}$$

for $k \in \{1, 2, \ldots, T\}$ and $i \in N$.

In addition, $\xi^i(k, x_k^*; \underline{\bar{u}}_{k-}^*)$ in (2.8) also satisfies group optimality throughout the cooperation duration because

$$\begin{aligned} \sum_{i=1}^n \xi^i(k, x_k^*; \underline{\bar{u}}_{k-}^*) &= \sum_{i=1}^n \frac{V^i(k, x_k^*; \underline{\bar{u}}_{k-}^*)}{\sum_{j=1}^n V^j(k, x_k^*; \underline{\bar{u}}_{k-}^*)} W(k, x_k^*; \underline{\bar{u}}_{k-}^*) \\ &= W(k, x_k^*; \underline{\bar{u}}_{k-}^*), \end{aligned} \tag{2.10}$$

for $k \in \{1, 2, \ldots, T\}$.

3. Subgame-Consistent Payoff Distribution

Crucial to the analysis is the derivation of a Payoff Distribution Procedure (PDP) leading to the realization of the agreed-upon payoffs

$\xi^i(k, x_k^*; \underline{\bar{u}}_{k-}^*)$ in (2.8) in all periods $k \in \{1, 2, \ldots, T\}$ throughout the cooperation duration. If this is satisfied, no nations will have an incentive to deviate from the cooperation scheme along its cooperative state trajectory.

3.1. *Payoff distribution procedure*

To do this, we have to design a Payoff Distribution Procedure (PDP) that leads to the realization of (2.8) in all periods $k \in \{1, 2, \ldots, T\}$. We use $\beta_k^i(x_k^*; \underline{\bar{u}}_{k-}^*)$ to denote the payment that nation i receives in period k under the cooperative agreement along the cooperative trajectory with strategies executed previously but still in effect. The payment scheme $\beta_k^i(x_k^*; \underline{\bar{u}}_{k-}^*)$ constitutes a payoff distribution scheme in the sense that the payoff to nation i over the periods from k to $T+1$ satisfies the following condition:

$$\xi^i(k, x_k^*; \underline{\bar{u}}_{k-}^*) = \beta_k^i(x_k^*; \underline{\bar{u}}_{k-}^*) + \left[\sum_{\zeta=k}^{T} \beta_\zeta^i(x_\zeta^*; \underline{\bar{u}}_{\zeta-}^*)\delta_1^\zeta + q_{T+1}^i(x_{T+1}^*; \underline{\bar{u}}_{(T+1)-}^*)\delta_1^{T+1}\right], \quad (3.1)$$

for $i \in N$ and $k \in \{1, 2, \ldots, T\}$.

Invoking Theorem III in Chapter 7, we obtain the values of $\beta_k^i(x_k^*; \underline{\bar{u}}_{k-}^*)$ in the following:

Theorem 3.1. *The agreed-upon payoff $\xi(k, x_k^*; \underline{\bar{u}}_{k-}^*)$, for $k \in \{1, 2, \ldots, T\}$ along the cooperative trajectory $\{x_k^*\}_{k=1}^T$, can be realized by a payment*

$$\beta_k^i(x_k^*; \underline{\bar{u}}_{k-}^*) = (\delta_1^k)^{-1}[\xi^i(k, x_k^*; \underline{\bar{u}}_{k-}^*) - \xi^i(k+1, f_k(x_k^*, \underline{u}_k^*, \underline{\bar{u}}_k^*; \underline{\bar{u}}_{k-}^*); \underline{\bar{u}}_{(k+1)-}^*)], \quad (3.2)$$

given to nation $i \in N$ at period $k \in \{1, 2, \ldots T\}$.

Proof. See the proof of Theorem III of Mathematical Solution Mechanisms given in Chapter 7. □

The payment scheme in (3.2) gives rise to the realization of the payoff guided by the agreed-upon optimality principle and constitutes a dynamically consistent payment scheme. More specifically, the payment of $\beta_k^i(x_k^*;\underline{\bar{u}}_{k-}^*)$ allotted to nation $i \in N$ in period $k \in \{1,2,\ldots,T\}$ will establish a cooperative plan that matches with the agreed-upon payoff to every nation along the cooperative path.

In the case where the gain-sharing optimality principle is to share the total cooperative payoff proportional to the nation's non-cooperative payoffs,

$$\xi^i(k,x_k^*;\underline{\bar{u}}_{k-}^*) = \frac{V^i(k,x_k^*;\underline{\bar{u}}_{k-}^*)}{\sum_{j=1}^n V^j(k,x_k^*;\underline{\bar{u}}_{k-}^*)} W(k,x_k^*;\underline{\bar{u}}_{k-}^*), \tag{3.3}$$

for $k \in \{1,2,\ldots,T\}$ and $i \in N$.

$\beta_k^i(x_k^*;\underline{\bar{u}}_{k-}^*)$ is the target allocation to nation $i \in N$. However, under cooperation, all nations would use the cooperative strategies and the gain that nation i will directly receive in period k along the cooperative trajectory equaling $\sum_{\varsigma=1}^{\Theta_i} w_k^{i(\varsigma)} g_k^{i(\varsigma)}(x_k^*,\underline{u}_k^*,\underline{\bar{u}}_k^*;\underline{\bar{u}}_{k-}^*)$. Therefore, a side payment

$$\pi_k^i(x_k^*;\underline{\bar{u}}_{k-}^*) = \beta_k^i(x_k^*;\underline{\bar{u}}_{k-}^*) - \sum_{\varsigma=1}^{\Theta_i} w_k^{i(\varsigma)} g_k^{i(\varsigma)}(x_k^*,\underline{u}_k^*,\underline{\bar{u}}_k^*;\underline{\bar{u}}_{k-}^*), \tag{3.4}$$

for $k \in \{1,2,\ldots,T\}$,

has to be given to nation $i \in N$ to yield the cooperative payoff $\xi^i(k,x_k;\underline{\bar{u}}_{k-})$, for $k \in \{1,2,\ldots,T\}$.

3.2. *Other gain-sharing optimality principle considerations*

While the gain sharing optimality principle of sharing the total cooperative payoff proportional to the nation's non-cooperative payoffs in (2.8) is a practicable alternative, there are considerations that the minor nations may deserve a fairer share of the cooperative gains. On the other extreme is sharing the excess of cooperative gains over the sum of the non-cooperative equally, which leads to a payoff:

$$\begin{aligned}\tilde{\xi}^i(k,x_k^*;\underline{\bar{u}}_{k-}^*) &= V^i(k,x_k^*;\underline{\bar{u}}_{k-}^*)\\ &\quad + \frac{1}{n}\left[W(k,x_k^*;\underline{\bar{u}}_{k-}^*) - \sum_{j=1}^n V^j(k,x_k^*;\underline{\bar{u}}_{k-}^*)\right],\end{aligned} \tag{3.5}$$

for $k \in \{1,2,\ldots,T\}$ and $i \in N$.

One of the ways to lean more favorably toward the minor (poorer) nations is to consider a gain-sharing optimality principle which is a weighted average of $\tilde{\xi}^i(k, x_k^*; \underline{\bar{u}}_{k-}^*)$ and $\xi^i(k, x_k^*; \underline{\bar{u}}_{k-}^*)$, that is,

$$\widehat{\xi}^{i}(k, x_k^*; \underline{\bar{u}}_{k-}^*) = \alpha\xi^i(k, x_k^*; \underline{\bar{u}}_{k-}^*) + (1-\alpha)\tilde{\xi}^i(k, x_k^*; \underline{\bar{u}}_{k-}^*), \quad (3.6)$$

for $k \in \{1, 2, \ldots, T\}$ and $i \in N$,

where the weight α is greater than zero and less than one.

Another sharing optimality principle may involve an adaptation fund being provided to participating nations that are less developed (or being significantly damaged) through a global adaptation fund with contributions from wealthier nations. We let $N^w \subset N$ be the set of wealthier nations that would make extra contributions, and $i^w \in N^w$ be a contributing nation. We let $N^r \subset N$ be the set of less-well-off nations that would receive adaptation funds, and $i^r \in N^r$ be a receiving nation. The payoff of wealthier nation $i^w \in N^w$ can be agreed to be

$$\begin{aligned}\bar{\xi}^{i^w}(k, x_k^*; \underline{\bar{u}}_{k-}^*) &= \xi^{i^w}(k, x_k^*; \underline{\bar{u}}_{k-}^*) - \bar{A}_k^{i^w} \\ &= \frac{V^{i^w}(k, x_k^*; \underline{\bar{u}}_{k-}^*)}{\sum_{j=1}^n V^j(k, x_k^*; \underline{\bar{u}}_{k-}^*)} W(k, x_k^*; \underline{\bar{u}}_{k-}^*) - \bar{A}_k^{i^w}, \quad (3.7)\end{aligned}$$

for $k \in \{1, 2, \ldots, T\}$.

The payoff of less-well-off nation $i^r \in N^r$ can be agreed to be

$$\begin{aligned}\hat{\xi}^{i^r}(k, x_k^*; \underline{\bar{u}}_{k-}^*) &= \xi^{i^r}(k, x_k^*; \underline{\bar{u}}_{k-}^*) + \hat{B}_k^{\hat{i}^r} \\ &= \frac{V^{i^r}(k, x_k^*; \underline{\bar{u}}_{k-}^*)}{\sum_{j=1}^n V^j(k, x_k^*; \underline{\bar{u}}_{k-}^*)} W(k, x_k^*; \underline{\bar{u}}_{k-}^*) + \hat{B}_k^{\hat{i}^r}, \quad (3.8)\end{aligned}$$

for $k \in \{1, 2, \ldots, T\}$.

$\bar{A}_k^{i^w}$ is an arbitrary amount of adaptation fund that nation $i^w \in N^w$ is willing to contribute, and $\hat{B}_k^{\hat{i}^r}$ is an arbitrary amount that $i^r \in N^r$ will receive by the Pact. In particular,

$$\xi^{i^w}(k, x_k^*; \underline{\bar{u}}_{k-}^*) - \bar{A}_k^{i^w} > V^{i^w}(k, x_k^*; \underline{\bar{u}}_{k-}^*), \quad \text{for } i^w \in N^w.$$

Moreover, $\sum_{i^w \in N^w} \bar{A}_k^{i^w} = \sum_{i^r \in N^r} \hat{B}_k^{\hat{i}^r}$.

4. Price of Anarchy and Cost of Delay

A measure that evaluates how the efficiency of an international collaboration degrades due to self-seeking optimization is the Price of Anarchy (PoA) (see Koutsoupias and Papadimitriou, 2009; Dubey, 1986). The essential idea of PoA is to see how poor an outcome one can get in a non-cooperative situation when the nations act in their own interests (fulfilling individuality only) versus the Pareto optimum that can be obtained for the same situation assuming that the nations can cooperate (fulfilling optimality collectively).

The formal definition of the price of anarchy (PoA) is the ratio of the most efficient (Pareto optimal) sum of nations' payoffs to the sum of the nations' payoffs in a non-cooperative equilibrium, that is, $\sum_{j=1}^{n} V^j(k, x_k^*; \underline{u}_{k-}^*)$. The measure of PoA can then be obtained as

$$PoA(k, x_k^*; \bar{\underline{u}}_{k-}^*) = \frac{W(k, x_k^*; \bar{\underline{u}}_{k-}^*)}{\sum_{j=1}^{n} V^j(k, x_k^*; \underline{u}_{k-}^*)}. \tag{4.1}$$

In addition, the percentage of efficiency loss in relation to PoA can be expressed as

$$\frac{W(k, x_k^*; \bar{\underline{u}}_{k-}^*) - \sum_{j=1}^{n} V^j(k, x_k^*; \underline{u}_{k-}^*)}{W(k, x_k^*; \bar{\underline{u}}_{k-}^*)} = 1 - (PoA)^{-1}. \tag{4.2}$$

The percentage of efficiency loss indicates the potential percentage loss of efficiency brought about by self-seeking optimization.

Given a subgame-consistent cooperation solution with consented payoff distribution optimality condition, we can measure the individual Price of Anarchy (iPoA) as follows:

$$iPoA(k, x_k^*; \bar{\underline{u}}_{k-}^*) = \frac{\xi^i(k, x_k^*; \bar{\underline{u}}_{k-}^*)}{V^i(k, x_k^*; \underline{u}_{k-}^*)}, \tag{4.3}$$

for $i \in N$.

In particular, iPoA measures the price that a nation has to pay for non-cooperation. The sum of iPoA equals PoA.

Next, we consider the case when the nations defer the starting of their cooperative scheme from period 1 to period $\tau + 1$. The loss in payoffs due to the delay in cooperation is measured by the cost of delay:

$$\begin{aligned}
&CoD(1, \tau+1) \\
&\quad = W(1, x_1^*; \underline{\bar{u}}_{1-}^*) - W(\tau, x_\tau^*; \underline{\bar{u}}_{\tau-}^*) \\
&\quad\quad - \left[\sum_{j=1}^{n} V^j(1, x_1^{**}; \underline{\bar{u}}_{1-}^{**}) - \sum_{j=1}^{n} V^j(\tau, x_\tau^{**}; \underline{\bar{u}}_{\tau-}^{**}) \right] \\
&\quad\quad + W(\tau+1, x_{\tau+1}^*; \underline{\bar{u}}_{(\tau+1)-}^*) - W(\tau+1, x_{\tau+1}^{**}; \underline{\bar{u}}_{(\tau+1)-}^{**}), \quad (4.4)
\end{aligned}$$

where

$W(1, x_1^*; \underline{\bar{u}}_{1-}^*) - W(\tau x_\tau^*; \underline{\bar{u}}_{\tau-}^*)$ is the optimal joint payoff from period 1 to period τ,

$[\sum_{j=1}^{n} V^j(1, x_1^{**}; \underline{\bar{u}}_{1-}^{**}) - \sum_{j=1}^{n} V^j(\tau, x_\tau^{**}; \underline{\bar{u}}_{\tau-}^{**})]$ is the sum of the nations' payoff from period 1 to period τ under a feedback Nash equilibrium,

$W(\tau+1, x_{\tau+1}^*; \underline{u}_{(\tau+1)-}^*)$ is the optimal cooperative joint payoff in period $\tau+1$ given that the vector of states is $x_{\tau+1}^*$ and the previously executed (but still effective) strategies are $\underline{u}_{(\tau+1)-}^*$,

and

$W(\tau+1, x_{\tau+1}^{**}; \underline{\bar{u}}_{(\tau+1)-}^{**})$ is the optimal cooperative joint payoff in period $\tau+1$ given that the vector of states is $x_{\tau+1}^{**}$ and the previously executed (but still effective) strategies are $\underline{\bar{u}}_{(\tau+1)-}^{**}$.

In particular, the term

$$\begin{aligned}
&W(1, x_1^*; \underline{\bar{u}}_{1-}^*) \\
&\quad - W(\tau, x_\tau^*; \underline{\bar{u}}_{\tau-}^*) - \left[\sum_{j=1}^{n} V^j(1, x_1^{**}; \underline{\bar{u}}_{1-}^{**}) - \sum_{j=1}^{n} V^j(\tau, x_\tau^{**}; \underline{\bar{u}}_{\tau-}^{**}) \right]
\end{aligned}$$

yields the loss in payoff from period 1 to period τ for choosing a feedback equilibrium outcome rather than a Pareto optimal solution. The term

$$W(\tau+1, x_{\tau+1}^*; \underline{\bar{u}}_{(\tau+1)-}^*) - W(\tau+1, x_{\tau+1}^{**}; \underline{\bar{u}}_{(\tau+1)-}^{**})$$

reflects the loss in the cooperative joint payoff from period $\tau+1$ to the end of the game if cooperation starts in period $\tau+1$ rather than period 1.

Hence, $COD(1, \tau + 1)$ gives the loss in payoffs due to the delay in cooperation.

5. Explication in Global Climate Agreement

In this section, we explicate the problem of cooperation and coordination with an analysis of global climate agreement.

5.1. *Existing challenges in global climate agreement*

After several decades of rapid technological advancement and economic growth, alarming levels of pollution and environmental degradation are emerging all over the world. Due to the geographical diffusion of pollutants and cross-national effects of ecosystem degradation, the unilateral response of one nation or region is often ineffective. Self-seeking optimization by individual nations can produce devastating outcomes. Reports are portraying the situation as an advanced civilization on the verge of suicide, destroying its environmental conditions of existence with people being held as prisoners on a runaway catastrophe-bound train. Although global cooperation in environmental control holds the best promise of effective action, success has been limited. This is the result of many hurdles, ranging from agreeable individual commitments, cost-sharing, and disparities in future developments under cooperation to the expected effectiveness of the policy instruments. It is hard to be convinced that multinational joint initiatives like the Paris Agreement can offer a long-term solution because there is no guarantee that participants will always be better off within the entire cooperation duration. To ensure stability in dynamic environmental cooperation over time, a stringent condition is required: the specific agreed-upon optimality principle must be maintained at any instant of time throughout the game along the optimal state trajectory. It is precisely due to the lack of this kind of foresight that current cooperative schemes would fail to provide an effective means to avert disaster. This is a 'classic' game-theoretic problem.

Existing international environmental cooperation arrangements focus mainly on emissions reduction, permit trading, and regulations. The focus on pollution emissions only leaves out the needs of two important elements in alleviating anthropogenic eco-degradation. The first is the improvement (and the reduction in damages) of the ecosystems which can produce desirable impacts on pollution accumulation, global warming, or climate change. Altering the ecosystems in the processes of satisfying the needs of society can cause detrimental effects on the survival of human beings. The second is the development of environmentally clean technology into efficient and affordable means of production which plays a key role in effectively solving the continually worsening global industrial pollution problem and meeting the industrial growth needs.

A number of crucial issues remain unresolved in the existing cooperation schemes. These crucial issues include the following:

(i) The inadequacy of the policy instruments of pollutant emissions in resolving the current environmental problem globally.
(ii) The lack of guarantee that the participating nations would not be worse off (compared to their non-cooperative well-being) throughout the cooperation duration.
(iii) The economic viability of the policy for all participants after joining the cooperation scheme is not ensured.
(iv) The original positions that nations are entitled to hold in the initial stage may not be maintained in all the subsequent stages of the cooperation scheme.
(v) The cooperation scheme does not yield Pareto optimal efficiency.
(iv) Cooperative development of clean technology as a long-term effective remedy has not been seriously deliberated.

The development of an effective plan to solve the catastrophe-bound global environmental problem is very much in order. Given that most existing clean techniques are in general more expensive cooperative schemes involving pollution reduction, and the adoption of clean technology may not be able to meet the needs of industrial growth. Collaborative development of environmentally clean techniques into efficient and affordable means of production plays a key role in effectively solving the continually worsening global industrial pollution problem and meeting industrial growth needs. In addition, subgame consistency, group optimality, individual rationality, and economic viability throughout the cooperation duration must be maintained to make the cooperative scheme sustainable. We provide an effective and efficient means for resolving the catastrophe-bound global environmental problem in the sequel.

5.2. *Modelling global climate agreement*

The development of an effective plan to solve the catastrophe-bound global environmental problem is very much in order. Since most existing clean techniques are in general more expensive cooperative schemes involving pollution reduction and the adoption of clean technology may not be able to meet the needs of industrial growth. Collaborative development of environmentally clean techniques into efficient and affordable means of production plays a key role in effectively solving the continually worsening global industrial pollution problem and meeting industrial growth needs. In addition, subgame consistency, group optimality, individual rationality, and economic viability throughout the cooperation duration must be maintained to make the cooperative scheme sustainable. To provide an effective and efficient means for resolving the catastrophe-bound global environmental problem, we follow the non-uncoordinated version of environmental degradation in Section 4.3 of Chapter 3.

5.2.1. *The set of target controls*

Under cooperation, the entire set of target controls of all nations can be manipulated by the joint decision-making unit. The set of non-durable target controls $\underline{u}_k$ include food production, manufactured products that cause one-shot environmental damages or no environmental damages, resources extraction, clean transportation, clean productions, government spending, taxes, subsidies, pollution abatement activities, environmental improvement activities, and policy instruments. The set of durable target controls $\bar{\underline{u}}_k$ include deforestation, irrigation, genetic engineering, waste disposal, fishery, coal mining and use, electricity generation, nuclear power production, petroleum extraction and use, resources extraction, transportation by land-sea-air, energy consumption, greenhouse gas emissions, oil spills, toxic waste, loss of bio-diversification, over-farming, disposal of solid waste, chemical waste, radioactive materials, plastics waste, and investment in green technology.

5.2.2. *State variables and evolution dynamics*

Major state variables include air and water pollution, the condition of the ecosystems, the level of green technology, the level of biodiversity, government debt, and public capital. In particular, we can express the vector of states as $x_k = (x_k^1, x_k^2, \ldots, x_k^m)$.

The evolution dynamics of the state can be expressed as follows:

$$\begin{pmatrix} x_{k+1}^1 = f_k^1(x_k, \underline{u}_k, \bar{u}_k; \bar{\underline{u}}_{k-}) \\ x_{k+1}^2 = f_k^2(x_k, \underline{u}_k, \bar{u}_k; \bar{\underline{u}}_{k-}) \\ \vdots \\ x_{k+1}^m = f_k^m(x_k, \underline{u}_k, \bar{u}_k; \bar{\underline{u}}_{k-}) \end{pmatrix}. \tag{5.1}$$

However, the evolution of the states in (5.1) may or may not be the same as the case of an uncoordinated scenario depending on whether cooperative actions, like joint R&D, lead to changes in these functions.

5.2.3. *Government objectives and payoffs*

The objectives of the cooperation scheme include (i) air quality improvement, (ii) improving waste management, (iii) minimizing toxic exposure, (iv) good water quality, (v) reducing noise pollution, (vi) resource conservation, (vii) efficiency in energy use, (viii) sustainable ecosystem, (ix) biodiversity, (x) maintaining employment and income, (xi) government revenue, and (xii) economic growth. Concerns will be given to all participating nations with respect to the monetary weights of their objectives. The joint payoff of the nations can be expressed as

$$\sum_{j=1}^{n} \left[\sum_{k=1}^{T} \sum_{\varsigma=1}^{\Theta_j} w_k^{j(\varsigma)} g_k^{j(\varsigma)}(x_k, \underline{u}_k, \bar{u}_k; \bar{\underline{u}}_{k-}) \delta_1^k \right. \\ \left. + q_{T+1}^j(x_{T+1}; \bar{\underline{u}}_{(T+1)-}) \delta_1^{T+1} \right]. \tag{5.2}$$

The set of Pareto optimal target controls $(\underline{u}_k^*, \bar{\underline{u}}_k^*)$ can be obtained by solving the joint maximization problem:

$$\max_{\substack{u_k^j, \bar{u}_k^j, j\in N \\ k\in\{1,2,\ldots,T\}}} \left\{ \sum_{j=1}^{n} \left[\sum_{k=1}^{T} \sum_{\varsigma=1}^{\Theta_j} w_k^{j(\varsigma)} g_k^{j(\varsigma)}(x_k, \underline{u}_k, \bar{\underline{u}}_k; \bar{\underline{u}}_{k-}) \delta_1^k \right. \right.$$
$$\left. \left. + \ q_{T+1}^j(x_{T+1}; \bar{\underline{u}}_{(T+1)-}) \delta_1^{T+1} \right] \right\} \tag{5.3}$$

subject to dynamics (5.1).

A set of comprehensive cross-national policy instruments includes levies on industrial pollution, royalties on natural resources, coordinated pollutant abatement efforts, cooperative development of environmentally sustainable technology, and transfer payments to businesses that invest in clean production technologies, trans-nation rehabilitation of the world ecosystem, cross-boundary pollution abatement, public medical care/education, international pollution abatement, moral suasion, multinational NGO formation, public–private partnership, cross-national welfare payments, and joint green technology R&D. leading to the realization of the optimal target controls $(\underline{u}_k^*, \bar{\underline{u}}_k^*)$.

The evolution of the environmental states under cooperation can be obtained as follows:

$$x_{k+1} = f(x_k, \underline{u}_k^*, \bar{\underline{u}}_k^*; \bar{\underline{u}}_{k-}^*). \tag{5.4}$$

We use $\{x_k^*\}_{k=1}^{T+1}$ to denote the Pareto optimal trajectory of the state path. The maximal value of the joint payoff along the cooperative trajectory $\{x_k^*\}_{k=1}^{T+1}$ can be expressed as follows:

$$W(k, x_k^*; \bar{\underline{u}}_{k-}^*), \text{ for } k \in \{1, 2, \ldots, T\}. \tag{5.5}$$

5.2.4. *Sustainable technology*

It is the effective joint development of environmentally sustainable technology and an optimal technology adoption pattern of conventional and clean technologies over time. The collaborated development of environmentally sustainable technology captures the positive externalities of R&D and innovation of such technology which is not possible under development by individual nations.

Consider a durable green technology investment $\bar{u}_k^{i(\ell)}$, invoking Theorem 2.1 its optimal choice requires

$$\sum_{j=1}^{n}\sum_{\varsigma=1}^{\Theta_j} w_k^{j(\varsigma)} \frac{\partial}{\partial \bar{u}_k} g_k^{j(\varsigma)}(x, \underline{u}_k, \bar{u}_k; \bar{\underline{u}}_{k-})\delta_1^k$$
$$+ \frac{\partial W}{\partial x_{k+1}^{\ell}} \frac{\partial f_k^{\ell}(x, \underline{u}_k, \bar{u}_k; \bar{\underline{u}}_{k-})}{\partial \bar{u}_k^{i(\ell)}} + \frac{\partial W}{\partial \bar{\underline{u}}_k^{i(\ell)}} = 0. \tag{5.6}$$

Note that the optimality condition (5.6) captures the effects of $\bar{u}_k^{i(\ell)}$ on

(i) the objectives of all nations that are affected by $\bar{u}_k^{i(\ell)}$,
(ii) the change in the future joint payoff induced by the change in green technology level x_k^{ℓ} caused by $\bar{u}_k^{i(\ell)}$, and
(iii) the change in the future joint payoff directly caused by $\bar{u}_k^{i(\ell)}$.

Hence the positive externalities in green technology investment are captured and lead to higher investment and joint gains.

5.2.5. *Cooperative curtailment of pollution*

Consider a pollution-generating output $\bar{u}_k^{i(\ell)}$ that would lead to a higher pollution stock and damage the ecosystem over time. Invoking Theorem 2.1, its optimal output choice under cooperation requires

$$\sum_{j=1}^{n}\sum_{\varsigma=1}^{\Theta_j} w_k^{j(\varsigma)} \frac{\partial}{\partial \bar{u}_k} g_k^{j(\varsigma)}(x, \underline{u}_k, \bar{u}_k; \bar{\underline{u}}_{k-})\delta_1^k + \frac{\partial W}{\partial x_{k+1}^{1}} \frac{\partial f_k^{i}(x, \underline{u}_k, \bar{u}_k; \bar{\underline{u}}_{k-})}{\partial \bar{u}_k^{i(\ell)}}$$
$$+ \frac{\partial W}{\partial x_{k+1}^{2}} \frac{\partial f_k^{2}(x, \underline{u}_k, \bar{u}_k; \bar{\underline{u}}_{k-})}{\partial \bar{u}_k^{i(l)}} + \frac{\partial W}{\partial \bar{u}_k^{i(\ell)}} = 0, \tag{5.7}$$

where $f_k^1(x_k, \underline{u}_k, \bar{u}_k; \bar{\underline{u}}_{k-}^*)$ is the increment of the pollution stock and $f_k^2(x_k, \underline{u}_k, \bar{u}_k; \bar{\underline{u}}_{k-}^*)$ is the damage to the condition of the ecosystem.

In particular, $\frac{\partial f_k^1(x, \underline{u}_k, \bar{u}_k; \bar{\underline{u}}_{k-})}{\partial \bar{u}_k^{i(\ell)}} > 0$ and $\frac{\partial f_k^2(x, \underline{u}_k, \bar{u}_k; \bar{\underline{u}}_{k-})}{\partial \bar{u}_k^{i(\ell)}} < 0$.

Note that the optimality condition (5.7) captures the effects of $\bar{u}_k^{i(\ell)}$ on

(i) the objectives of all nations that are affected by $\bar{u}_k^{i(\ell)}$,
(ii) the change in the future joint payoff induced by the change in the pollution stock x_k^1 caused by $\bar{u}_k^{i(\ell)}$,
(iii) the change in the future joint payoff induced by the damage to the ecosystem x_k^2 caused by $\bar{u}_k^{i(\ell)}$,
(iv) the change in the future joint payoff directly caused by $\bar{u}_k^{i(\ell)}$.

Hence the negative externalities in polluting production to all nations are captured and hence less pollution-generating output will be produced under cooperation.

6. Subgame-Consistent Cooperation

In this section, we consider subgame-consistent cooperation among nations in tackling the global environmental degradation problem.

6.1. *Gain-distribution optimality principle*

Now consider the case when the nations agree to distribute the cooperative payoff among themselves according to an optimality principle. We use $\xi^i(k, x_k; \underline{\bar{u}}_{k-})$ to denote the agreed-upon payoff that nation $i \in N$ conditional on the state x_k and the previously executed durable strategies $\underline{\bar{u}}_{k-}$. As stated in Section 2, if the nations agree to share the total cooperative payoff proportional to their non-cooperative payoffs, the payoff to nation i at the start of the cooperation scheme is

$$\xi^i(1, x_1^0; \underline{\bar{u}}_{1-}^0) = \frac{V^i(1, x_1^0; \underline{\bar{u}}_{1-}^0)}{\sum_{j=1}^n V^j(1, x_1^0; \underline{\bar{u}}_{1-}^0)} W(1, x_1^0; \underline{\bar{u}}_{1-}^0), \tag{6.1}$$

for $i \in N$.

To satisfy subgame consistency, the condition in (6.1) has to be maintained throughout the cooperation duration along the cooperative trajectory $\{x_k^*\}_{k=1}^{T+1}$, that is,

$$\xi^i(k, x_k^*; \underline{\bar{u}}_{k-}^*) = \frac{V^i(k, x_k^*; \underline{\bar{u}}_{k-}^*)}{\sum_{j=1}^n V^j(k, x_k^*; \underline{\bar{u}}_{k-}^*)} W(k, x_k^*; \underline{\bar{u}}_{k-}^*), \tag{6.2}$$

for $k \in \{1, 2, \ldots, T\}$ and $i \in N$.

However, to ensure the economic viability of the policy for all participants after joining the cooperation scheme, a sharing optimality principle involving adaptation funds is provided to participating nations that are less developed with contributions from wealthier nations. The fund will help less developed nations in technology transfer, removal of environmentally damaging production techniques, and clean technology innovation. We use $N^w \subset N$ to denote the set of wealthier nations that would make contributions and $i^w \in N^w$ to denote a contributing nation. We use $N^r \subset N$ to denote the set of less-well-off nations that would receive adaptation funds and $i^r \in N^r$ to denote a receiving nation. The payoff of wealthier nation $i^w \in N^w$ can be agreed to be

$$\begin{aligned}\bar{\xi}^{i^w}(k, x_k^*; \underline{\bar{u}}_{k-}^*) &= \xi^{i^w}(k, x_k^*; \underline{\bar{u}}_{k-}^*) - \bar{A}_k^{i^w} \\ &= \frac{V^{i^w}(k, x_k^*; \underline{\bar{u}}_{k-}^*)}{\sum_{j=1}^{n} V^j(k, x_k^*; \underline{\bar{u}}_{k-}^*)} W(k, x_k^*; \underline{\bar{u}}_{k-}^*) - \bar{A}_k^{i^w}, \end{aligned} \tag{6.3}$$

for $k \in \{1, 2, \ldots, T\}$.

The payoff of less-well-off nation $i^r \in N^r$ can be agreed to be

$$\begin{aligned}\hat{\xi}^{i^r}(k, x_k^*; \underline{\bar{u}}_{k-}^*) &= \xi^{i^r}(k, x_k^*; \underline{\bar{u}}_{k-}^*) + \hat{B}_k^{\hat{i}^r} \\ &= \frac{V^{i^r}(k, x_k^*; \underline{\bar{u}}_{k-}^*)}{\sum_{j=1}^{n} V^j(k, x_k^*; \underline{\bar{u}}_{k-}^*)} W(k, x_k^*; \underline{\bar{u}}_{k-}^*) + \hat{B}_k^{\hat{i}^r}, \end{aligned} \tag{6.4}$$

for $k \in \{1, 2, \ldots, T\}$.

In particular, $\bar{A}_k^{i^w}$ is an arbitrary amount of adaptation fund that nation $i^w \in N^w$ is willing to contribute, and $\hat{B}_k^{\hat{i}^r}$ is an arbitrary amount that $i^r \in N^r$ will receive by the Pact. In particular,

$$\xi^{i^w}(k, x_k^*; \underline{\bar{u}}_{k-}^*) - \bar{A}_k^{i^w} > V^{i^w}(k, x_k^*; \underline{\bar{u}}_{k-}^*), \text{ for } i^w \in N^w.$$

Moreover, $\sum_{i^w \in N^w} \bar{A}_k^{i^w} = \sum_{i^r \in N^r} \bar{B}_k^{\hat{i}^r}$.

We use $N^o \subset N$ to denote the set of nations that are neither contributors nor receivers and $i^o \in N^o$ to denote a member in N^o. The payoffs of nation $i^o \in N^o$ will be

$$\xi^{i^o}(k, x_k^*; \underline{\bar{u}}_{k-}^*) = \frac{V^{i^o}(k, x_k^*; \underline{\bar{u}}_{k-}^*)}{\sum_{j=1}^{n} V^j(k, x_k^*; \underline{\bar{u}}_{k-}^*)} W(k, x_k^*; \underline{\bar{u}}_{k-}^*), \tag{6.5}$$

for $k \in \{1, 2, \ldots, T\}$.

This economically viable scheme provides each participating nation with a cooperative payoff higher than its non-cooperative payoff.

Existing climate funds, like the UNFCCC multilateral climate funds, also hold critical political significance, reflecting an acknowledgment by developed countries for historical greenhouse gas emissions and in line with the commitments made by developed countries to support developing countries mitigate and adapt to climate change. Such climate funds play an important role in supporting countries to adopt low-emission, climate-resilient development trajectories. They have a role in capacity building, research, piloting and demonstrating new approaches and technologies, and removing barriers to other climate finance flows. Another source of funding — the Official Development Assistance (ODA) — is defined by the Organization for Economic Cooperation and Development (OECD) as "flows of official financing administered with the promotion of the economic development and welfare of developing countries as the main objective, and which are concessional with a grant element of at least 25 percent. By convention, ODA flows comprise contributions of donor government agencies, at all levels, to developing countries, and to multilateral institutions. ODA receipts comprise disbursements by bilateral donors and multilateral institutions". Developed countries had announced pledges totaling USD 10.1 billion during the initial resource mobilization of the Green Climate Fund. A group of 11 countries announced contributions totaling USD 248 million to the Least Developed Countries Fund, hosted by the Global Environment Facility. The Adaptation Fund has received pledges amounting to almost USD 75 million from Germany, Sweden, Italy, and the Walloon Region of Belgium.

6.2. *Subgame-consistent payoff distribution*

Crucial to the analysis is the derivation of a payoff distribution scheme leading to the realization of the agreed-upon payoffs in (6.3)–(6.5) in all periods $k \in \{1, 2, \ldots, T\}$ throughout the cooperation duration. If this is satisfied, no nations will have an incentive to deviate from the cooperation scheme along its cooperative state trajectory. To do this, we have to design a Payoff Distribution Procedure

(PDP) that leads to the realization of (6.3)–(6.5) in all periods $k \in \{1, 2, \ldots, T\}$.

(i) The non-contributing and non-receiving nations

We use $\beta_k^{i^o}(x_k^*; \underline{\bar{u}}_{k-}^*)$ to denote the payment that non-contributing and non-receiving nation $i^o \in N^o$ receives in period k under the cooperative agreement along the cooperative trajectory with strategies executed previously but still in effect.

Invoking Theorem 3.1, we can obtain the value of $\beta_k^{i^o}(x_k^*; \underline{\bar{u}}_{k-}^*)$ as follows:

$$\beta_k^{i^o}(x_k^*; \underline{\bar{u}}_{k-}^*)$$
$$= (\delta_1^k)^{-1} \left[\xi^{i^o}(k, x_k^*; \underline{\bar{u}}_{k-}^*) - \xi^{i^o}(k+1, f_k(x_k^*, \underline{u}_k^*, \underline{\bar{u}}_k^*; \underline{\bar{u}}_{k-}^*); \underline{\bar{u}}_{(k+1)-}^*) \right], \tag{6.6}$$

where $\xi^{i^o}(k, x_k^*; \underline{\bar{u}}_{k-}^*) = \frac{V^{i^o}(k, x_k^*; \underline{\bar{u}}_{k-}^*)}{\sum_{j=1}^n V^j(k, x_k^*; \underline{\bar{u}}_{k-}^*)} W(k, x_k^*; \underline{\bar{u}}_{k-}^*)$.

(ii) The contributing nations

We use $\bar{\beta}_k^{i^w}(x_k^*; \underline{\bar{u}}_{k-}^*)$ to denote the payment that the contributing nation $i^w \in N^w$ receives in period k under the cooperative agreement along the cooperative trajectory with strategies executed previously but still in effect.

Invoking Theorem 3.1, we can obtain the value of $\bar{\beta}_k^{i^w}(x_k^*; \underline{\bar{u}}_{k-}^*)$ as follows:

$$\bar{\beta}_k^{i^w}(x_k^*; \underline{\bar{u}}_{k-}^*)$$
$$= (\delta_1^k)^{-1} \left[\bar{\xi}^{i^w}(k, x_k^*; \underline{\bar{u}}_{k-}^*) - \bar{\xi}^{i^w}(k+1, f_k(x_k^*, \underline{u}_k^*, \underline{\bar{u}}_k^*; \underline{\bar{u}}_{k-}^*); \underline{\bar{u}}_{(k+1)-}^*) \right], \tag{6.7}$$

where $\bar{\xi}^{i^w}(k, x_k^*; \underline{\bar{u}}_{k-}^*) = \frac{V^{i^w}(k, x_k^*; \underline{\bar{u}}_{k-}^*)}{\sum_{j=1}^n V^j(k, x_k^*; \underline{\bar{u}}_{k-}^*)} W(k, x_k^*; \underline{\bar{u}}_{k-}^*) - \bar{A}_k^{i^w}$

and $\bar{A}_k^{i^w}$ is contribution made by nation i^w.

(iii) The receiving nations

We use $\hat{\beta}_k^{i^r}(x_k^*; \underline{\bar{u}}_{k-}^*)$ to denote the payment that contributing nation $i^r \in N^r$ receives in period k under the cooperative agreement along the cooperative trajectory with strategies executed previously but still in effect.

Invoking Theorem 3.1, we can obtain the value of $\bar{\beta}_k^{i^r}(x_k^*; \underline{\bar{u}}_{k-}^*)$ as follows:

$$\bar{\beta}_k^{i^r}(x_k^*; \underline{\bar{u}}_{k-}^*)$$
$$= (\delta_1^k)^{-1} \left[\hat{\xi}^{i^r}(k, x_k^*; \underline{\bar{u}}_{k-}^*) - \hat{\xi}^{i^r}(k+1, f_k(x_k^*, \underline{u}_k^*, \underline{\bar{u}}_k^*; \underline{\bar{u}}_{k-}^*); \underline{\bar{u}}_{(k+1)-}^*)\right], \tag{6.8}$$

where $\hat{\xi}^{i^r}(k, x_k^*; \underline{\bar{u}}_{k-}^*) = \frac{V^{i^r}(k, x_k^*; \underline{\bar{u}}_{k-}^*)}{\sum_{j=1}^{n} V^j(k, x_k^*; \underline{\bar{u}}_{k-}^*)} W(k, x_k^*; \underline{\bar{u}}_{k-}^*) + \hat{B}_k^{i^r}$

and $\hat{B}_k^{i^r}$ is the amount received by nation i^r.

The payment scheme in (6.6), (6.7), and (6.8) gives rise to the realization of the payoff guided by the agreed-upon optimality principle and constitutes a dynamically consistent payment scheme. More specifically, the payment scheme establishes a cooperative plan that matches the agreed-upon payoff to every nation along the cooperative path.

7. A Computational Illustration

To illustrate the computational details of the analysis in this chapter, we present a collaborative scheme of the $n-$nation international political encounter in environmental collaboration.

7.1. *Problem formulation*

Consider the T-period international political encounter in Chapter 3 in which the n nations consider to collaborate and increase their payoffs. They also agree to share the cooperative gains proportional to their non-cooperative payoffs. In order to achieve group optimality, they would act cooperatively to maximize their joint payoff:

$$\sum_{i=1}^{n} \left\{ \sum_{k=1}^{T} \left[\alpha_k^i \bar{q}_k^i - m_k^i (\bar{q}_k^i)^2 - c_k^i \bar{q}_k^i + 2\varpi_k^i (\vartheta_k q_k^i)^{\frac{1}{2}} \right.\right.$$
$$\left. - \varepsilon_k^i q_k^i + \chi_k^i \bar{w}_k^i - (\bar{w}_k^i)^2 - c_k^{(a)i} (u_k^i)^2 - h_k^i x_k \right] \delta^{k-1}$$
$$\left. + (h_{T+1}^i x_{T+1} + y_{T+1}^i \vartheta_{T+1} + Q_{T+1}^i) \delta^T \right\}, \tag{7.1}$$

subject to the pollution accumulation dynamics

$$
\begin{aligned}
x_{k+1} &= x_k + \sum_{i=1}^{n} a_k^i \bar{q}_k^i + \sum_{i=1}^{n} a_k^{(1)i} \bar{q}_{k-1}^i \\
&\quad + \sum_{i=1}^{n} a_k^{(2)i} \bar{q}_{k-2}^i - \sum_{i=1}^{n} b_k^i u_k^i (x_k)^{\frac{1}{2}} - \gamma_k x_k, \\
x_1 &= x_1^0,
\end{aligned} \tag{7.2}
$$

and the green technology evolution dynamics

$$
\begin{aligned}
\vartheta_{k+1} &= \vartheta_k + \sum_{i=1}^{n} z_k^i \bar{w}_k^i + \sum_{i=1}^{n} z_k^{(1)i} \bar{w}_k^i - \lambda_k \vartheta_k, \\
\vartheta_1 &= \vartheta_1^0.
\end{aligned} \tag{7.3}
$$

7.2. *Collaborative outcome*

We first obtain the joint payoffs of the nations under collaboration and the optimal target controls.

Proposition 7.1. (1) *The joint payoffs of the nations under collaboration are as follows:*
$W(k, x, \vartheta; \underline{\bar{q}}_{k-}, \underline{\bar{w}}_{k-}) = (A_k x_k + B_k \vartheta_k + C_k)\delta^{k-1}$, *where*

$$
\begin{aligned}
A_k &= \sum_{i=1}^{n} \left[-c_k^{(a)i} \left(\frac{b_k^j}{2c_k^{(a)j}} A_{k+1}\delta \right)^2 - h_k^i \right] \\
&\quad + A_{k+1} \left[(1-\gamma_k) + \sum_{j=1}^{n} \frac{(b_k^j)^2}{2c_k^{(a)j}} A_{k+1}\delta \right] \delta, \\
B_k &= \sum_{i=1}^{n} \frac{(\varpi_k^i)^2}{\varepsilon_k^i} + B_{k+1}(1-\lambda_k)\delta,
\end{aligned}
$$

C_k *is a function involving strategies executed before period* k, *namely,* $\bar{q}_{k-1}^j$, $\bar{q}_{k-2}^j$, *and* $\bar{w}_{k-1}^j$ *for* $j \in N$ *and* $k \in \{1, 2, \ldots, T-2\}$,

$$\tag{7.4}$$

$$A_{T-1} = \sum_{i=1}^{n} \left[-c_{T-1}^{(a)i} \left(\frac{b_{T-1}^{j}}{2c_{T-1}^{(a)j}} A_T \delta \right)^2 - h_{T-1}^{i} \right]$$

$$+ A_T \left[(1-\gamma_{T-1}) + \sum_{j-1}^{n} \frac{(b_{T-1}^{j})^2}{2c_{T-1}^{(a)j}} A_{T-1}\delta \right] \delta,$$

$$B_{T-1} = \sum_{i=1}^{n} \frac{(\varpi_{T-1}^{i})^2}{\varepsilon_{T-1}^{i}} + B_T(1-\lambda_{T-1})\delta,$$

C_{T-1} is a function involving strategies executed before period $T-1$, namely, $\bar{q}_{T-2}^{j}$, $\bar{q}_{T-3}^{j}$, and $\bar{w}_{T-2}^{j}$, for $j \in N$,

$$(7.5)$$

$$A_T = \sum_{i=1}^{n} \left[-c_T^{(a)i} \left(\frac{b_T^{i}}{2c_T^{(a)i}} A_{T+1} \delta \right)^2 - h_T^{i} \right]$$

$$+ A_{T+1} \left[(1-\gamma_T) + \sum_{j=1}^{n} \frac{(b_T^{j})^2}{2c_T^{(a)j}} A_{T+1} \right] \delta,$$

$$B_T = \sum_{i=1}^{n} \frac{(\varpi_T^{i})^2}{\varepsilon_T^{i}} + B_{T+1}(1-\lambda_T)\delta,$$

C_T is a function involving strategies executed before period T, namely, $\bar{q}_{T-1}^{j}$, $\bar{q}_{T-2}^{j}$, and $\bar{w}_{T-1}^{j}$, for $j \in N$.

$$(7.6)$$

(2) *The optimal target controls under collaboration are*

$$\bar{q}_T^{i*} = \frac{\alpha_T^{i} - c_T^{i} + A_{T+1}\delta a_T^{i}}{2m_T^{i}}, \quad q_T^{i*} = \frac{(\varpi_T^{i})^2 \vartheta_T}{(\varepsilon_T^{i})^2},$$

$$u_T^{i*} = -\frac{b_T^{i}}{2c_T^{(a)i}} A_{T+1}\delta (x_T)^{1/2}, \quad \bar{w}_T^{i*} = \frac{\chi_T^{i} + B_{T+1}\delta z_T^{i}}{2}, \quad \text{for } i \in N,$$

$$\bar{q}_{T-1}^{i*} = \frac{\alpha_{T-1}^{i} - c_{T-1}^{i} + A_T\delta a_{T-1}^{i} + A_{T+1}\delta^2 a_{T-1}^{(1)i}}{2m_{T-1}^{i}},$$

$$
\begin{aligned}
q_{T-1}^{i*} &= \frac{(\varpi_{T-1}^{i})^2 \vartheta_{T-1}}{(\varepsilon_{T-1}^{i})^2}, \\
u_{T-1}^{i*} &= -\frac{b_{T-1}^{i}}{2c_{T-1}^{(a)i}} A_T \delta (x_{T-1})^{1/2}, \\
\bar{w}_{T-1}^{i*} &= \frac{\chi_{T-1}^{i} + B_T \delta z_{T-1}^{i} + B_{T+1} \delta^2 z_{T-1}^{(1)i}}{2}, \text{ for } i \in N, \\
\bar{q}_k^{i*} &= \frac{\alpha_k^i - c_k^i + A_{k+1} \delta a_k^i + A_{k+2} \delta^2 a_k^{(1)i} + A_{k+3} \delta^3 a_k^{(2)i}}{2m_k^i}, \\
q_k^{i*} &= \frac{(\varpi_k^i)^2 \vartheta_k}{(\varepsilon_k^i)^2}, \\
u_k^{i*} &= -\frac{b_k^i}{2c_k^{(a)i}} A_{k+1} \delta (x_k)^{1/2}, \quad \bar{w}_k^{i*} = \frac{\chi_k^i + B_k \delta z_k^i + B_{k+2} \delta^2 z_k^{(1)i}}{2},
\end{aligned}
\tag{7.7}
$$

for $i \in N$ *and* $k \in \{T-2, T-3, \ldots, 1\}$.

Proof. See Appendix B in the Mathematical Appendices. □

Comparing Proposition 7.1 in this chapter and Proposition 5.1 in Chapter 3 shows the negative magnitude of A_k is stronger than the negative magnitude A_k^i which leads to a lower level of polluting outputs under cooperation. Similarly, $B_k > B_k^i$ and leads to a higher level of green technology R&D. Hence the level of green output is larger under cooperation.

Substituting the optimal target controls in (7.7) into (7.2) and (7.3), we obtain the pollution accumulation dynamics and the green technology evolution dynamics under collaboration. We denote the solutions to the dynamics by $\{x_k^*, \vartheta_k^*\}_{k=1}^{T+1}$.

7.3. *Global policy instruments and implementation*

In this section, we consider the nations' collaborative policy instruments in fulfilling the optimal target controls.

(i) The Polluting Industrial Sector

In the pollution-generating sector, the sum of all the nation's net economic profits is $\sum_{i=1}^{n} [\alpha_k^i \bar{q}_k^i - m_k^i(\bar{q}_k^i)^2 - c_k^i \bar{q}_k^i]$. The market's profit maximization with respect to $\bar{q}_k^i$ yields the first-order condition

$$\alpha_k^i - 2\beta_k^i \bar{q}_k^i - c_k^i = 0,$$

leading to the industry's profit-maximizing output $\bar{q}_k^i = \frac{\alpha_k^i - c_k^i}{2m_k^i}$.

However, the optimal target controls under collaboration in the polluting industrial sector are

$$\bar{q}_k^{i*} = \frac{\alpha_k^i - c_k^i + A_{k+1}\delta a_k^i + A_{k+2}\delta^2 a_k^{(1)i} + A_{k+3}\delta^3 a_k^{(2)i}}{2m_k^i},$$

for $k \in \{1, 2, \ldots, T-2\}$.

A Pigouvian tax equaling $\frac{A_{k+1}\delta a_k^i + A_{k+2}\delta^2 a_k^{(1)i} + A_{k+3}\delta^3 a_k^{(2)i}}{2m_k^i}$ per unit of output has to be levied on the polluting output to achieve the optimal target control in period $k \in \{1, 2, \ldots, T-2\}$.

Similarly, a Pigouvian tax $\frac{A_T \delta a_{T-1}^i + A_{T+1}\delta^2 a_{T-1}^{(1)i}}{2m_{T-1}^i}$ in period $T-1$ and a Pigouvian tax $\frac{A_{T+1}\delta a_T^i}{2m_T^i}$ in period T have to be imposed.

(ii) The Green Sector

The net economic profit in the green sector is $2\varpi_k(\vartheta_k q_k^i)^{1/2} - \varepsilon_k^i q_k^i$. Maximizing economic profit with respect to q_k^i yields the first-order condition

$$\varpi_k(\vartheta_k)^{1/2}(q_k^i)^{-1/2} - \varepsilon_k^i = 0,$$

which leads to an industry's profit-maximizing output $q_k^i = \frac{(\varpi_k^i)^2 \vartheta_k}{(\varepsilon_k^i)^2}$.

The optimal target control under collaboration in the green industry is

$$q_k^{i*} = \frac{(\varpi_k^i)^2 \vartheta_k}{(\varepsilon_k^i)^2},$$

which corresponds to that industry's profit maximizing output. Hence there is no need for a government intervention.

(iii) Green Technology R&D Sector

The net economic profit in the green technology R&D sector is $\chi_k \bar{w}_k^i - (\bar{w}_k^i)^2$. Maximizing economic profit with respect to $\bar{w}_k^i$ yields the first-order condition

$$\chi_k - 2\bar{w}_k^i = 0,$$

which leads to an industry's profit-maximizing R&D $\bar{w}_k^i = \frac{\chi_k^i}{2}$.

However, the optimal target control of R&D under collaboration is

$$\bar{w}_k^{i*} = \frac{\chi_k^i + B_k^i \delta z_k^i + B_{k+2}^i \delta^2 z_k^{(1)i}}{2}, \text{ for } k \in \{1, 2, \ldots, T-1\}.$$

A Pigouvian subsidy equaling $\frac{B_k^i \delta z_k^i + B_{k+2}^i \delta^2 z_k^{(1)i}}{2}$ per unit of output has to be provided to the green technology R&D sector to achieve the optimal target control in period $k \in \{1, 2, \ldots, T-1\}$.

Similarly, a Pigouvian subsidy $\frac{B_{T+1}^i \delta z_T^i}{2}$ has to be provided in period T.

(iv) The Government Sector for Pollution Abatement

The pollution abatement sector is a government sector in which the government would spend abatement up to the level when the marginal abatement cost equals the marginal benefit from reducing pollution, that is,

$$2c_k^{(a)i} u_k^i = -A_{k+1} b_k^i u_k^i (x_k)^{1/2} \delta,$$

which leads to the target control $u_k^{i*} = -\frac{b_k^i}{2c_k^{(a)i}} A_{k+1} \delta (x_k)^{1/2}$.

If the tax revenue collected from the polluting industrial sectors of all nations is higher than the total subsidy offered to the green technology R&D sectors, that is,

$$\sum_{i=1}^{n} \frac{A_{k+1}\delta a_k^i + A_{k+2}\delta^2 a_k^{(1)i} + A_{k+3}\delta^3 a_k^{(2)i}}{2m_k^i} \bar{q}_k^{i*}$$

$$\geq \sum_{i=1}^{n} \frac{B_k \delta z_k^i + B_{k+2}\delta^2 z_k^{(1)i}}{2} \frac{B_k \delta z_k^i + B_{k+2}\delta^2 z_k^{(1)i}}{2} \bar{w}_k^{i*},$$

the collaborating governments can keep the balance as reserves or distribute it to the private sector.

On the other hand, if the tax revenue collected from the polluting industrial sector of all nations is less than the total subsidy offered to the green technology R&D sector, that is,

$$\sum_{i=1}^{n} \frac{A_{k+1}\delta a_k^i + A_{k+2}\delta^2 a_k^{(1)i} + A_{k+3}\delta^3 a_k^{(2)i}}{2m_k^i} \bar{q}_k^{i*}$$
$$< \frac{B_k \delta z_k^i + B_{k+2}\delta^2 z_k^{(1)i}}{2} \bar{w}_k^{i*},$$

the collaborating governments can bear the loss or impose a general tax proportional to the collaborative payoffs of the nations or according to an agreed-upon sharing formula.

7.4. *Subgame-consistent payoff distribution among nations*

The nations agree to share their collaborative gains according to the relative proportions of their non-collaborative payoffs.

Hence nation $i \in N$ will receive a payoff under collaboration equaling

$$\begin{aligned}
&\xi^i(k, x_k^*, \vartheta_k^*; \underline{q}_{k-}, \underline{\bar{w}}_{k-}) \\
&= W(k, x_k^*, \vartheta_k^*; \underline{q}_{k-}, \underline{\bar{w}}_{k-}) \frac{V^i(k, x_k^*, \vartheta_k^*; \underline{q}_{k-}, \underline{\bar{w}}_{k-})}{\sum_{j=1}^{n} V^j(k, x_k^*, \vartheta_k^*; \underline{q}_{k-}, \underline{\bar{w}}_{k-})} \\
&= (A_k x_k^* + B_k \vartheta_k^* + C_k) \frac{(A_k^i x_k^* + B_k^i \vartheta_k^* + C_k^i)}{\sum_{j=1}^{n}(A_k^j x_k^* + B_k^j \vartheta_k^* + C_k^j)} \delta^{k-1}, \quad (7.8)
\end{aligned}$$

$k \in \{1, 2, \ldots, T\}$.

In order for $\xi^i(k, x, \vartheta; \underline{q}_{k-}, \underline{\bar{w}}_k)$ to hold throughout all period $k \in \{1, 2, \ldots, T\}$, a payoff distribution procedure has to be formulated.

Proposition 7.2. *Invoking Theorem* III *in Chapter* 7 *on subgame-consistent solution mechanisms, we can maintain the condition* *in* (7.8) *with a payment equaling*

$$\beta_k^i(x_k^*, \vartheta_k^*; \underline{\bar{q}}_{k-}, \underline{\bar{w}}_{k-})$$

$$= \delta^{1-k} \Bigg[W(k, x_k^*, \vartheta_k^*; \underline{\bar{q}}_{k-}, \underline{\bar{w}}_{k-}) \frac{V^i(k, x_k^*, \vartheta_k^*; \underline{\bar{q}}_{k-}, \underline{\bar{w}}_{k-})}{\sum_{j=1}^n V^j(k, x_k^*, \vartheta_k^*; \underline{\bar{q}}_{k-}, \underline{\bar{w}}_{k-})}$$

$$- W(k+1, x_{k+1}^*, \vartheta_{k+1}^*; \underline{\bar{q}}_{(k+1)-}, \underline{\bar{w}}_{(k+1)-})$$

$$\times \frac{V^i(k+1, x_{k+1}^*, \vartheta_{k+1}^*; \underline{\bar{q}}_{(k+1)-}, \underline{\bar{w}}_{(k+1)-})}{\sum_{j=1}^n V^j(k+1, x_{k+1}^*, \vartheta_{k+1}^*; \underline{\bar{q}}_{(k+1)-}, \underline{\bar{w}}_{(k+1)-})} \Bigg]$$

$$= (A_k x_k^* + B_k, \vartheta_k^* + C_k) \frac{(A_k^i x_k^* + B_k^i \vartheta_k^* + C_k^i)}{\sum_{j=1}^n (A_k^j x_k^* + B_k^j \vartheta_k^* + C_k^j)}$$

$$- (A_{k+1} x_{k+1}^* + B_{k+1} \vartheta_{k+1}^* + C_{k+1})$$

$$\times \frac{(A_{k+1}^i x_{k+1}^* + B_{k+1}^i \vartheta_{k+1}^* + C_{k+1}^i)}{\sum_{j=1}^n (A_{k+1}^j x_{k+1}^* + B_{k+1}^j \vartheta_{k+1}^* + C_{k+1}^j)} \delta \tag{7.9}$$

given to nation $i \in N$ at period $k \in \{1, 2, \ldots, T\}$.

Proof. See proof of Theorem III in Chapter 6 on subgame-consistent solution mechanisms. □

7.5. *Other subgame-consistent payoff distributions*

The subgame-consistent payoff distribution in Section 6.4 can be applied to other gain-sharing optimality principles.

(i) Sharing the excess of cooperative gains over the sum of the non-cooperative equally:

A payment equaling

$$\tilde{\beta}_k^i(x_k^*, \vartheta_k^*; \underline{\bar{q}}_{k-}, \underline{\bar{w}}_{k-}) = (A_k^i x_k^* + B_k^i \vartheta_k^* + C_k^i)$$

$$+ \frac{1}{n} \left[(A_k x_k^* + B_k \vartheta_k^* + C_k) - \sum_{j=1}^n (A_{k+1}^j x_{k+1}^* + B_{k+1}^j \vartheta_{k+1}^* + C_{k+1}^j) \right]$$

$$- \Bigg[(A_{k+1}^i x_{k+1}^* + B_{k+1}^i \vartheta_{k+1}^* + C_{k+1}^i)$$

$$+\frac{1}{n}\left[(A_{k+1}x_{k+1}^{*}+B_{k+1}\vartheta_{k+1}^{*}+C_{k+1})\right.$$

$$\left.\left.-\sum_{j=1}^{n}(A_{k+1}^{j}x_{k+1}^{*}+B_{k+1}^{j}\vartheta_{k+1}^{*}+C_{k+1}^{j})\right]\right]\delta \tag{7.10}$$

will be given to nation $i \in N$ at period $k \in \{1, 2, \ldots, T\}$.

(ii) A gain sharing optimality principle which is a weighted average of $\xi^i(k, x_k^*, \vartheta_k^*; \underline{\bar{q}}_{k-}, \underline{\bar{w}}_{k-})$ *and* $\tilde{\xi}^i(k, x_k^*, \vartheta_k^*; \underline{\bar{q}}_{k-}, \underline{\bar{w}}_{k-})$, *that is,*

$$\tilde{\xi}^i(k, x_k^*, \vartheta_k^*; \underline{\bar{q}}_{k-}, \underline{\bar{w}}_{k-}) = \alpha\xi^i(k, x_k^*, \vartheta_k^*; \underline{\bar{q}}_{k-}, \underline{\bar{w}}_{k-})$$
$$+ (1-\alpha)\tilde{\xi}^i(k, x_k^*, \vartheta_k^*; \underline{\bar{q}}_{k-}, \underline{\bar{w}}_{k-}):$$

A payment equaling

$$\widehat{\beta}_k^i(x_k^*, \vartheta_k^*; \underline{\bar{q}}_{k-}, \underline{\bar{w}}_{k-}) = \alpha\beta_k^i(x_k^*, \vartheta_k^*; \underline{\bar{q}}_{k-}, \underline{\bar{w}}_{k-})$$
$$+ (1-\alpha)\tilde{\beta}_k^i(x_k^*, \vartheta_k^*; \underline{\bar{q}}_{k-}, \underline{\bar{w}}_{k-}) \tag{7.11}$$

will be given to nation $i \in N$ at period $k \in \{1, 2, \ldots, T\}$.

(iii) A gain-sharing optimality principle with Adaptation Fund:

A payment equaling

$$\bar{\beta}_k^{i^*}(x_k^*, \vartheta_k^*; \underline{\bar{q}}_{k-}, \underline{\bar{w}}_{k-}) = \hat{\beta}_k^{i^w}(x_k^*, \vartheta_k^*; \underline{\bar{q}}_{k-}, \underline{\bar{w}}_{k-}) - \bar{A}_k^{i^w}$$

will be given to a contributing nation $i^w \in N^w$ at period $k \in \{1, 2, \ldots, T\}$.

A payment equaling

$$\hat{\beta}_k^{i^*}(x_k^*, \vartheta_k^*; \underline{\bar{q}}_{k-}, \underline{\bar{w}}_{k-}) = \hat{\beta}_k^{i^r}(x_k^*, \vartheta_k^*; \underline{\bar{q}}_{k-}, \underline{\bar{w}}_{k-}) + \hat{B}_k^{i^r}$$

will be given to a receiving nation $i^r \in N^r$ at period $k \in \{1, 2, \ldots, T\}$.

8. Chapter Notes

This chapter presents a general class of cooperative dynamic environmental games involving anthropogenic eco-degradation and green technology development under durable strategy lags. Cooperative differential games in environmental control had been studied by Dockner and Long (1993), Jørgensen and Zaccour (2001), Fredj *et al.* (2004), Breton *et al.* (2005, 2006), Petrosyan and Zaccour (2003), Yeung (2007, 2008), and Yeung and Petrosyan (2008). Cooperative differential games that have identified dynamically consistent solutions can be found in Jørgensen and Zaccour (2001), Petrosyan and Zaccour (2003), Yeung (2007), and Yeung and Petrosyan (2004, 2005, 2006a, 2006b, 2008), and Yeung *et al.* (2017, 2021). Yeung (2014) presented a dynamic cooperative game of environmental management in which discrete choices of production techniques are available. Yeung and Petrosyan (2016b) provided the first cooperative dynamic environmental game with technology switching and clean technology development. This chapter enriches the Yeung and Petrosyan (2016b) game with the incorporation of natural resources and their effects on the evolution of the pollution stock. Studies on using green technology to alleviate environmental damages had been presented in Billatos and Basaly (1997), Ane (2000), Eis *et al.* (2016), Yeung and Petrosyan (2016b), and Yeung *et al.* (2017). The derivation of $\beta_k^i(x_k^*; \underline{\bar{u}}_{k-}^*)$ is provided in Yeung and Petrosyan (2010, 2016a, 2019).

The question of dynamic stability in differential games has been rigorously explored in the past three decades. Haurie (1976) raised the problem of instability when the Nash bargaining solution is extended to differential games. Petrosyan (1977) formalized the notion of dynamic stability in solutions of differential games. Petrosyan and Danilov (1985) introduced the notion of "payoff distribution procedure" for cooperative solution. Tolwinski *et al.* (1986) investigated cooperative equilibria in differential games in which memory-dependent strategies and threats are introduced to maintain the agreed-upon control path. Petrosyan (1993) and Petrosyan and Zenkevich (1996) presented a detailed analysis of dynamic stability in cooperative differential games, in which the method of regularization was introduced to construct time-consistent solutions.

Yeung and Petrosyan (2001) designed time-consistent solutions in differential games and characterized the conditions that the allocation-distribution procedure must satisfy. Petrosyan (2003) employed the regularization method to construct time-consistent bargaining procedures. Yeung *et al.* (2023) developed subgame-consistent Shapley value allocation in a dynamic network game.

References

Ane, R. P. (2000). Stimulating innovation in green technology. Policy alternatives and opportunities. *American Behavioral Scientist*, 44(2), 188–212.

Billatos, S. & Basaly, N. (1997). *Green Technology and Design for the Environment.* Washington, DC: CRC Press, 312 p.

Breton, M., Zaccour, G., & Zahaf, M. (2005). A differential game of joint implementation of environmental projects. *Automatica*, 41(10), 1737–1749.

Breton, M., Zaccour, G., & Zahaf, M. (2006). A game-theoretic formulation of joint implementation of environmental projects. *European Journal of Operational Research*, 168(1), 221–239.

Dockner, E. & Long, N. V. (1993). International pollution control: Cooperative versus noncooperative strategies. *Journal of Environmental Economics and Management*, 24, 13–29.

Dubey, P. (1986). Inefficiency of Nash equilibria. *Mathematics of Operations Research*, 11(1), 1–8.

Eis, J., Bishop, R., & Gradwell, P. (2016). Galvanising low-carbon innovation. *A New Climate Economy Working Paper for Seizing the Global Opportunity: Partnerships for Better Growth and a Better Climate.* New Climate Economy, London, 1–28 pp.

Fredj, K., Martín-Herrán, G., & Zaccour, G. (2004). Slowing deforestation pace through subsidies: A differential game. *Automatica*, 40, 301–309. DOI: 10.1016/j.automatica.2003.10.020.

Haurie, A. (1976). A note on nonzero-sum differential games with bargaining solutions. *Journal of Optimization Theory and Applications*, 18, 31–39.

Jørgensen, S. & Zaccour, G. (2001). Time consistent side payments in a dynamic game of downstream pollution. *Journal of Economic Dynamics and Control*, 25, 1973–1987.

Koutsoupias, E. & Papadimitriou, C. (2009). Worst-case equilibria. *Computer Science Review*, 3(2), 65–69. Archived from the original on 2016-03-13. Retrieved 2010-09-12.

Petrosyan, L. A. (1977). Stability of solutions of nonzero sum differential games with many participants. *Viestnik (Transactions) of Leningrad University*, 19, 6–52.

Petrosyan, L. A. (1993). *Differential Games of Pursuit*. World Scientific, Singapore, London, 326 p.

Petrosyan, L. A. (1997). Agreeable solutions in differential games. *Game Theory and Applications III*. Nova Science Publishers, Commack, pp. 91–103.

Petrosyan, L. A. (2003). *Bargaining in Dynamic Games*. In: Petrosyan, L. A., Yeung, D. W. K. (eds.) ICM Millennium Lectures on Games. Springer, Berlin, Heidelberg.

Petrosyan, L. A. & Danilov, N. N. (1985). *Cooperative Differential Games and Their Applications*. Tomskogo Gosudarstvennogo Universiteta, Tomsk, 276 p.

Petrosyan, L. A. & Yeung, D. W. K. (2020a). Cooperative dynamic games with durable controls: Theory and application. *Dynamic Games and Applications*, 10, 872–896. DOI: 10.1007/s13235-019-00336-w.

Petrosyan, L. A. & Zaccour, G. (2003). Time-consistent Shapley value allocation of pollution cost reduction. *Journal of Economic Dynamics and Control*, 27, 381–398.

Petrosyan, L. A. & Zenkevich, N. A. (1996). *Game Theory*. World Scientific: New Jersey, 380 p.

Tolwinski, B., Haurie, A., & Leitmann, G. (1986). Cooperative equilibria in differential games. *Journal of Mathematical Analysis and Applications*, 119, 182–202.

Yeung, D. W. K. (2005). Subgame consistent dormant-firm cartel. In A. Haurie and G. Zaccour (eds.), *Dynamic Games and Applications*. Springer, Berlin, pp. 255–271.

Yeung, D. W. K. (2006). Solution mechanisms for cooperative stochastic differential games. *International Game Theory Review*, 8(2), 309–326.

Yeung, D. W. K. (2007). Dynamically consistent cooperative solution in a differential game of transboundary industrial pollution. *Journal of Optimization Theory and Applications*, 134, 143–160.

Yeung, D. W. K. (2008). Dynamically consistent solution for a pollution management game in collaborative abatement with uncertain future payoffs. Special Issue on Frontiers in Game Theory: In Honour of John F. Nash. *International Game Theory Review*, 10(4), 517–538.

Yeung, D. W. K. (2010). Subgame consistent Shapley value imputation for cost-saving joint ventures. *Mathematical Game Theory and Applications*, 2(3), 137–149.

Yeung, D. W. K. (2014). Dynamically consistent collaborative environmental management with production technique choices. *Annals of Operations Research*, 220(1), 181–204. DOI: 10.1007/s10479-011-0844-0.

Yeung, D. W. K. & Petrosyan, L. A. (2001). Proportional time-consistent solution in differential games. In Yanovskaya, E. B. (ed.), *International Conference on Logic, Game Theory and Social Choice*. St Petersburg State University, pp. 254–256.

Yeung, D. W. K. & Petrosyan, L. A. (2004). Subgame consistent cooperative solutions in stochastic differential games. *Journal of Optimization Theory and Applications*, 120(3), 651–666.

Yeung, D. W. K. & Petrosyan, L. A. (2005). Subgame consistent solution of a cooperative stochastic differential game with nontransferable payoffs. *Journal of Optimization Theory and Applications*, 124(3), 701–724.

Yeung, D. W. K. & Petrosyan, L. A. (2006a). Dynamically stable corporate joint ventures. *Automatica*, 42, 365–370.

Yeung, D. W. K. & Petrosyan, L. A. (2006b). *Cooperative Stochastic Differential Games*. Springer-Verlag, New York, 242 p.

Yeung, D. W. K. & Petrosyan, L. A. (2007a). Managing catastrophe-bound industrial pollution with game-theoretic algorithm: The St Petersburg Initiative. In Petrosyan, L. A. & Zenkevich, N. A. (eds.), *Contributions to Game Theory and Management*. St. Petersburg State University, pp. 524–538.

Yeung, D. W. K. & Petrosyan, L. A. (2007b). The crux of dynamic economic cooperation: Subgame consistency and equilibrating transitory compensation. *Game Theory and Applications*, 11, 207–221.

Yeung, D. W. K. & Petrosyan, L. A. (2007c). The tenet of transitory compensation in dynamically stable cooperation. *International Journal of Tomography and Statistics*, 7, 60–65.

Yeung, D. W. K. & Petrosyan, L. A. (2008). A cooperative stochastic differential game of transboundary industrial pollution. *Automatica*, 44(6), 1532–1544.

Yeung, D. W. K. & Petrosyan, L. A. (2010). Subgame consistent solutions for cooperative stochastic dynamic games. *Journal of Optimization Theory and Applications*, 145, 579–596.

Yeung, D. W. K. & Petrosyan, L. A. (2013). Subgame-consistent cooperative solutions in randomly furcating stochastic dynamic games. *Mathematical and Computer Modelling*, 57(3–4), 976–991.

Yeung, D. W. K. & Petrosyan, L. A. (2014). Subgame consistent cooperative solutions for randomly furcating stochastic dynamic games with uncertain horizon. *International Game Theory Review*, 16(2), 1440012.01–1440012.29.

Yeung, D. W. K. & Petrosyan, L. A. (2016a). *Subgame Consistent Cooperation: A comprehensive treatise.* Singapore: Springer, 520 p.

Yeung, D. W. K. & Petrosyan, L. A. (2016b). A cooperative dynamic environmental game of subgame consistent clean technology development. *International Game Theory Review*, 18(2), 164008.01–164008.23.

Yeung, D. W. K. & Petrosyan, L. A. (2019). Cooperative dynamic games with control lags. *Dynamic Games and Applications*, 9(2), 550–567. https://doi.org/10.1007/s13235-018-0266-6.

Yeung, D. W. K. & Petrosyan, L. A. (2021). Asynchronous horizons durable-strategies dynamic games and tragedy of cross-generational environmental commons. *International Game Theory Review*, 23(4). DOI: 10.1142/s0219198921500201.

Yeung, D. W. K. & Petrosyan, L. A. (2022). *Durable-Strategies Dynamic Games: Theory, Solution Techniques and Applications.* Springer, Switzerland, 303 p.

Yeung, D. W. K., Petrosyan, L. A., & Zhang, Y. X. (2023). A dynamic network game of the fintech industry. *Journal of the Operations Research Society of China.* https://doi.org/10.1007/s40305-022-00434-4.

Yeung, D. W. K., Zhang, Y. X., Bai, H. T., & Islam, S. (2021). Collaborative environmental management for transboundary air pollution problems: A differential levies game. *Journal of Industrial and Management Optimization*, 17(2), 517–531. DOI: 10.3934/jimo.2019121.

Yeung, D. W. K., Petrosyan, L. A., Zhang, Y. X., & Cheung, F. (2017). *BEST SCORES Solution to the Catastrophe Bound Environment*, Nova Science, New York, 184 pp., ISBN 978-1 53610-924-5.

Chapter 5

Coalition and Blocs: Mathematical Theory and Explication in Trade Networks

"One of the things the United States does well is building coalitions. What the U.S. knows is that if you don't have a coalition with you, you will have a coalition against you." — Shimon Peres

In many real-life scenarios, groups or nations with common interests form coalition blocs by agreement for mutual support and joint actions. Coalition blocs are formed by players with common interests to enhance their gains through cooperation. In coalitions, the participating players — whether in international political unions, trade networks, military blocs, climate management groups, or medical research joint ventures — would retain their distinctive identity and interests. Yet the purpose of collaboration is ultimately the same: to aggregate actors' strengths to achieve some shared goal that none could achieve individually. The organization of this chapter is as follows. Coalition and bloc formation is discussed in Section 1. The mathematical foundation of GPE with coalitions is presented in Section 2. Grand coalition cooperation is examined in Section 3. An explication of trade networks involving China's Belt-Road Initiative (BRI) and G7 countries' Build Back Better World (B3W) Initiative is given in Section 4. A grand coalition cooperation that would improve the payoffs of BRI and B3W is provided in Section 5.

A computational illustration is given in Section 6 and chapter notes are given in Section 7.

1. Coalition and Bloc Formation

International coalition blocs are groups of nations that share similar aims and interests and that act together over some issues. The term bloc can also refer to the alignment of nations during the Cold War when the world was divided into two main camps: the Western Bloc, led by the United States and its allies, and the Eastern Bloc, led by the former Soviet Union and its allies.

Some of the factors that influence the formation of coalition blocs are geography, ideology, culture, religion, security, and economic interests. Coalition blocs can cooperate or compete with each other on various issues, such as trade, diplomacy, human rights, security, and development. Some examples of coalition blocs in the contemporary world include the following:

(i) **The European Union (EU):** It is a political and economic union of 27 member states that are located primarily in Europe. The EU is an internal single market through a standardized system of laws. The EU policies aim to ensure the free movement of people, capital, and goods and services within the internal market; enact common legislation in justice and home affairs; and maintain common policies on trade, agriculture, fisheries, and regional development.

(ii) **The North Atlantic Treaty Organization (NATO):** It is a military alliance of 30 European and North American countries that was established in 1949 to counter the threat of the former Soviet Union expansionism. NATO constitutes a system of collective defense whereby its independent member states agree to mutual defense in response to an attack by any external party.

(iii) **The Association of Southeast Asian Nations (ASEAN):** It is a regional intergovernmental organization comprising 10 countries in Southeast Asia, which promotes intergovernmental cooperation and facilitates economic, political, security, military, educational, and sociocultural integration among its members and other countries in Asia. Its motto is "One Vision, One Identity, One Community".

(iv) **The African Union (AU):** It is a continental union consisting of 55 member states located on the continent of Africa. The AU was launched on 9 July 2002 in South Africa. The main objectives of the AU are to achieve greater unity and solidarity between the African countries and Africans; to defend the sovereignty, territorial integrity, and independence of its member states; to accelerate the political and social-economic integration of the continent; to promote and defend African common positions on issues of interest to the continent and its peoples; to encourage international cooperation; and to advance the development of the continent by promoting research in all fields.

(v) **BRICS:** It is an intergovernmental organization comprising Brazil, Russia, India, China, and South Africa. It was formed in 2010 by the addition of South Africa to its predecessor BRIC. In August 2023, Egypt, Ethiopia, Iran, Saudi Arabia, and the United Arab Emirates were invited to join the organization. Full membership was scheduled to take effect on 1 January 2024. The BRICS were originally identified to highlight investment opportunities and had not been a formal intergovernmental organization. Since 2009, they have increasingly formed into a more cohesive geopolitical bloc, with their governments meeting annually at formal summits and coordinating multilateral policies. Bilateral relations among BRICS are conducted mainly on the basis of non-interference, equality, and mutual benefit. The BRICS nations are considered to be a geopolitical rival to the G7 bloc of leading advanced economies, implementing competing initiatives, such as the New Development Bank, the BRICS Contingent Reserve Arrangement, the BRICS payment system, the BRICS Joint Statistical Publication, and the BRICS basket reserve currency.

2. Mathematical Foundation of GPE with Coalitions

In this section, we provide the mathematical foundation for analyzing a global political economy with coalition blocs. We characterize a GPE with coalitions, the coalitional equilibrium, and a grand coalition cooperation outcome.

2.1. Characterization of GPE with coalitions

We first characterize the scenario of a global political and economic arena involving n coalitions with a planning horizon of T periods. We use S^j to denote the set of nations belonging to Coalition j, and there are n^j nations in Coalition j. In particular, the number of nations in a coalition can be 1. We use $i_j \in \hat{S}^j$ to denote the index of nation i_j in Coalition j, where $\hat{S}^j$ is the set of nation index of Coalition S^j.

The constituent elements are as follows:

(i) The set of non-durable target controls

We use $u_k^{i_j}$ to denote the set of non-durable target controls of nation $i_j \in \hat{S}^j$. We use $u_k^j = \cup_{i_j \in \hat{S}^j} u_k^{i_j}$ to denote the sets of non-durable target controls of Coalition j. Finally, we use $\underline{u}_k = (u_k^1, u_k^2, \ldots, u_k^n)$ to denote the sets of non-durable target controls of Coalitions 1 to n.

(ii) The set of durable target controls

We use $\bar{u}_k^{i_j}$ to denote the set of durable target controls of nation $i_j \in \hat{S}^j$. We use $\bar{u}_k^j = \cup_{i_j \in \hat{S}^j} \bar{u}_k^{i_j}$ to denote the sets of durable target controls of Coalition j. Finally, we use $\underline{\bar{u}}_k = (\bar{u}_k^1, \bar{u}_k^2, \ldots, \bar{u}_k^n)$ to denote the sets of durable target controls of Coalitions 1 to n.

(iii) The set of effective durable target controls previously executed

We use $\bar{u}_{k-}^{i_j}$ to denote the set of durable target controls executed before period k but still have effects of nation $i_j \in \hat{S}^j$. We use $\bar{u}_{k-}^j = \cup_{i_j \in \hat{S}^j} \bar{u}_{k-}^{i_j}$ to denote the sets of durable target controls executed before period k but still have effects of Coalition j. Finally, we use $\underline{u}_{k-} = (\bar{u}_{k-}^1, \bar{u}_{k-}^2, \ldots, \bar{u}_{k-}^n)$ to denote the sets of durable target controls executed before period k but still have effects of Coalitions 1 to n.

(iv) State variables and evolution dynamics

The vector of state variables of the economy at period k is denoted by $x_k \subset X$. The evolution dynamics of the vector of state variables are affected by $\underline{u}_k$, $\bar{u}_k$, $\underline{\bar{u}}_{k-}$, and x_k. We denote the evolution dynamics by the vector difference equation:

$$x_{k+1} = f_k(x_k, \underline{u}_k, \bar{u}_k; \underline{\bar{u}}_{k-}), \quad x_1 = x_1^0.$$

(v) Single-stage gain of nation in period k

The gain of nation $i_j \in \hat{S}^j$ in period k depends on $\underline{u}_k$, $\bar{u}_k$, $\underline{\bar{u}}_{k-}$, and x_k, and it is expressed as $\sum_{\varsigma=1}^{\Theta_{i_j}} w_k^{i_j(\varsigma)} g_k^{i_j(\varsigma)}(x_k, \underline{u}_k, \bar{u}_k; \underline{\bar{u}}_{k-})$. The terminal payoff of nation $i_j \in \hat{S}^j$ at period $T+1$ is $q_{T+1}^{i_j}(x_{T+1}, \underline{\bar{u}}_{(T+1)-})$.

(vi) The payoffs of the nations

The payoff of nation $i_j \in \hat{S}^j$ over the period 1 to period T can be expressed as follows:

$$\sum_{k=1}^{T} \sum_{\varsigma=1}^{\Theta_{i_j}} w_k^{i_j(\varsigma)} g_k^{i_j(\varsigma)}(x_k, \underline{u}_k, \bar{u}_k; \underline{\bar{u}}_{k-}) \delta_1^k + q_{T+1}^{i_j}(x_{T+1}, \underline{\bar{u}}_{(T+1)-}) \delta_1^{T+1}.$$

(vii) The payoffs of the coalitions

The payoff of Coalition S^j is the sum of the payoffs of all its member nations $i_j \in \hat{S}^j$:

$$\sum_{i_j \in \hat{S}^j} \left[\sum_{k=1}^{T} \sum_{\varsigma=1}^{\Theta_{i_j}} w_k^{i_j(\varsigma)} g_k^{i_j(\varsigma)}(x_k, \underline{u}_k, \bar{u}_k; \underline{\bar{u}}_{k-}) \delta_1^k \right.$$
$$\left. + q_{T+1}^{i_j}(x_{T+1}, \underline{\bar{u}}_{(T+1)-}) \delta_1^{T+1} \right].$$

Note that the influences of durable and non-durable strategies, previously executed durable strategies, and the current state of the nation's objectives and terminal payoffs are known realities. The influence on the evolution of the state by durable and non-durable strategies and previously executed durable strategies, and the current state are commonly observed. The relational contents exhibited in the system indeed reflect real-life phenomena in the global political economy. Coalitions can be competitive or complementary. In addition, there can be only one coalition and two or more coalitions. In addition, coalitions can be competitors and collaborators.

2.2. Coalitional equilibrium

The above scenario of a dynamic coalition economy involving n coalitions can be modeled as a durable-strategies dynamic game (see Yeung and Petrosyan, 2022) as follows:

$$\max_{\substack{u_k^j,\bar{u}_k^j \\ k\in\{1,2,\ldots,T\}}} \left\{ \sum_{i_j\in\hat{S}^j} \left[\sum_{k=1}^{T} \sum_{\varsigma=1}^{\Theta_{i_j}} w_k^{i_j(\varsigma)} g_k^{i_j(\varsigma)}(x_k, \underline{u}_k, \bar{u}_k; \underline{\bar{u}}_{k-})\delta_1^k \right.\right.$$
$$\left.\left. + q_{T+1}^{i_j}(x_{T+1}, \underline{\bar{u}}_{(T+1)-})\delta_1^{T+1} \right] \right\}, \tag{2.1}$$

for $j \in \{1, 2, \ldots, n\}$,
subject to

$$x_{k+1} = f_k(x_k, \underline{u}_k, \bar{u}_k; \underline{\bar{u}}_{k-}), \quad x_1 = x_1^0. \tag{2.2}$$

A theorem to characterize an equilibrium of the 2-coalition dynamic game (2.1)–(2.2) is provided as follows.

Theorem 2.1. *A set target controls $(\underline{u}_k^{**}, \underline{\bar{u}}_k^{**})$, for $k \in \{1, 2, \ldots, T\}$, provides a coalition equilibrium to the game (2.1)–(2.2) if there exist functions $W^j(k, x; \underline{\bar{u}}_{k-}^{**})$, for $k \in \{1, 2, \ldots, T\}$ and $j \in \{1, 2, \ldots, n_j\}$, such that the following recursive relations are satisfied:*

$$W^j(T+1, x; \underline{\bar{u}}_{(T+1)-}^{**}) = \sum_{i_j\in S^j} q_{T+1}^{i_j}(x_{T+1}, \underline{\bar{u}}_{(T+1)-}^{**})\delta_1^{T+1}, \tag{2.3}$$

$$W^j(k, x; \underline{\bar{u}}_{k-}^{**})$$
$$= \max_{u_k^j,\bar{u}_k^j} \left\{ \sum_{i_j\in\hat{S}^j} \left[\sum_{\varsigma=1}^{\Theta_{i_j}} w_k^{i_j(\varsigma)} g_k^{i_j(\varsigma)}(x_k, u_k^j, \bar{u}_k^j, \underline{u}_k^{(\neq S^j)^{**}}, \underline{\bar{u}}_k^{(\neq S^j)^{**}}; \underline{\bar{u}}_k^{**})\delta_1^k \right] \right.$$
$$+ W^j \Big[k+1, f_k(x, u_k^j, \bar{u}_k^j, \underline{u}_k^{(\neq S^j)^{**}}, \underline{\bar{u}}_k^{(\neq S^j)^{**}}; \underline{\bar{u}}_{k-}^{**});$$
$$\left. \times u_k^j, \bar{u}_k^j, \underline{\bar{u}}_{(k+1)-}^{**} \cap \underline{\bar{u}}_{k-}^{**} \Big] \right\}, \quad \text{for } k\in\{1, 2, \ldots, T\}, \tag{2.4}$$

where $\underline{u}_k^{(\neq S^j)**} = \underline{u}_k^{**} \backslash u_k^{j**}$ is the set of game equilibrium non-durable strategies of all nations excluding the set of non-durable strategies of

Coalition S^j and $\underline{\bar{u}}_k^{(\neq S^j)**} = \underline{\bar{u}}_k^{**} \backslash \bar{u}_k^{j**}$ is the set of game equilibrium durable strategies of all nations excluding the set of durable strategies of Coalition S^j.

Proof. Since nations belonging to the same coalition would cooperate and maximize the coalition payoff and act like one participant. Therefore, the coalitional game is analogous to a game with j coalitions. The proof of Theorem 2.1 can be found in Theorem II of Mathematical Solution Mechanisms given in Chapter 7. □

$W^j(k, x; \underline{\bar{u}}_{k-}^{**})$ is the game equilibrium payoffs of Coalition S^j.

The coalition equilibrium dynamics becomes

$$x_{k+1} = f_k(x_k, \underline{u}_k^{**}, \underline{\bar{u}}_k^{**}; \underline{\bar{u}}_{k-}^{**}), \quad x_1 = x_1^0. \tag{2.5}$$

We use $\{x_k^{**}\}_{k=1}^{T+1}$ to denote the solution to (2.7).

Since nations within a coalition (in general) share some common interests, forming a coalition would yield a coalition payoff which is larger than the sum of the nations' payoffs if they act individually. The characterization of the coalitional equilibrium above allows one to derive the equilibrium target controls. Policy instruments have to be implemented by the coalitions to achieve their target controls.

2.3. *Intra-coalition payoff distribution*

A coalition maximizes the joint payoffs of its member nations and shares the cooperative gains among themselves according to their agreed-upon optimality principle of gain sharing. We use $\xi^{i_j}(k, x_k; \underline{\bar{u}}_{k-})$ to denote the agreed-upon payoff that nation $i_j \in \hat{S}^j$. In particular, group optimality requiring the sum of the payoffs allocated to the member nations to be equal to the coalition payoff must be satisfied, that is,

$$\sum_{i_j \in \hat{S}^j} \xi^{i_j}(k, x_k; \underline{\bar{u}}_{k-}) = W^j(k, x_k; \underline{\bar{u}}_{k-}).$$

In addition, individual rationality requiring the payoff allocated to a member nation to be no less than the payoff when they are acting independently must also be satisfied, that is,

$$\xi^{i_j}(k, x_k; \underline{\bar{u}}_{k-}) \geq V^{i_j}(k, x_k; \underline{\bar{u}}_{k-}),$$

where $V^{i_j}(k, x_k; \underline{\bar{u}}_{k-})$ is the payoff of nation $i_j \in \hat{S}^j$ when the nations in Coalition S^j act independently. The corresponding game can be expressed as follows:

$$\max_{\substack{u_k^\ell, \bar{u}_k^\ell \\ k\in\{1,2,\ldots,T\}}} \left\{ \sum_{i_l \in \hat{S}^j} \left[\sum_{k=1}^{T} \sum_{\varsigma=1}^{\Theta_{i_l}} w_k^{i_l(\varsigma)} g_k^{i_l(\varsigma)}(x_k, \underline{u}_k, \bar{u}_k; \underline{\bar{u}}_{k-}) \delta_1^k \right.\right.$$
$$\left.\left. + q_{T+1}^{i_l}(x_{T+1}, \underline{\bar{u}}_{(T+1)-}) \delta_1^{T+1} \right] \right\}, \tag{2.6}$$

for $\ell \in \{1, 2, \ldots, \ell-1, \ell+1, \ldots, n\}$,

$$\max_{\substack{u_k^{i_j}, \bar{u}_k^{i_j} \\ k\in\{1,2,\ldots,T\}}} \left\{ \sum_{k=1}^{T} \sum_{\varsigma=1}^{\Theta_{i_l}} w_k^{i_l(\varsigma)} g_k^{i_l(\varsigma)}(x_k, \underline{u}_k, \bar{u}_k; \underline{\bar{u}}_{k-}) \delta_1^k \right.$$
$$\left. + q_{T+1}^{i_j}(x_{T+1}, \underline{\bar{u}}_{(T+1)-}) \delta_1^{T+1} \right\}, \tag{2.7}$$

for $i_j \in \hat{S}^j$ and $j \in \{1, 2, \ldots, n\} \equiv N$, subject to dynamics (2.2).

The system (2.2) and (2.6)–(2.7) is a game with $(n-1)+n_j$ players of which the solution can be characterized as a similar version of Theorem 2.1. (See also Theorem II of Mathematical Solution Mechanisms given in Chapter 7.)

Any $\xi^{i_j}(k, x_k; \underline{\bar{u}}_{k-})$ satisfying group optimality and individual rationality can be adopted. Among the possible sharing optimality principles given in Chapter 4, if the nations within a coalition would agree to share the coalition gain proportional to the relative size of their non-cooperative payoffs, then the share of nation $i_j \in \hat{S}^j$ along the coalition equilibrium trajectory becomes

$$\xi^{i_j}(k, x_k^{**}; \underline{\bar{u}}_{k-}^{**}) = \frac{V^{i_j}(k, x_k^{**}; \underline{\bar{u}}_{k-}^{**})}{\sum_{\varsigma \in \hat{S}^j} V^{\varsigma}(k, x_k^{**}; \underline{\bar{u}}_{k-}^{**})} W^j(k, x_k^{**}; \underline{\bar{u}}_{k-}^{**}), \tag{2.8}$$

for $i_j \in \hat{S}^j$ and $j \in N$.

To ensure stability in the dynamic cooperation of the nations in a coalition over time, subgame consistency has to be maintained in that the specific agreed-upon optimality principle must be maintained at any instant of time throughout the game along the optimal state trajectory.

$$\xi^{i_j}(k, x_k^{**}; \underline{\bar{u}}_{k-}^{**}) \text{ must be maintained in all periods}$$
$$k \in \{1, 2, \ldots, T\}, \quad \text{for } i_j \in \hat{S}^j. \tag{2.9}$$

This leads to the derivation of a payoff distribution procedure leading to the realization of the agreed-upon imputations. If this is satisfied, all the nations within a coalition will have no incentive to deviate from the cooperation scheme along its cooperative state trajectory.

A general theorem of a Payoff Distribution Procedure (PDP) leading to a specific agreed-upon payoff-sharing optimality principle $\xi^{i_j}(k, x_k^{**}; \underline{\bar{u}}_{k-}^{**})$ is provided as follows.

Theorem 2.2. *The agreed-upon payoff* $\xi^{i_j}(k, x_k^{**}; \underline{\bar{u}}_{k-}^{**})$ *for* $i_j \in \hat{S}^j$ *along the coalition equilibrium trajectory* $\{x_k^{**}\}_{k=1}^T$ *can be realized by a payment*

$$\begin{aligned}\beta_k^{i_j}(x_k^{**}; \underline{\bar{u}}_{k-}^{**}) = (\delta_1^k)^{-1}\{&\xi^{i_j}(k, x_k^{**}; \underline{\bar{u}}_{k-}^{**})\\ &- \xi^{i_j}[k+1, f_k(x_k^{**}, \underline{u}_k^{**}, \underline{\bar{u}}_k^{**}; \underline{\bar{u}}_{k-}^{**}); \underline{\bar{u}}_{(k+1)-}^{**}]\},\\ &\text{for } i_j \in \hat{S}^j \text{ and } j \in N,\end{aligned} \tag{2.10}$$

given to nation $i_j \in \hat{S}^j$ *at period* $k \in \{1, 2, \ldots, T\}$.

Proof. See the proof of Theorem III of Mathematical Solution Mechanisms given in Chapter 7. □

The payment scheme in (2.10) gives rise to the realization of the payoff guided by the agreed-upon optimality principle and constitutes a dynamically consistent payment scheme.

$\beta_k^{i_j}(x_k^{**}; \underline{\bar{u}}_{k-}^{**})$ is the target allocation to nation $i_j \in \hat{S}^j$. However, in the coalition equilibrium, the payment that nation $i_j \in \hat{S}^j$ will directly receive in period k along the cooperative trajectory is

$$\sum_{\varsigma=1}^{\Theta_{i_j}} w_k^{i_j(\varsigma)} g_k^{i_j(\varsigma)}(x_k^{**}, \underline{u}_k^{**}, \underline{\bar{u}}_k^{**}; \underline{\bar{u}}_{k-}^{**}).$$

Therefore, a side payment

$$\pi_k^{i_j}(x_k^{**};\underline{\bar{u}}_{k-}^{**}) = \beta_k^{i_j}(x_k^{**};\underline{\bar{u}}_{k-}^{**}) - \sum_{\varsigma=1}^{\Theta_{i_j}} w_k^{i_j(\varsigma)} g_k^{i_j(\varsigma)}(x_k^{**}, \underline{u}_k^{**}, \underline{\bar{u}}_k^{**}; \underline{\bar{u}}_{k-}^{**}),$$

$$\text{for } k \in \{1, 2, \ldots, T\}, \tag{2.11}$$

has to be given to nation $i_j \in \hat{S}^j$ to yield the equilibrium coalition payoff $\xi^{i_j}(k, x_k^{**}; \underline{\bar{u}}_{k-}^{**})$ in period $k \in \{1, 2, \ldots, T\}$.

In coalitions, a considerable portion of the side payment may be paid by payment-in-kind, like food, technology, infrastructure, natural resources, weapons, manufactured goods, capital, and scientific knowledge.

3. Grand Coalition Cooperation

Coalitions often have conflicting interests and pursue antagonizing strategies. Commonly observed antagonizing encounters include arms race, technology competition, trade sanctions, and defense security. A grand coalition cooperation of the coalitional blocs would improve the payoffs of both coalitions substantially. As proposed by Secretary-General António Guterres, the world can develop common solutions to common problems — grounded in goodwill, trust, and the rights shared by every human being — and work as a coalition of the world.

3.1. *Pareto optimal target controls*

The Pareto optimal target controls under the grand coalition can be obtained by solving the joint maximization problem:

$$\max_{\substack{u_k^j, \bar{u}_k^j, j \in N \\ k \in \{1,2,\ldots,T\}}} \left\{ \sum_{j \in N} \left[\sum_{i_j \in \hat{S}^j} \left[\sum_{k=1}^{T} \sum_{\varsigma=1}^{\Theta_{i_j}} w_k^{i_j(\varsigma)} g_k^{i_j(\varsigma)}(x_k, \underline{u}_k, \underline{\bar{u}}_k; \underline{\bar{u}}_{k-})\delta_1^k \right.\right.\right.$$
$$\left.\left.\left. + q_{T+1}^{i_j}(x_{T+1}, \underline{\bar{u}}_{(T+1)-})\delta_1^{T+1} \right]\right]\right\}, \tag{3.1}$$

subject to

$$x_{k+1} = f_k(x_k, \underline{u}_k, \underline{\bar{u}}_k; \underline{\bar{u}}_{k-}), \quad x_1 = x_1^0. \tag{3.2}$$

A theorem for characterizing the solution of the grand coalitional optimization problem (3.1)–(3.2) is provided as follows.

Theorem 3.1. *A set target controls* $(\underline{u}_k^*, \underline{\bar{u}}_k^*)$, *for* $k \in \{1, 2, \ldots, T\}$, *provides an optimal solution of the grand coalition cooperation* (3.1)–(3.2), *if there exists function* $W(k, x; \underline{\bar{u}}_{k-}^*)$, *for* $k \in \{1, 2, \ldots, T\}$, *such that the following recursive relations are satisfied:*

$$W(T+1, x; \underline{\bar{u}}_{(T+1)-}^*) = \sum_{j \in N} \sum_{i_j \in S^j} q_{T+1}^{i_j}(x_{T+1}, \underline{\bar{u}}_{(T+1)-}^{**})\delta_1^{T+1}, \tag{3.3}$$

$$W(k, x; \underline{\bar{u}}_{k-}^*)$$
$$= \max_{\underline{u}_k, \underline{\bar{u}}_k} \left\{ \sum_{j \in N} \left[\sum_{i_j \in \hat{S}^j} \left[\sum_{\varsigma=1}^{\Theta_{i_J}} w_k^{i_j(\varsigma)} g_k^{i_j(\varsigma)} \left(x_k, u_k^j, \bar{u}_k^j, \underline{u}_k^{(\neq S^j)**}, \right. \right. \right. \right.$$
$$\left. \left. \left. \underline{\bar{u}}_k^{(\neq S^j)**}; \underline{\bar{u}}_{k-}^{**} \right) \delta_1^k \right] \right] + W[k+1, f_k(x, \underline{u}_k, \underline{\bar{u}}_k;$$
$$\left. \underline{\bar{u}}_{k-}^{**}); \underline{u}_k, \underline{\bar{u}}_k, \underline{\bar{u}}_{(k+1)-}^{**} \cap \underline{\bar{u}}_{k-}^{**}] \right\}, \tag{3.4}$$

for $k \in \{1, 2, \ldots, T\}$.

Proof. See the proof of Theorem I of Mathematical Solution Mechanisms given in Chapter 7. □

The grand coalition state dynamics becomes

$$x_{k+1} = f_k(x_k, \underline{u}_k^*, \underline{\bar{u}}_k^*; \underline{\bar{u}}_{k-}^*), \quad x_1 = x_1^0. \tag{3.5}$$

We use $\{x_k^*\}_{k=1}^{T+1}$ to denote the solution to (3.5).

3.2. *Coalition payoff distribution*

Let $\psi^j(k, x_k^*; \underline{\bar{u}}_{k-}^*)$ be the payoff that the grand coalition would allocate Coalition S^j, for $j \in N$. In particular, group optimality requiring the sum of the payoffs allocated to the n coalitions to be equal to the coalition payoff must be satisfied, that is,

$$\sum_{j\in N} \psi^j(k, x_k; \underline{\bar{u}}_{k-}) = W(k, x_k; \underline{\bar{u}}_{k-}).$$

In addition, individual rationality requiring the payoff allocated to the coalitions to be no less than the payoff when the coalitions are acting independently must also be satisfied, that is,

$$\psi^j(k, x_k; \underline{\bar{u}}_{k-}) \geq W^j(k, x_k; \underline{\bar{u}}_{k-}).$$

Any $\psi^j(k, x_k; \underline{\bar{u}}_{k-})$ satisfying group optimality and individual rationality can be adopted. Among the possible sharing optimality principles as given in Chapter 4, if the grand coalition agrees to share the grand coalition payoff $W(k, x; \underline{\bar{u}}^*_{k-})$ proportional to the relative size of the non-cooperative payoffs of the coalitions, then the share of Coalition j along the Pareto optimal trajectory becomes

$$\psi^j(k, x^{**}_k; \underline{\bar{u}}^{**}_{k-}) = \frac{W^j(k, x^{**}_k; \underline{\bar{u}}^{**}_{k-})}{\sum_{\varsigma\in N} W^\varsigma(k, x^{**}_k; \underline{\bar{u}}^{**}_{k-})} W(k, x^{**}_k; \underline{\bar{u}}^{**}_{k-}), \tag{3.6}$$

for $j \in N$.

To ensure stability in dynamic cooperation between the coalitions over time, subgame consistency has to be maintained in that the specific agreed-upon optimality principle must be maintained at any instant of time throughout the game along the optimal state trajectory, that is,

$$\psi^j(k, x^*_k; \underline{\bar{u}}^*_{k-}) \text{ must be maintained in all periods } k \in \{1, 2, \ldots, T\}, \tag{3.7}$$

for $j \in N$.

This leads to the derivation of a payoff distribution procedure leading to the realization of the agreed-upon payoff $\psi^j(k, x^*_k; \underline{\bar{u}}^*_{k-})$ in all periods $k \in \{1, 2, \ldots, T\}$. If this is satisfied, both coalitions will have no incentive to deviate from the cooperation scheme along its cooperative state trajectory.

A general theorem of a Payoff Distribution Procedure (PDP) leading to the realization of condition (3.7) is provided as follows.

Theorem 3.2. *The agreed-upon payoff* $\psi^j(k, x_k^*; \underline{\bar{u}}_{k-}^*)$ *for* $j \in N$ *along the grand coalition optimal trajectory* $\{x_k^*\}_{k=1}^T$*, can be realized by a payment*

$$\begin{aligned}\hat{\beta}_k^j(x_k^*; \underline{\bar{u}}_{k-}^*) = (\delta_1^k)^{-1}\{&\psi^j(k, x_k^*; \underline{\bar{u}}_{k-}^*) \\ &- \psi^j[k+1, f_k(x_k^*, \underline{u}_k^*, \underline{\bar{u}}_k^*; \underline{\bar{u}}_{k-}^*); \underline{\bar{u}}_{(k+1)-}^*]\},\end{aligned} \quad (3.8)$$

for $j \in N$*, given to Coalition* $j \in N$ *at period* $k \in \{1, 2, \ldots, T\}$.

Proof. See the proof of Theorem III of Mathematical Solution Mechanisms given in Chapter 7. □

Let $\hat{\xi}^{i_j}(k, x_k^*; \underline{\bar{u}}_{k-}^*)$ denote the payoff that nation $i_j \in \hat{S}^j$ would receive under the grand coalition optimization. Again, for the requirement of subgame consistency,

$$\begin{aligned}&\hat{\xi}^{i_j}(k, x_k^*; \underline{\bar{u}}_{k-}^*) \text{ has to be maintained for } k \in \{1, 2, \ldots, T\}, \\ &i_j \in \hat{S}^j \text{ and } j \in N.\end{aligned} \quad (3.9)$$

A general theorem of a Payoff Distribution Procedure (PDP) leading to the realization of condition (3.9) is provided as follows.

Theorem 3.3. *The agreed-upon payoff* $\hat{\xi}^{i_j}(k, x_k^*; \underline{\bar{u}}_{k-}^*)$ *for* $k \in \{1, 2, \ldots, T\}$ *along the grand coalition optimal state trajectory* $\{x_k^*\}_{k=1}^T$ *can be realized by a payment*

$$\begin{aligned}\hat{\beta}_k^{i_j}(x_k^*; \underline{\bar{u}}_{k-}^*) = (\delta_1^k)^{-1}\{&\hat{\xi}^{i_j}(k, x_k^*; \underline{\bar{u}}_{k-}^*) \\ &- \hat{\xi}^{i_j}[k+1, f_k(x_k^*, \underline{u}_k^*, \underline{\bar{u}}_k^*; \underline{\bar{u}}_{k-}^*); \underline{\bar{u}}_{(k+1)-}^*]\},\end{aligned} \quad (3.10)$$

given to nation $i_j \in \hat{S}^j$ *at period* $k \in \{1, 2, \ldots, T\}$.

Proof. See the proof of Theorem III of Mathematical Solution Mechanisms given in Chapter 7. □

4. Explication in Trade Networks

In this section, we explicate the theoretical analysis and mathematical theory of coalition in trade networks with China's Belt-Road Initiative (BRI) and G7 countries' Build Back Better World (B3W) Initiative.

4.1. *China's Belt-Road Initiative*

In 2013, China unveiled the Silk Road Economic Belt and the 21st-Century Maritime Silk Road Initiative. These initiatives evolved into the Belt and Road Initiative (BRI). Many government bilateral agreements, MOUs, joint declarations, and the like have been officially signed and various commercial undertakings have been established. China's BRI is a transnational development initiative to further the growth of its economy and the development of participating nations. China also strives to increase participation in global affairs and governance through this trade connection. From a development strategy perspective, the economic logic behind the BRI appears to be highly sound. China's economy, which after decades of high economic growth due to the unprecedented increase in foreign direct investment, experiences a decline of the marginal efficiency of investment.

China strives to expand and improve its infrastructure to enhance its overall production efficiency by building connections with trading partners over land and sea to facilitate trade and investment costs. China also wants to secure its supply of energy and raw materials and ensure secured and unimpeded trade routes. It is against such background that the BRI was formulated. The BRI is therefore important to China's internal development needs as well as growing integration into the global economy. The BRI indeed calls for transnational cooperation, public and private sector partnerships, and massive support from the global and regional financial communities in general.

In 2023, the number of countries that have joined the Belt and Road Initiative (BRI) by signing a Memorandum of Understanding (MoU) with China is 148. The countries of the Belt and Road Initiative (BRI) are spread across all continents: 44 countries are in Sub-Saharan Africa, 35 BRI countries are in Europe and Central Asia,

25 BRI countries are in East Asia and Pacific (including China), 21 BRI countries are in Latin America and Caribbean, 18 BRI countries in the Middle East and North Africa, and 6 countries are in South East Asia. The BRI may be the largest trade network in the world. The sustainability of the BRI requires all the nations at any time to see that their interests are not to be compromised, and they will gain from cooperation from both the national interest and economic perspectives.

In trade blocs and networks, the most commonly used non-durable strategies $\underline{u}_k$ include various industrial productions, natural resource extraction, manpower application, agricultural production, aids, and subsidies. Durable strategies $\bar{\underline{u}}_k$ include transnational financing, trans-border infrastructure building, foreign trade and investment, innovation and technology transfer, international joint venture, rural development, urbanization, and capital investment.

The state vector x_k includes financial assets, technology level, environmental state, trans-infrastructure, number of international joint ventures, state of rural development, urbanization, and level of capital stocks.

Major objectives of BRI — like (i) infrastructure development, (ii) strengthening of regional political cooperation, (iii) unimpeded trade, (iv) financial integration, and (v) people and business cultural exchange — would appear in the vector of objectives

$$\begin{pmatrix} g_k^{i_j(1)}(x_k, \underline{u}_k, \bar{\underline{u}}_k; \bar{\underline{u}}_{k-}) \\ g_k^{i_j(2)}(x_k, \underline{u}_k, \bar{\underline{u}}_k; \bar{\underline{u}}_{k-}) \\ \vdots \\ g_k^{i_j(\Theta_{i_j})}(x_k, \underline{u}_k, \bar{\underline{u}}_k; \bar{\underline{u}}_{k-}) \end{pmatrix} \quad \text{for } i_j \in \hat{S}^j,$$

where $\hat{S}^j$ is the number of nations in the BRI and i_j is the i_j-th nation in BRI.

China obviously understands that the BRI will not succeed unless all the participating countries find that the cooperation project is consistent with their own national interest and economic development needs.

The coalition payoff of BRI can be modeled as

$$\sum_{i_j \in \hat{S}^j} \left[\sum_{k=1}^{T} \sum_{\varsigma=1}^{\Theta_{i_j}} w_k^{i_j(\varsigma)} g_k^{i_j(\varsigma)}(x_k, \underline{u}_k, \underline{\bar{u}}_k; \underline{\bar{u}}_{k-}) \delta_1^k \right.$$

$$\left. + q_{T+1}^{i_j}(x_{T+1}, \underline{\bar{u}}_{(T+1)-}) \delta_1^{T+1} \right]. \tag{4.1}$$

4.2. *Build Back Better World (B3W) Initiative*

Considered as a counter-balance to China's Belt and Road Initiative, the G7 countries launched the Build Back Better World (B3W) initiative in 2021. It acts as an alternative channel to BRI for infrastructure development of low- and middle-income countries. The B3W initiative seeks to address the $40 trillion worth of infrastructure needed by developing countries by 2035. It aims to catalyze funding for quality infrastructure from the private sector and will encourage private-sector investments that support "climate, health and health security, digital technology, and gender equity and equality". The building of B3W is influenced by the founding of Blue Dot Network (BDN) in 2019. The BDN is a joint project of the United States, Japan, and Australia that supports investment in high-quality infrastructure projects around the world, relating to the environment and climate, labor and social safeguards, financing, construction, anticorruption, and other areas. It was founded with $60 billion in initial funding. B3W is committed to concrete actions to help low- and middle-income nations for infrastructure formation. It proposes a value-driven, high-standard, and transparent infrastructure partnership led by major democracies to help narrow the forty trillion+ dollars in infrastructure need in the developing world. Through B3W, the G7 and other like-minded partners will coordinate in mobilizing private-sector capital in four areas of focus: climate, health and health security, digital technology, and gender equity and equality.

The White House Statement and Releases (12 June 2021) "FACT SHEET: President Biden and G7 Leaders Launch Build Back Better World (B3W) Partnership" stated that together with leaders of the G7, the Biden Administration fully endorses the guiding principles of B3W:

Values-Driven: Infrastructure development carried out in a transparent and sustainable manner — financially, environmentally, and socially — will lead to a better outcome for recipient countries and communities. We will offer countries a positive vision and a sustainable, transparent source of financing to meet their infrastructure needs.

Good Governance and Strong Standards: High standards have become ever more important at a time when governments are grappling with complex decisions on how to tackle climate change, build back local economies, direct scarce financing, and boost employment in an inclusive way. We are committed to providing citizens of recipient communities with the long-run benefits they expect and deserve from infrastructure projects. Our efforts will be guided by high standards and principles, such as those promoted by the updated Blue Dot Network, relating to the environment and climate, labor and social safeguards, transparency, financing, construction, anti-corruption, and other areas.

Climate-Friendly: The investments will be made in a manner consistent with achieving the goals of the Paris Climate Agreement.

Strong Strategic Partnerships: Infrastructure that is developed in partnership with those whom it benefits will last longer and generate more development impact. Infrastructure created under the B3W will be developed through consultation with communities and assessing local needs as true partners. The US attempts to establish a task force together as a G7, and with others, to coordinate, harmonize our efforts, and increase our impact and reach.

Mobilize Private Capital Through Development Finance: *Status quo* funding and financing approaches are inadequate to address the tremendous infrastructure gap in low- and middle-income countries. We are committed to augmenting the development finance tools at our disposal to support and catalyze a significant increase in private capital to address infrastructure needs. Infrastructure investment by a responsible and market-driven private sector, paired with high standards and transparency in public funding, is crucial for long-run development effectiveness and sustainability.

Enhancing the Impact of Multilateral Public Finance: Multilateral development banks and other international financial institutions (IFIs) have developed rigorous standards for project planning, implementation, social and environmental safeguards, and analytical capability. The United States will incorporate these standards and safeguards to help ensure that U.S. taxpayer resources are used appropriately and effectively. We will work with the IFIs to enhance their catalytic impact and increase the mobilization of capital — both public and private — needed for impactful and sustainable infrastructure investment.

Major objectives of Build Back Better World (B3W) — like (i) values-driven infrastructure development, (ii) good governance and strong standards, (iii) climate control (iv) strong strategic partnerships, (v) private capital through development finance, and (vi) impact of multilateral public finance — would appear in the B3W's vector of objectives

$$\begin{pmatrix} g_k^{i_\ell(1)}(x_k, \underline{u}_k, \bar{u}_k; \bar{\underline{u}}_{k-}) \\ g_k^{i_\ell(2)}(x_k, \underline{u}_k, \bar{u}_k; \bar{\underline{u}}_{k-}) \\ \vdots \\ g_k^{i_\ell(\Theta_{i_\ell})}(x_k, \underline{u}_k, \bar{u}_k; \bar{\underline{u}}_{k-}) \end{pmatrix} \quad \text{for } i_l \in \hat{S}^l,$$

where $\hat{S}^\ell$ is the number of nations in the B3W and i_ℓ the i_ℓ-th nation in B3W.

The coalition payoff of B3W can be modeled as

$$\sum_{i_l \in \hat{S}^l} \left[\sum_{k=1}^{T} \sum_{\varsigma=1}^{\Theta_{i_l}} w_k^{i_l(\varsigma)} g_k^{i_l(\varsigma)}(x_k, \underline{u}_k, \bar{u}_k; \bar{\underline{u}}_{k-}) \delta_1^k \right.$$
$$\left. + q_{T+1}^{i_\ell}(x_{T+1}, \bar{\underline{u}}_{(T+1)-}) \delta_1^{T+1} \right]. \tag{4.2}$$

4.3. *Coalitional equilibrium and intra-coalition payoff distribution*

In this section, we first characterize the coalition equilibrium of BRI and B3W. Then we derive the intra-coalition of payoff distribution to member nations in each of these coalitions.

4.3.1. *Coalition equilibrium*

The above scenario of a dynamic coalition economy involving two coalitions can be modeled as a durable-strategies dynamic game:

$$\max_{\substack{u_k^j, \bar{u}_k^j \\ k \in \{1,2,\ldots,T\}}} \left\{ \sum_{i_j \in \hat{S}^j} \left[\sum_{k=1}^{T} \sum_{\varsigma=1}^{\Theta_{i_j}} w_k^{i_j(\varsigma)} g_k^{i_j(\varsigma)}(x_k, \underline{u}_k, \bar{u}_k; \bar{\underline{u}}_{k-}) \delta_1^k \right.\right.$$
$$\left.\left. + q_{T+1}^{i_j}(x_{T+1}, \bar{\underline{u}}_{(T+1)-}) \delta_1^{T+1} \right] \right\}, \tag{4.3}$$

$$\max_{\substack{u_k^\ell, u_k^\ell \\ k \in \{1,2,\ldots,T\}}} \left\{ \sum_{i_l \in \hat{S}^l} \left[\sum_{k=1}^{T} \sum_{\varsigma=1}^{\Theta_{i_l}} w_k^{i_l(\varsigma)} g_k^{i_l(\varsigma)}(x_k, \underline{u}_k, \bar{u}_k; \bar{\underline{u}}_{k-}) \delta_1^k \right.\right.$$
$$\left.\left. + q_{T+1}^{i_\ell}(x_{T+1}, \bar{\underline{u}}_{(T+1)-}) \delta_1^{T+1} \right] \right\}, \tag{4.4}$$

subject to

$$x_{k+1} = f_k(x_k, \underline{u}_k, \bar{u}_k; \bar{\underline{u}}_{k-}), \quad x_1 = x_1^0. \tag{4.5}$$

Using Theorem 2.1, we can derive the set of game equilibrium strategies $(\underline{u}_k^{**}, \bar{u}_k^{**})$, for $k \in \{1, 2, \ldots, T\}$, and the game equilibrium payoff of BRI $W^j(k, x; \bar{\underline{u}}_{k-}^{**})$ and the game equilibrium payoff of B3W $W^\ell(k, x; \bar{\underline{u}}_{k-}^{**})$. The equilibrium dynamics becomes

$$x_{k+1} = f_k(x_k, \underline{u}_k^{**}, \bar{u}_k^{**}; \bar{\underline{u}}_{k-}^{**}), \quad x_1 = x_1^0. \tag{4.6}$$

We use $\{x_k^{**}\}_{k=1}^{T+1}$ to denote the solution to (4.6).

4.3.2. *Intra-coalition payoff distribution*

A coalition maximizes the joint payoffs of its member nations and shares the cooperative gains among themselves according to their agreed-upon optimality principle of gain sharing. We use $\xi^{i_j}(k, x_k; \bar{\underline{u}}_{k-})$ to denote the agree-upon payoff to nation $i_j \in \hat{S}^j$ in BRI and $\xi^{i_\ell}(k, x_k; \bar{\underline{u}}_{k-})$ to denote the agree-upon payoff to nation $i_\ell \in S^\ell$ in B3W.

In particular, group optimality requiring the sum of the payoffs allocated to the member nations to be equal to the coalition payoff must be satisfied, that is,

$$\sum_{i_j \in \hat{S}^j} \xi^{i_j}(k, x_k; \underline{\bar{u}}_{k-}) = W^j(k, x_k; \underline{\bar{u}}_{k-})$$

and

$$\sum_{i_\ell \in \hat{S}^\ell} \xi^{i_j}(k, x_k; \underline{\bar{u}}_{k-}) = W^\ell(k, x_k; \underline{\bar{u}}_{k-}). \tag{4.7}$$

In addition, individual rationality requiring the payoff allocated to a member nation to be no less than the payoff when they are acting independently must also be satisfied, that is,

$$\xi^{i_j}(k, x_k; \underline{\bar{u}}_{k-}) \geq V^{i_j}(k, x_k; \underline{\bar{u}}_{k-}), \text{ for } i_j \in \hat{S}^j, \text{ and}$$

$$\xi^{i_\ell}(k, x_k; \underline{\bar{u}}_{k-}) \geq V^{i_\ell}(k, x_k; \underline{\bar{u}}_{k-}), \text{ for } i_\ell \in \hat{S}^\ell, \tag{4.8}$$

where $V^{i_j}(k, x_k; \underline{\bar{u}}_{k-})$ is the payoff of nation $i_j \in \hat{S}^j$ when the nations in coalition S^j acts independently; and

$V^{i_\ell}(k, x_k; \underline{\bar{u}}_{k-})$ is the payoff of nation $i_\ell \in \hat{S}^\ell$ when the nations in coalition S^ℓ act independently.

Any $\xi^{i_j}(k, x_k; \underline{\bar{u}}_{k-})$ and $\xi^{i_\ell}(k, x_k; \underline{\bar{u}}_{k-})$ satisfying group optimality and individual rationality can be adopted. If the nations within BRI agree to share the coalition gain proportional to the relative size of their non-cooperative payoffs, then the share of nation $i_j \in \hat{S}^j$ along the coalition equilibrium trajectory becomes

$$\xi^{i_j}(k, x_k^{**}; \underline{\bar{u}}_{k-}^{**}) = \frac{V^{i_j}(k, x_k^{**}; \underline{\bar{u}}_{k-}^{**})}{\sum_{\varsigma \in \hat{S}^j} V^{\varsigma}(k, x_k^{**}; \underline{\bar{u}}_{k-}^{**})} W^j(k, x_k^{**}; \underline{\bar{u}}_{k-}^{**}),$$

$$\text{for } i_j \in \hat{S}^j. \tag{4.9}$$

Similarly, if the nations within B3W agree to share the coalition gain proportional to the relative size of their non-cooperative payoffs, then the share of nation $i_\ell \in \hat{S}^\ell$ along the coalition equilibrium trajectory becomes

$$\xi^{i_\ell}(k, x_k^{**}; \underline{\bar{u}}_{k-}^{**}) = \frac{V^{i_\ell}(k, x_k^{**}; \underline{\bar{u}}_{k-}^{**})}{\sum_{\varsigma \in \hat{S}^\ell} V^{\varsigma}(k, x_k^{**}; \underline{\bar{u}}_{k-}^{**})} W^\ell(k, x_k^{**}; \underline{\bar{u}}_{k-}^{**}),$$

$$\text{for } i_\ell \in \hat{S}^\ell. \tag{4.10}$$

To ensure stability in the dynamic cooperation of the nations in a coalition over time, subgame consistency has to be maintained in that the specific agreed-upon optimality principle must be maintained at any instant of time throughout the game along the optimal state trajectory.

$$\xi^{i_j}(k, x_k^{**}; \underline{\bar{u}}_{k-}^{**}), \text{ for } i_j \in \hat{S}^j, \text{ and } \xi^{i_\ell}(k, x_k^{**}; \underline{\bar{u}}_{k-}^{**}), \text{ for } i_\ell \in \hat{S}^\ell, \tag{4.11}$$

must be maintained in all periods $k \in \{1, 2, \ldots, T\}$.

Invoking Theorem 2.2, the agreed-upon payoff $\xi^{i_j}(k, x_k^{**}; \underline{\bar{u}}_{k-}^{**})$ in (4.9) will be realized by a payment

$$\begin{aligned}\beta_k^{i_j}(x_k^{**}; \underline{\bar{u}}_{k-}^{**}) &= (\delta_1^k)^{-1}\xi^{i_j}(k, x_k^{**}; \underline{\bar{u}}_{k-}^{**}) \\ &\quad - \xi^{i_j}[k+1, f_k(x_k^{**}, \underline{u}_k^{**}, \underline{\bar{u}}_k^{**}; \underline{\bar{u}}_{k-}^{**}); \underline{\bar{u}}_{(k+1)-}^{**}],\end{aligned} \tag{4.12}$$

given to nation $i_j \in \hat{S}^j$ at period $k \in \{1, 2, \ldots, T\}$.

Similarly, the agreed-upon payoff $\xi^{i_\ell}(k, x_k^{**}; \underline{\bar{u}}_{k-}^{**})$ in (4.10) will be realized by a payment

$$\begin{aligned}\beta_k^{i_\ell}(x_k^{**}; \underline{\bar{u}}_{k-}^{**}) &= (\delta_1^k)^{-1}\{\xi^{i_\ell}(k, x_k^{**}; \underline{\bar{u}}_{k-}^{**}) \\ &\quad - \xi^{i_\ell}[k+1, f_k(x_k^{**}, \underline{u}_k^{**}, \underline{\bar{u}}_k^{**}; \underline{\bar{u}}_{k-}^{**}); \underline{\bar{u}}_{(k+1)-}^{**}]\},\end{aligned} \tag{4.13}$$

given to nation $i_\ell \in \hat{S}^\ell$ at period $k \in \{1, 2, \ldots, T\}$.

However, in the coalition equilibrium, the payment that nation $i_j \in \hat{S}^j$ will directly receive in period k along the cooperative trajectory is

$$\sum_{\varsigma=1}^{\Theta_{i_j}} w_k^{i_j(\varsigma)} g_k^{i_j(\varsigma)}(x_k^{**}, \underline{u}_k^{**}, \underline{\bar{u}}_k^{**}; \underline{\bar{u}}_{k-}^{**}).$$

Similarly, the payment that nation $i_\ell \in \hat{S}^\ell$ will directly receive in period k along the cooperative trajectory is

$$\sum_{\varsigma=1}^{\Theta_{i_\ell}} w_k^{i_\ell(\varsigma)} g_k^{i_\ell(\varsigma)}(x_k^{**}, \underline{u}_k^{**}, \underline{\bar{u}}_k^{**}; \underline{\bar{u}}_{k-}^{**}).$$

Therefore, a side payment

$$\pi_k^{i_j}(x_k^{**};\underline{\bar{u}}_{k-}^{**}) = \beta_k^{i_j}(x_k^{**};\underline{\bar{u}}_{k-}^{**}) - \sum_{\varsigma=1}^{\Theta_{i_j}} w_k^{i_j(\varsigma)} g_k^{i_j(\varsigma)}(x_k^{**}, \underline{u}_k^{**}, \underline{\bar{u}}_k^{**}; \underline{\bar{u}}_{k-}^{**}) \tag{4.14}$$

has to be given to nation $i_j \in \hat{S}^j$ in BRI in period $k \in \{1, 2, \ldots, T\}$.

A side payment

$$\pi_k^{i_\ell}(x_k^{**};\underline{\bar{u}}_{k-}^{**}) = \beta_k^{i_\ell}(x_k^{**};\underline{\bar{u}}_{k-}^{**}) - \sum_{\varsigma=1}^{\Theta_{i_\ell}} w_k^{i_\ell(\varsigma)} g_k^{i_\ell(\varsigma)}(x_k^{**}, \underline{u}_k^{**}, \underline{\bar{u}}_k^{**}; \underline{\bar{u}}_{k-}^{**}) \tag{4.15}$$

has to be given to nation $i_\ell \in \hat{S}^\ell$ in B3W in period $k \in \{1, 2, \ldots, T\}$.

A considerable portion of the side payment may be paid by payment-in-kind, like food, technology, infrastructure, natural resources, weapons, manufactured goods, capital, and scientific knowledge.

5. Grand Coalition Cooperation Analysis

A grand coalition cooperation of the trade blocs would improve the payoffs of BRI and B3W substantially through the positive externality in international trade and cross-national investments.

5.1. *Pareto optimal target controls*

The Pareto optimal target controls under the grand coalition can be obtained by solving the joint maximization problem:

$$\max_{\substack{u_k^j, \bar{u}_k^j, u_k^\ell, \bar{u}_k^\ell \\ j \in N \\ k \in \{1,2,3,\ldots,T\}}} \left\{ \sum_{i_j \in \hat{S}^j} \left[\sum_{k=1}^{T} \sum_{\varsigma=1}^{\Theta_{i_j}} w_k^{i_j(\varsigma)} g_k^{i_j(\varsigma)}(x_k, \underline{u}_k, \underline{\bar{u}}_k; \underline{\bar{u}}_{k-}) \delta_1^k \right.\right.$$

$$+ q^{i_j}_{T+1}(x_{T+1}, \underline{\bar{u}}_{(T+1)-})\delta_1^{T+1}\Bigg] + \sum_{i_l \in \hat{S}^l}\left[\sum_{k=1}^{T}\sum_{\varsigma=1}^{\Theta_{i_l}} w_k^{i_l(\varsigma)}\right.$$

$$\left.\left. \times\, g_k^{i_l(\varsigma)}(x_k, \underline{u}_k, \bar{u}_k; \underline{\bar{u}}_{k-})\delta_1^k + q^{i_\ell}_{T+1}(x_{T+1}, \underline{\bar{u}}_{(T+1)-})\delta_1^{T+1}\right]\right\}, \tag{5.1}$$

subject to

$$x_{k+1} = f_k(x_k, \underline{u}_k, \bar{u}_k; \underline{\bar{u}}_{k-}), \quad x_1 = x_1^0. \tag{5.2}$$

Using Theorem 3.1, we can derive the set of optimal cooperative strategies $(\underline{u}_k^*, \bar{u}_k^*)$, for $k \in \{1, 2, \ldots, T\}$; and the optimal joint payoff in the grand coalition $W(k, x; \underline{\bar{u}}_{k-}^*)$. The grand coalition state dynamics becomes

$$x_{k+1} = f_k(x_k, \underline{u}_k^*, \bar{u}_k^*; \underline{\bar{u}}_{k-}^*), \quad x_1 = x_1^0. \tag{5.3}$$

We use $\{x_k^*\}_{k=1}^{T+1}$ to denote the solution to (5.3).

5.2. *Coalition payoff distribution*

Let $\psi^j(k, x_k^*; \underline{\bar{u}}_{k-}^*)$ be the payoff that the grand coalition would allocate to BRI and $\psi^l(k, x_k^*; \underline{\bar{u}}_{k-}^*)$ be the payoff that the grand coalition would allocate to B3W. Group optimality requires the sum of the payoffs allocated to BRI and B3W be the same as grand coalition optimal joint payoff $W(k, x; \underline{\bar{u}}_{k-}^*)$, that is

$$\psi^j(k, x_k^*; \underline{\bar{u}}_{k-}^*) + \psi^\ell(k, x_k^*; \underline{\bar{u}}_{k-}^*) = W(k, x_k; \underline{\bar{u}}_{k-}). \tag{5.4}$$

Individual rationality requires the payoff allocated to the BRI and B3W be no less than their payoffs when they are acting independently, that is,

$$\begin{aligned} \psi^j(k, x_k^*; \underline{\bar{u}}_{k-}^*) &\geq W^j(k, x_k; \underline{\bar{u}}_{k-}) \text{ and} \\ psi^\ell(k, x_k^*; \underline{\bar{u}}_{k-}^*) &\geq W^\ell(k, x_k; \underline{\bar{u}}_{k-}). \end{aligned} \tag{5.5}$$

If the grand coalition agrees to share the grand coalition payoff $W(k, x; \underline{\bar{u}}_{k-}^{*})$ proportional to the relative size of the non-cooperative payoffs of BRI and B3W, then the share of BRI along the Pareto optimal trajectory becomes

$$\psi^{j}(k, x_k^*; \underline{\bar{u}}_{k-}^{*}) = \frac{W^{j}(k, x_k^*; \underline{\bar{u}}_{k-}^{*})}{W^{j}(k, x_k^*; \underline{\bar{u}}_{k-}^{*}) + W^{\ell}(k, x_k^*; \underline{\bar{u}}_{k-}^{*})} W(k, x_k^*; \underline{\bar{u}}_{k-}^{*}). \tag{5.6}$$

Correspondingly, the share of B3W along the Pareto optimal trajectory becomes

$$\psi^{\ell}(k, x_k^{**}; \underline{\bar{u}}_{k-}^{**}) = \frac{W^{\ell}(k, x_k^*; \underline{\bar{u}}_{k-}^{*})}{W^{j}(k, x_k^*; \underline{\bar{u}}_{k-}^{*}) + W^{\ell}(k, x_k^*; \underline{\bar{u}}_{k-}^{*})} W(k, x_k^*; \underline{\bar{u}}_{k-}^{*}). \tag{5.7}$$

Invoking Theorem 3.2, the payoffs of BRI and B3W in (5.6)–(5.7) will be realized with a payment

$$\begin{aligned}\hat{\beta}_k^{j}(x_k^*; \underline{\bar{u}}_{k-}^{*}) = (\delta_1^k)^{-1}\{&\psi^{j}(k, x_k^*; \underline{\bar{u}}_{k-}^{*}) \\ &- \psi^{j}[k+1, f_k(x_k^*, \underline{u}_k^*, \underline{\bar{u}}_k^*; \underline{\bar{u}}_{k-}^{*}); \underline{\bar{u}}_{(k+1)-}^{*}]\},\end{aligned}$$

given to BRI at period $k \in \{1, 2, \ldots, T\}$, and a payment

$$\begin{aligned}\hat{\beta}_k^{\ell}(x_k^*; \underline{\bar{u}}_{k-}^{*}) = (\delta_1^k)^{-1}\{&\psi^{\ell}(k, x_k^*; \underline{\bar{u}}_{k-}^{*}) \\ &- \psi^{l}[k+1, f_k(x_k^*, \underline{u}_k^*, \underline{\bar{u}}_k^*; \underline{\bar{u}}_{k-}^{*}); \underline{\bar{u}}_{(k+1)-}^{*}]\},\end{aligned}$$

given to B3W at period $k \in \{1, 2, \ldots, T\}$.

5.3. *Intra-coalition payoff distribution*

Recall that the nations in BRI and B3W agree to share the coalition gain proportional to the relative size of their non-cooperative payoffs. Let $\hat{\xi}^{i_j}(k, x_k^*; \underline{\bar{u}}_{k-}^{*})$ denote the payoff that nation $i_j \in S^j$ would receive under the grand coalition optimization, then the share of nation

$i_j \in \hat{S}^j$ in BRI becomes

$$\hat{\xi}^{i_j}(k, x_k^{**}; \underline{\bar{u}}_{k-}^{**}) = \frac{V^{i_j}(k, x_k^{**}; \underline{\bar{u}}_{k-}^{**})}{\sum_{\varsigma \in \hat{S}^j} V^{\varsigma}(k, x_k^{**}; \underline{\bar{u}}_{k-}^{**})} \psi^j(k, x_k^{**}; \underline{\bar{u}}_{k-}^{**}), \text{ for } i_j \in \hat{S}^j. \tag{5.8}$$

Correspondingly, the share of nation $i_\ell \in \hat{S}^\ell$ in B3W becomes

$$\hat{\xi}^{i_\ell}(k, x_k^{**}; \underline{\bar{u}}_{k-}^{**}) = \frac{V^{i_\ell}(k, x_k^{**}; \underline{\bar{u}}_{k-}^{**})}{\sum_{\varsigma \in \hat{S}^\ell} V^{\varsigma}(k, x_k^{**}; \underline{\bar{u}}_{k-}^{**})} \psi^\ell(k, x_k^{**}; \underline{\bar{u}}_{k-}^{**}), \text{ for } i_\ell \in \hat{S}^\ell. \tag{5.9}$$

Invoking Theorem 3.3, the payoffs of nation $i_j \in \hat{S}^j$ in BRI and nation $i_\ell \in S^\ell$ in B3W specified in (5.8)–(5.9) will be realized with a payment

$$\begin{aligned}\hat{\beta}_k^{i_j}(x_k^*; \underline{\bar{u}}_{k-}^*) = (\delta_1^k)^{-1}\{&\hat{\xi}^{i_j}(k, x_k^*; \underline{\bar{u}}_{k-}^*) \\ &- \hat{\xi}^{i_j}[k+1, f_k(x_k^*, \underline{u}_k^*, \underline{\bar{u}}_k^*; \underline{\bar{u}}_{k-}^*); \underline{\bar{u}}_{(k+1)-}^*]\},\end{aligned}$$

given to nation $i_j \in S^j$ in BRI at period $k \in \{1, 2, \ldots, T\}$, a payment

$$\begin{aligned}\hat{\beta}_k^{i_\ell}(x_k^*; \underline{\bar{u}}_{k-}^*) = (\delta_1^k)^{-1}\{&\hat{\xi}^{i_\ell}(k, x_k^*; \underline{\bar{u}}_{k-}^*) \\ &- \hat{\xi}^{i_\ell}[k+1, f_k(x_k^*, \underline{u}_k^*, \underline{\bar{u}}_k^*; \underline{\bar{u}}_{k-}^*); \underline{\bar{u}}_{(k+1)-}^*]\},\end{aligned}$$

given to nation $i_\ell \in S^\ell$ in B3W at period $k \in \{1, 2, \ldots, T\}$.

Again, a considerable portion of the side payment may be paid by payment-in-kind, like technology transfer and exchange of scientific knowledge on top of food, infrastructure, natural resources, manufactured goods, and capital.

6. A Computational Illustration

Consider a T-period dynamic game involving a political economic arena with two coalitions with conflicting political stands and a group of unallied nations. Coalition 1 has n_1 member nations and Coalition 2 has n_2 member nations. The set of nations in Coalition 1 is denoted by S^1, and the set of nations in Coalition 2 is denoted by S^2. There are n_3 unallied nations. The two coalitions are concerned about their security/defense level and income levels. The set of unallied is denoted by S^3. The unallied nations have a neutral stand and are concerned with income levels only. We use x_k to denote the defense level of Coalition 1 in period k, and y_k to denote the defense level of Coalition 2 in period k. The level of activities enhancing Coalition 1's defense level by nation $i \in S^1$ is denoted by $\bar{u}_k^{(1)i}$. In particular, $\bar{u}_k^{(1)i}$ is a durable strategy that lasts for two periods. The level of activities enhancing Coalition 2's defense level by nation $j \in S^2$ is denoted by $\bar{u}_k^{(2)i}$. Again, $\bar{u}_k^{(2)i}$ is a durable strategy that lasts for two periods.

The dynamics of the defense level of Coalition 1 is governed by

$$x_{k+1} = x_k + \sum_{\ell=1}^{n_1} \gamma_k^{(1)\ell} \bar{u}_k^{(1)\ell} + \sum_{\ell=1}^{n_1} \sigma_k^{(1)\ell} \bar{u}_{k-1}^{(1)\ell} - \lambda_k^{(1)} x_k,$$

$$x_1 = x_1^0, \tag{6.1}$$

where $\gamma_k^{(1)\ell} \bar{u}_k^{(1)\ell}$ is the increment in Coalition 1's defense level brought about by the defense activity $\bar{u}_k^{(1)\ell}$ in period k and $\gamma_k^{(1)\ell} \bar{u}_k^{(1)\ell}$ is the increment in Coalition 1's defense level brought about by the defense activity $\bar{u}_k^{(1)\ell}$ in period k, $\sigma_k^{(1)\ell} \bar{u}_{k-1}^{(1)\ell}$ is the increment in Coalition 1's defense level brought about by the defense activity $\bar{u}_{k-1}^{(1)\ell}$ in period k, and $\lambda_k^{(1)}$ is the rate of degradation and obsolescence of the defense level of Coalition 1.

The dynamics of the defense level of Coalition 2 is governed by

$$y_{k+1} = y_k + \sum_{\ell=1}^{n_1} \gamma_k^{(2)\ell} \bar{u}_k^{(2)\ell} + \sum_{\ell=1}^{n_2} \sigma_k^{(2)\ell} \bar{u}_{k-1}^{(2)\ell} - \lambda_k^{(2)} x_k,$$

$$y_1 = y_1^0, \tag{6.2}$$

where $\gamma_k^{(2)\ell} \bar{u}_k^{(2)\ell}$ is the increment in Coalition 2's defense level brought about by the defense activity $\bar{u}_k^{(2)\ell}$ in period k and $\gamma_k^{(2)\ell} \bar{u}_k^{(2)\ell}$ is the increment in Coalition 2's defense level brought about by the defense activity $\bar{u}_k^{(2)\ell}$ in period k, $\sigma_k^{(2)\ell} \bar{u}_{k-1}^{(2)\ell}$ is the increment in Coalition 2's defense level brought about by the defense activity $\bar{u}_{k-1}^{(2)\ell}$ in period k, and $\lambda_k^{(2)}$ is the rate of degradation and obsolescence of the defense level of Coalition 2.

Nations in Coalition 1 prefer x_k and against y_k, while nations in Coalition 2 prefer y_k and against x_k. This is one of the incentives for nations in S^1 to form a coalition and nations in S^2 to form another coalition. The level of income enhancing activities by nation $i \in S^1$ is denoted by $u_k^{(1)i}$, and the level of income enhancing activities by nation $j \in S^2$ is denoted by $u_k^{(2)j}$. The cost of $u_k^{(1)i}$ is $c_k^{(1)i}(u_k^{(1)i})^2$ and The cost of $u_k^{(2)j}$ is $c_k^{(2)j}(u_k^{(2)j})^2$. The cost of defense-enhancing activities $\bar{u}_k^{(1)i}$ is $w_k^{(1)i}(\bar{u}_k^{(1)i})^2$, and the cost of defense-enhancing activities $\bar{u}_k^{(2)i}$ is $w_k^{(2)i}(\bar{u}_k^{(2)i})^2$.

The net income from the economic activities of nation $i \in S^1$ is

$$\left[p_k^{(1)i} - b_k^{(1)i(i)}u_k^{(1)i} + \sum_{\substack{\ell \in S^1 \\ \ell \neq i}} \varpi_k^{(1)i(\ell)}u_k^{(1)\ell}\right]u_k^{(1)i} - c_k^{(1)i}(u_k^{(1)i})^2. \tag{6.3}$$

Note that the income of nation $i \in S^1$ is positively related to the economic activities of nations in Coalition 1 because $\varpi_k^{(1)i(\ell)} > 0$. This is another incentive for nations in S^1 to form a coalition.

The net income from economic activities of nation $j \in S^2$ is

$$\left[p_k^{(2)j} - b_k^{(2)j(j)}u_k^{(2)j} + \sum_{\substack{\ell \in S^2 \\ \ell \neq j}} \varpi_k^{(2)j(\ell)}u_k^{(2)\ell}\right]u_k^{(2)j} - c_k^{(2)j}(u_k^{(2)j})^2. \tag{6.4}$$

Note that the income of nation $j \in S^2$ is positively related to the economic activities of nations in Coalition 2 because $\varpi_k^{(2)j(\ell)} > 0$. This is another incentive for nations in S^2 to form a coalition.

The level of income-enhancing activities by nation $\varsigma \in S^3$ is denoted by $u_k^{(3)\varsigma}$, and the cost of $u_k^{(3)\varsigma}$ is $c_k^{(3)\varsigma}(u_k^{(3)\varsigma})^2$. The net income from the economic activities of nation $\varsigma \in S^3$ is

$$\left[p_k^{(3)\varsigma} - b_k^{(3)\varsigma(\varsigma)} u_k^{(3)\varsigma} + \sum_{\substack{\ell \in S^3 \\ \ell \neq \varsigma}} \varpi_k^{(3)\varsigma(\ell)} u_k^{(3)\varsigma}\right] u_k^{(3)\varsigma} - c_k^{(3)\varsigma} (u_k^{(3)\varsigma})^2. \quad (6.5)$$

Note that the income of nation $\varsigma \in S^3$ is related to the economic activities of other unallied nations. In particular, $\varpi_k^{(3)\varsigma(\ell)}$ can be positive, negative, or zero, and these relationships between the unallied nations may not be reciprocal. Therefore, there is no incentives for nations in S^3 to form a coalition. In addition, the neutral nations are either neutral to or mildly dislike defense build-ups.

The scenario of the above political-economic arena can be characterized by the following dynamic game:

$$\max_{\substack{u_t^{(1)i}, \bar{u}_t^{(1)i} \\ i \in s^1 \\ t \in \{1,2,\ldots,T\}}} \left\{ \sum_{i \in S^1} \left\{ \sum_{t=1}^{T} \left[\left(p_t^{(1)i} - b_t^{(1)i(i)} u_t^{(1)i} + \sum_{\substack{\ell \in S^1 \\ \ell \neq i}} \varpi_t^{(1)i(\ell)} u_t^{(1)\ell} \right) u_t^{(1)i} \right.\right.\right.$$

$$\left. - c_t^{(1)i} (u_t^{(1)i})^2 - w_t^{(1)i} (\bar{u}_t^{(1)i})^2 + h_t^{(1)i} x_t - \hat{h}_t^{(1)i} y_t \right] \delta^{t-1}$$

$$\left.\left. + \sum_{i \in S^1} (h_{T+1}^{(1)i} x_{T+1} - \hat{h}_{T+1}^{(1)i} y_{T+1} + Q_{T+1}^{(1)i}) \delta^T \right\}\right\}, \quad (6.6a)$$

$$\max_{\substack{u_t^{(2)j}, \bar{u}_t^{(2)j} \\ j \in S^2 \\ t \in \{1,2,\ldots,T\}}} \left\{ \sum_{j \in S^2} \left\{ \sum_{t=1}^{T} \left[\left(p_t^{(2)j} - b_t^{(2)j(j)} u_t^{(2)j} + \sum_{\substack{\ell \in S^2 \\ \ell \neq j}} \varpi_t^{(2)j(\ell)} u_t^{(2)\ell} \right) u_t^{(2)j} \right.\right.\right.$$

$$\left. - c_t^{(2)j} (u_t^{(2)j})^2 - w_t^{(2)j} (\bar{u}_t^{(2)j})^2 + h_t^{(2)j} y_t - \hat{h}_t^{(2)j} x_t \right] \delta^{t-1}$$

$$\left.\left. + \sum_{j \in S^2} (h_{T+1}^{(2)j} y_{T+1} - \hat{h}_{T+1}^{(2)j} x_{T+1} + Q_{T+1}^{(2)j}) \delta^T \right\}\right\}, \quad (6.6b)$$

$$\max_{\substack{u_t^{(3)\varsigma} \\ t\in\{1,2,\ldots,T\}}} \left\{ \sum_{t=1}^{T} \left[\left(p_t^{(3)\varsigma} - b_t^{(3)\varsigma(\varsigma)} u_t^{(3)\varsigma} + \sum_{\substack{\ell\in S^3 \\ \ell\neq\varsigma}} \varpi_t^{(3)\varsigma(\ell)} u_t^{(3)\varsigma} \right) u_t^{(3)\varsigma} \right.\right.$$

$$\left. - c_t^{(3)\varsigma} (u_t^{(3)\varsigma})^2 - h_t^{(3)\varsigma} x_t - \hat{h}_t^{(3)\varsigma} y_t \right] \delta^{t-1}$$

$$\left. + (Q_{T+1}^{(3)\varsigma} - h_{T+1}^{(3)\varsigma} x_{T+1} - \hat{h}_{T+1}^{(3)\varsigma} y_{T+1}) \delta^T \right\}, \quad \text{for } \varsigma \in S^3, \tag{6.6c}$$

subject to dynamics (6.1)—(6.2),
where $\sum_{i\in S^1}(h_{T+1}^{(1)i} x_{T+1} - \hat{h}_{T+1}^{(1)i} y_{T+1} + Q_{T+1}^{(1)i})\delta^T$, $\sum_{j\in S^2}(h_{T+1}^{(2)j} x_{T+1} - \hat{h}_{T+1}^{(2)j} y_{T+1} + Q_{T+1}^{(2)j})\delta^T$, and $(Q_{T+1}^{(3)\varsigma} - h_{T+1}^{(3)\varsigma} x_{T+1} - \hat{h}_{T+1}^{(3)\varsigma} y_{T+1})\delta^T$ are terminal payoffs.

6.1. *Game equilibrium*

Invoking Theorem 2.1, we have the following Corollary.

Corollary 6.1. *Let* $\underline{u}_k^{**} = (\underline{u}_k^{(1)**}, \underline{u}_k^{(2)**}, \underline{u}_k^{(3)**})$ *and* $\bar{\underline{u}}_k^{**} = (\bar{\underline{u}}_k^{(1)**}, \bar{\underline{u}}_k^{(2)**})$*, where* $u_k^{(z)**} = (u_k^{(z)1**}, u_k^{(z)2**}, \ldots, u_k^{(z)n_z**})$*, for* $z \in \{1,2\}$*,* $u_k^{(3)**} = (u_k^{(3)1**}, u_k^{(3)2**}, \ldots, u_k^{(3)n_3**})$*,* $\bar{u}_k^{(z)**} = (\bar{u}_k^{(z)1**}, \bar{u}_k^{(z)2**}, \ldots, \bar{u}_k^{(z)n_z**})$*, for* $z \in \{1,2\}$*, be the set of coalition equilibrium target controls and* $W^z(k,x,y;\bar{\underline{u}}_{k-}^{**})$ *be the equilibrium payoff of coalition* $z \in \{1,2\}$*, and* $W^{3(\varsigma)}(k,x,y;\bar{\underline{u}}_{k-}^{**})$*, for* $\varsigma \in S^3$*, be the equilibrium payoffs of the* n_3 *unallied nations at period* k *of the coalition game (6.6), then the functions* $W^z(k,x,y;\bar{\underline{u}}_{k-}^{**})$*, for* $z \in \{1,2\}$*, and* $W^{3(\varsigma)}(k,x,y;\bar{\underline{u}}_{k-}^{**})$*, for* $\varsigma \in S^3$*, satisfy the following recursive equations:*

$$W^1(T+1,x,y;\bar{\underline{u}}_{(T+1)-}^{**})$$
$$= \sum_{i\in S^1}(h_{T+1}^{(1)i} x_{T+1} - \hat{h}_{T+1}^{(1)i} y_{T+1} + Q_{T+1}^{(1)i})\delta^T,$$

$$\begin{aligned}
&W^2(T+1,x,y;\underline{\bar{u}}^{**}_{(T+1)-})\\
&\quad=\sum_{j\in S^2}(h^{(2)j}_{T+1}y_{T+1}-\hat{h}^{(2)j}_{T+1}x_{T+1}+Q^{(2)j}_{T+1})\delta^T,\\
&W^{3(\varsigma)}(T+1,x,y;\underline{\bar{u}}^{**}_{(T+1)-})\\
&\quad=(Q^{(3)\varsigma}_{T+1}-h^{(3)\varsigma}_{T+1}x_{T+1}-\hat{h}^{(3)\varsigma}_{T+1}y_{T+1})\delta^T,\quad for\ \varsigma\in S^3,
\end{aligned}\tag{6.7}$$

$$\begin{aligned}
&W^1(k,x,y;\underline{\bar{u}}^{**}_{k-})\\
&=\max_{u^{(1)}_k,\bar{u}^{(1)}_k}\Bigg\{\sum_{i\in S^1}\Bigg[\Bigg(p^{(1)i}_k-b^{(1)i(i)}_k u^{(1)i}_k+\sum_{\substack{l\in S^1\\ l\neq i}}\varpi^{(1)i(l)}_k u^{(1)l}_k\Bigg)u^{(1)i}_k\\
&\quad-c^{(1)i}_k(u^{(1)i}_k)^2-w^{(1)i}_k(\bar{u}^{(1)i}_k)^2+h^{(1)i}_k x_k-\hat{h}^{(1)i}_k y_k\Bigg]\delta^{k-1}\\
&\quad+W^1[k+1,x^{(\neq i)**}_{k+1},y^{**}_{k+1};u^{(1)}_k,\bar{u}^{(1)}_k,\underline{\bar{u}}^{**(\neq 1)}_k,\underline{\bar{u}}^{**}_{(k+1)-}\cap\underline{\bar{u}}^{**}_{k-}]\Bigg\},
\end{aligned}\tag{6.8}$$

where

$$\begin{aligned}
x^{(\neq 1)**}_{k+1}&=x_k+\sum_{\ell=1}^{n_1}\gamma^{(1)\ell}_k\bar{u}^{(1)\ell}_k+\sum_{\ell=1}^{n_1}\sigma^{(1)\ell}_k\bar{u}^{(1)\ell**}_{k-1}-\lambda^{(1)}_k x_k,\\
y^{**}_{k+1}&=y_k+\sum_{\ell=1}^{n_1}\gamma^{(2)\ell}_k\bar{u}^{(2)\ell**}_k+\sum_{\ell=1}^{n_2}\sigma^{(2)\ell}_k\bar{u}^{(2)\ell**}_{k-1}-\lambda^{(2)}_k x_k;
\end{aligned}$$

$$\begin{aligned}
&W^2(k,x,y;\underline{\bar{u}}^{**}_{k-})\\
&=\max_{u^{(2)}_k,\bar{u}^{(2)}_k}\Bigg\{\sum_{j\in S^2}\Bigg[\Bigg(p^{(2)j}_k-b^{(2)j(j)}_k u^{(2)j}_k+\sum_{\substack{l\in S^2\\ l\neq j}}\varpi^{(2)j(l)}_k u^{(2)l}_k\Bigg)u^{(1)j}_k
\end{aligned}$$

$$- c_k^{(2)j}(u_k^{(2)j})^2 - w_k^{(2)j}(\bar{u}_k^{(2)j})^2 + h_k^{(2)j}y_k - \hat{h}_k^{(2)j}x_k\Bigg]\delta^{k-1}$$

$$+ W^2[k+1, x_{k+1}^{**}, y_{k+1}^{(\neq 2)**}; u_k^{(2)}, \bar{u}_k^{(2)}, \underline{\bar{u}}_k^{**(\neq 2)}, \underline{\bar{u}}_{(k+1)-}^{**} \cap \underline{\bar{u}}_{k-}^{**}]\Bigg\}, \tag{6.9}$$

where

$$x_{k+1}^{**} = x_k + \sum_{\ell=1}^{n_1} \gamma_k^{(1)\ell}\bar{u}_k^{(1)\ell**} + \sum_{\ell=1}^{n_1} \sigma_k^{(1)\ell}\bar{u}_{k-1}^{(1)\ell**} - \lambda_k^{(1)}x_k,$$

$$y_{k+1}^{(\neq 2)**} = y_k + \sum_{\ell=1}^{n_1} \gamma_k^{(2)\ell}\bar{u}_k^{(2)\ell} + \sum_{\ell=1}^{n_2} \sigma_k^{(2)\ell}\bar{u}_{k-1}^{(2)\ell**} - \lambda_k^{(2)}x_k,$$

$$W^{(3)\varsigma}(k, x, y; \underline{\bar{u}}_{k-}^{**})$$

$$= \max_{u_k^{(3)\varsigma}} \Bigg\{\Bigg[\Bigg(-b_k^{(3)\varsigma(\varsigma)}u_k^{(3)\varsigma} + \sum_{\substack{l\in S^3 \\ l\neq\varsigma}} \varpi_k^{(3)\varsigma(l)}u_k^{(3)l**}\Bigg)u_k^{(3)\varsigma}$$

$$-c_k^{(3)\varsigma}(u_k^{(3)\varsigma})^2 - h_k^{(3)\varsigma}x_k - \hat{h}_k^{(3)\varsigma}y_k\Bigg]\delta^{k-1} + W^{(3)\varsigma}[k+1, x_{k+1}^{**}, y_{k+1}^{**};$$

$$u_k^{(3)\varsigma}, \underline{\bar{u}}_k^{**}, \underline{\bar{u}}_{(k+1)-}^{**} \cap \underline{\bar{u}}_{k-}^{**}]\Bigg\}, \quad \text{for } \varsigma \in \{1, 2, \ldots, n_3\}, \tag{6.10}$$

where

$$x_{k+1}^{**} = x_k + \sum_{\ell=1}^{n_1} \gamma_k^{(1)\ell}\bar{u}_k^{(1)\ell**} + \sum_{\ell=1}^{n_1} \sigma_k^{(1)\ell}\bar{u}_{k-1}^{(1)\ell**} - \lambda_k^{(1)}x_k,$$

$$y_{k+1}^{**} = y_k + \sum_{\ell=1}^{n_1} \gamma_k^{(2)\ell}\bar{u}_k^{(2)\ell**} + \sum_{\ell=1}^{n_2} \sigma_k^{(2)\ell}\bar{u}_{k-1}^{(2)\ell**} - \lambda_k^{(2)}x_k.$$

Performing the indicated maximization operator in Corollary 6.1 *and solving yield the the equilibrium payoffs of the two coalitions and the equilibrium payoffs of unallied nations plus the equilibrium target controls of all nations.*

Proposition 6.1. *The equilibrium payoff of Coalition* 1 *at period* $k \in \{1, 2, \ldots, T\}$ *can be obtained as follows:*

$$W^1(k, x, y; \underline{\bar{u}}^{**}_{k-}) = \left(A^{(1)}_k x_k + B^{(1)}_k y_k + C^{(1)}_k\right) \delta^{k-1},$$

where

$$A^{(1)}_k = \sum_{i \in S^1} h^{(1)i}_k + A^{(1)}_{k+1}(1 - \lambda^{(1)}_k)\delta,$$

$$B^{(1)}_k = -\sum_{i \in S^1} \hat{h}^{(1)i}_k + B^{(1)}_{k+1}(1 - \lambda^{(2)}_k)\delta, \text{ and}$$

$C^{(1)}_k$ *is a term which does not involve* (x, y) *but has previously executed controls* $(\bar{u}^{(1)}_{k-}, \bar{u}^{(2)}_{k-})$; *with*

$$A^{(1)}_{T+1} = \sum_{i \in S^1} h^{(1)i}_{T+1}, \quad B^{(1)}_{T+1} = -\sum_{i \in S^1} \hat{h}^{(1)i}_{T+1}, \quad C^{(1)}_{T+1} = \sum_{i \in S^1} Q^{(1)i}_{T+1}.$$

The equilibrium payoff of Coalition 2 *at period* $k \in \{1, 2, \ldots, T\}$ *can be obtained as follows:*

$$W^2(k, x, y; \underline{\bar{u}}^{**}_{k-}) = \left(A^{(2)}_k x_k + B^{(2)}_k y_k + C^{(2)}_k\right) \delta^{k-1},$$

where

$$A^{(2)}_k = -\sum_{j \in S^2} \hat{h}^{(2)j}_k + A^{(2)}_{k+1}(1 - \lambda^{(1)}_k)\delta,$$

$$B^{(2)}_k = \sum_{j \in S^2} h^{(2)j}_k + B^{(2)}_{k+1}(1 - \lambda^{(2)}_k)\delta, \text{ and}$$

$C^{(2)}_k$ *is a term which does not involve* (x, y) *but has previously executed controls* $(\bar{u}^{(1)}_{k-}, \bar{u}^{(2)}_{k-})$, *with* $A^{(2)}_{T+1} = -\sum_{j \in S^2} \hat{h}^{(2)j}_{T+1}$, $B^{(2)}_{T+1} = \sum_{j \in S^2} h^{(2)j}_{T+1}$, $C^{(2)}_{T+1} = \sum_{j \in S^2} Q^{(2)j}_{T+1}$.

The equilibrium payoff of the n_3 unallied nations at period $k \in \{1, 2, \ldots, T\}$ can be obtained as follows:

$$W^{(3)\varsigma}(k, x, y; \underline{\bar{u}}_{k-}^{**}) = \left(A_k^{(3)\varsigma} x_k - B_k^{(3)\varsigma} y_k + C_k^{(3)\varsigma}\right) \delta^{k-1},$$
$$\varsigma \in \{1, 2, \ldots, n_3\},$$

where

$$A_k^{(3)\varsigma} = -h_k^{(3)\varsigma} + A_{k+1}^{(3)\varsigma}(1 - \lambda_k^{(1)})\delta,$$
$$B_k^{(3)\varsigma} = -\hat{h}_k^{(3)\varsigma} + B_{k+1}^{(3)\varsigma}(1 - \lambda_k^{(2)})\delta, \text{ and}$$

$C_k^{(3)\varsigma}$ *is a term which does not involve (x, y) but has previously executed controls $(\bar{u}_{k-}^{(1)}, \bar{u}_{k-}^{(2)})$, with $A_{T+1}^{(3)\varsigma} = -h_{T+1}^{(3)\varsigma}$, $B_{T+1}^{(3)\varsigma} = -\hat{h}_{T+1}^{(3)\varsigma}$, $C_{T+1}^{(3)\varsigma} = Q_{T+1}^{(3)\varsigma}$.*

Proof. See Appendix C in the Mathematical Appendices. □

Proposition 6.2. *The equilibrium target controls of Coalition 1 are $(u_k^{(1)1**}, u_k^{(1)2**}, \ldots, u_k^{(1)n_1**})$ which is the solution of the linear equation system*

$$\frac{p_k^{(1)i}}{2b_k^{(1)i(i)} + 2c_k^{(1)i}} = u_k^{(1)i} - \sum_{\substack{\ell \in S^1 \\ \ell \neq i}} \frac{\varpi_k^{(1)i(\ell)} + \varpi_k^{(1)\ell(i)}}{2b_k^{(1)i(i)} + 2c_k^{(1)i}} u_k^{(1)\ell},$$

for $i \in \{1, 2, \ldots, n_1\}$, and

$$\bar{u}_T^{(1)i**} = \frac{A_{T+1}^{(1)} \gamma_T^{(1)i} \delta}{2w_T^{(1)i}},$$
$$\bar{u}_k^{(1)i**} = \frac{A_{k+1}^{(1)} \gamma_k^{(1)i} \delta + A_{k+2}^{(1)} \sigma_k^{(1)i} \delta^2}{2w_k^{(1)i}}, \text{ for } k \in \{1, 2, \ldots, T-1\} \text{ and}$$
$$i \in \{1, 2, \ldots, n_1\}. \tag{6.11}$$

The equilibrium target controls of Coalition 2 are $(u_k^{(2)1**}, u_k^{(2)2**}, \ldots, u_k^{(2)n_2**})$ which is the solution of the linear equation system

$$\frac{p_k^{(2)j}}{2b_k^{(2)j(j)} + 2c_k^{(2)j}} = u_k^{(2)j} - \sum_{\substack{\ell \in S^2 \\ \ell \neq j}} \frac{\varpi_k^{(2)j(\ell)} + \varpi_k^{(2)\ell(j)}}{2b_k^{(2)j(j)} + 2c_k^{(2)j}} u_k^{(2)\ell},$$

for $j \in \{1, 2, \ldots, n_2\}$, and

$$\bar{u}_T^{(2)j**} = \frac{B_{T+1}^{(2)} \gamma_T^{(2)j} \delta}{2w_T^{(2)k}},$$

$$\bar{u}_k^{(2)j**} = \frac{B_{k+1}^{(2)} \gamma_k^{(2)j} \delta + B_{k+2}^{(2)} \sigma_k^{(2)j} \delta^2}{2w_k^{(2)j}}, \text{ for } k \in \{1, 2, \ldots, T-1\} \text{ and}$$
$$j \in \{1, 2, \ldots, n_2\}. \tag{6.12}$$

The equilibrium target controls of unallied ς nations are $(u_k^{(3)1**}, u_k^{(3)2**}, \ldots, u_k^{(3)n_2**})$ which is the solution of the linear equation system

$$\frac{p_k^{(3)\varsigma}}{2b_k^{(3)\varsigma(\varsigma)} + 2c_k^{(3)\varsigma}} = u_k^{(3)\varsigma} - \sum_{\substack{\ell \in S^3 \\ \ell \neq \varsigma}} \frac{\varpi_k^{(3)\varsigma(\ell)}}{2b_k^{(3)\varsigma(\varsigma)} + 2c_k^{(3)\varsigma}} u_k^{(3)\ell},$$
$$\text{for } \varsigma \in \{1, 2, \ldots, n_3\}. \tag{6.13}$$

Proof. See Appendix C in the Mathematical Appendices. □

The equilibrium dynamics becomes

$$x_{k+1} = x_k + \sum_{\ell=1}^{n_1} \gamma_k^{(1)\ell} \bar{u}_k^{(1)\ell**} + \sum_{\ell=1}^{n_1} \sigma_k^{(1)\ell} \bar{u}_{k-1}^{(1)\ell**} - \lambda_k^{(1)} x_k,$$

$$x_1 = x_1^0,$$

$$y_{k+1} = y_k + \sum_{\ell=1}^{n_1} \gamma_k^{(2)\ell} \bar{u}_k^{(2)\ell**} + \sum_{\ell=1}^{n_2} \sigma_k^{(2)\ell} \bar{u}_{k-1}^{(2)\ell**} - \lambda_k^{(2)} x_k,$$

$$y_1 = y_1^0. \tag{6.14}$$

We use $\{x_k^{**}, y_k^{**}\}_{k=1}^{T+1}$ *to denote the solution to the linear difference equation system in* (6.14).

6.2. *Intra-coalition payoff distribution*

Within each coalition, the nations would adopt the optimal sharing of gains among nations within which a coalition is proportional to the relative size of their non-cooperative payoffs.

As a benchmark, we also consider the non-cooperative outcome of nations within a coalition. Assume the case that Coalition 2 is not formed and nations in S^2 act as an individual nation. The scenario becomes

$$\max_{\substack{u_t^{(1)i},\bar{u}_t^{(1)i}\\ i\in S^1\\ t\in\{1,2,\ldots,T\}}} \left\{\sum_{i\in S^1}\left\{\sum_{t=1}^{T}\left[\left(p_t^{(1)i} - b_t^{(1)i(i)}u_t^{(1)i} + \sum_{\substack{\ell\in S^1\\ \ell\neq i}} \varpi_t^{(1)i(\ell)}u_t^{(1)\ell}\right)u_t^{(1)i}\right.\right.\right.$$
$$\left. - c_t^{(1)i}(u_t^{(1)i})^2 - w_t^{(1)i}(\bar{u}_t^{(1)i})^2 + h_t^{(1)i}x_t - \hat{h}_t^{(1)i}y_t\right]\delta^{t-1}$$
$$\left.\left. + \sum_{i\in S^1}(h_{T+1}^{(1)i}x_{T+1} - \hat{h}_{T+1}^{(1)i}y_{T+1} + Q_{T+1}^{(1)i})\delta^T\right\}\right\}, \tag{6.15a}$$

$$\max_{\substack{u_t^{(2)j},\bar{u}_t^{(2)j}\\ t\in\{1,2,\ldots,T\}}} \left\{\sum_{t=1}^{T}\left[\left(p_t^{(2)j} - b_t^{(2)j(j)}u_t^{(2)j} + \sum_{\substack{\ell\in S^2\\ \ell\neq j}} \varpi_t^{(2)j(\ell)}u_t^{(2)\ell}\right)u_t^{(2)j}\right.\right.$$
$$- c_t^{(2)j}(u_t^{(2)j})^2 - w_t^{(2)j}(\bar{u}_t^{(2)j})^2 + h_t^{(2)j}y_t - \hat{h}_t^{(2)j}x_t]\delta^{t-1}$$
$$\left.\left. + (h_{T+1}^{(2)j}y_{T+1} - \hat{h}_{T+1}^{(2)j}x_{T+1} + Q_{T+1}^{(2)j})\delta^T\right]\right\}, \quad \text{for } j\in S^2, \tag{6.15b}$$

$$\max_{\substack{u_t^{(3)\varsigma}\\ t\in\{1,2,\ldots,T\}}} \left\{\sum_{t=1}^{T}\left[(p_t^{(3)\varsigma} - b_t^{(3)\varsigma(\varsigma)}u_t^{(3)\varsigma} + \sum_{\substack{\ell\in S^3\\ \ell\neq\varsigma}} \varpi_t^{(3)\varsigma(\ell)}u_t^{(3)\varsigma})u_t^{(3)\varsigma}\right.\right.$$

$$- c_t^{(3)\varsigma}(u_t^{(3)\varsigma})^2 - h_t^{(3)\varsigma}x_t - \hat{h}_t^{(3)\varsigma}y_t\Bigg]\delta^{t-1}$$

$$+ (Q_{T+1}^{(3)\varsigma} - h_{T+1}^{(3)\varsigma}x_{T+1} - \hat{h}_{T+1}^{(3)\varsigma}y_{T+1})\delta^T\Bigg\}, \text{ for } \varsigma \in S^3, \tag{6.15c}$$

subject to

$$x_{t+1} = x_t + \sum_{\ell=1}^{n_1}\gamma_t^{(1)\ell}\bar{u}_t^{(1)\ell} + \sum_{\ell=1}^{n_1}\sigma_t^{(1)\ell}\bar{u}_{t-1}^{(1)\ell} - \lambda_t^{(1)}x_t, \quad x_1 = x_1^0,$$

$$y_{t+1} = y_t + \sum_{\ell=1}^{n_1}\gamma_t^{(2)\ell}\bar{u}_t^{(2)\ell} + \sum_{\ell=1}^{n_2}\sigma_t^{(2)\ell}\bar{u}_{t-1}^{(2)\ell} - \lambda_k^{(2)}x_k, \quad y_1 = y_1^0. \tag{6.16}$$

The equilibrium payoffs of Coalition 1 and the equilibrium payoffs of unallied nations in S^2 and S^3 can be obtained as follows:

Proposition 6.3. *The equilibrium payoff of Coalition* 1 *at period* $k \in \{1, 2, \ldots, T\}$ *can be obtained as follows:*

$$W^1(k, x, y; \underline{\bar{u}}_{k-}^{**}) = (A_k^{(1)}x_k + B_k^{(1)}y_k + C_k^{(1)})\delta^{k-1},$$

where

$$A_k^{(1)} = \sum_{i\in S^1} h_k^{(1)i} + A_{k+1}^{(1)}(1 - \lambda_k^{(1)})\delta,$$

$$B_k^{(1)} = -\sum_{i\in S^1} \hat{h}_k^{(1)i} + B_{k+1}^{(1)}(1 - \lambda_k^{(2)})\delta, \text{ and}$$

$C_k^{(1)}$ *is a term which does not involve* (x, y) *but has previously executed controls* $(\bar{u}_{k-}^{(1)}, \bar{u}_{k-}^{(2)})$, *with*

$$A_{T+1}^{(1)} = \sum_{i\in S^1} h_{T+1}^{(1)i},$$

$$B_{T+1}^{(1)} = -\sum_{i\in S^1} \hat{h}_{T+1}^{(1)i},$$
$$C_{T+1}^{(1)} = \sum_{i\in S^1} Q_{T+1}^{(1)i}.$$

The equilibrium payoff of independent nation $j \in S^2$ at period $k \in \{1, 2, \ldots, T\}$ can be obtained as follows:

$$W^{(2)j}(k, x, y; \underline{\bar{u}}_{k-}^{**}) = (A_k^{(2)j} x_k + B_k^{(2)j} y_k + C_k^{(2)j})\delta^{k-1},$$

where

$$A_k^{(2)j} = -\hat{h}_k^{(2)j} + A_{k+1}^{(2)}(1 - \lambda_k^{(1)})\delta,$$
$$B_k^{(2)j} = h_k^{(2)j} + B_{k+1}^{(2)j}(1 - \lambda_k^{(2)})\delta,$$

and $C_k^{(2)j}$ is a term which does not involve (x, y) but has previously executed controls $(\bar{u}_{k-}^{(1)}, \bar{u}_{k-}^{(2)})$, with $A_{T+1}^{(2)j} = -\hat{h}_{T+1}^{(2)j}$, $B_{T+1}^{(2)j} = h_{T+1}^{(2)j}$, $C_{T+1}^{(2)j} = Q_{T+1}^{(2)j}$.

The equilibrium payoff of the n_3 unallied nations at period $k \in \{1, 2, \ldots, T\}$ can be obtained as follows:

$$W^{(3)\varsigma}(k, x, y; \underline{\bar{u}}_{k-}^{**})$$
$$= (A_k^{(3)\varsigma} x_k + B_k^{(3)\varsigma} y_k + C_k^{(3)\varsigma})\delta^{k-1}, \varsigma \in \{1, 2, \ldots, n_3\},$$

where

$$A_k^{(3)\varsigma} = -h_k^{(3)\varsigma} + A_{k+1}^{(3)\varsigma}(1 - \lambda_k^{(1)})\delta,$$
$$B_k^{(3)\varsigma} = -\hat{h}_k^{(3)\varsigma} + B_{k+1}^{(3)\varsigma}(1 - \lambda_k^{(2)})\delta, \text{ and}$$

$C_k^{(3)\varsigma}$ is a term which does not involve (x, y) but has previously executed controls $(\bar{u}_{k-}^{(1)}, \bar{u}_{k-}^{(2)})$; with

$$A_{T+1}^{(3)\varsigma} = -h_{T+1}^{(3)\varsigma}, \quad B_{T+1}^{(3)\varsigma} = -\hat{h}_{T+1}^{(3)\varsigma}, \quad C_{T+1}^{(3)\varsigma} = Q_{T+1}^{(3)\varsigma}. \tag{6.17}$$

Proof. Follow the proof of Proposition 6.1. □

The equilibrium target controls of Coalition 1 and the equilibrium target controls of unallied nations in S^2 and S^3 can be obtained as follows:

Proposition 6.4. *The equilibrium target controls of Coalition* 1 *are* $(u_k^{(1)1**}, u_k^{(1)2**}, \ldots, u_k^{(1)n_1**})$ *which is the solution of the linear equation system*

$$\frac{p_k^{(1)i}}{2b_k^{(1)i(i)} + 2c_k^{(1)i}} = u_k^{(1)i} - \sum_{\substack{\ell \in S^1 \\ \ell \neq i}} \frac{\varpi_k^{(1)i(\ell)} + \varpi_k^{(1)\ell(i)}}{2b_k^{(1)i(i)} + 2c_k^{(1)i}} u_k^{(1)\ell},$$

for $i \in \{1, 2, \ldots, n_1\}$, *and*

$$\bar{u}_T^{(1)i**} = \frac{A_{T+1}^{(1)} \gamma_T^{(1)i} \delta}{2w_T^{(1)i}},$$

$$\bar{u}_k^{(1)i**} = \frac{A_{k+1}^{(1)} \gamma_k^{(1)i} \delta + A_{k+2}^{(1)} \sigma_k^{(1)i} \delta^2}{2w_k^{(1)i}}, \text{ for } k \in \{1, 2, \ldots, T-1\}$$
$$\text{and } i \in \{1, 2, \ldots, n_1\}.$$

The equilibrium target controls of of independent nation $j \in S^2$ are $(u_k^{(2)1***}, u_k^{(2)2***}, \ldots, u_k^{(2)n_2***})$ which is the solution of the linear equation system

$$\frac{p_k^{(2)j}}{2b_k^{(2)j(j)} + 2c_k^{(2)j}} = u_k^{(2)j} - \sum_{\substack{\ell \in S^2 \\ \ell \neq j}} \frac{\varpi_k^{(2)j(\ell)}}{2b_k^{(2)j(j)} + 2c_k^{(2)j}} u_k^{(2)\ell},$$

for $j \in \{1, 2, \ldots, n_2\}$, and

$$\bar{u}_T^{(2)j***} = \frac{B_{T+1}^{(2)j} \gamma_T^{(2)j} \delta}{2w_T^{(2)k}},$$

$$\bar{u}_k^{(2)j***} = \frac{B_{k+1}^{(2)j} \gamma_k^{(2)j} \delta + B_{k+2}^{(2)j} \sigma_k^{(2)j} \delta^2}{2w_k^{(2)j}}, \text{ for } k \in \{1, 2, \ldots, T-1\}$$
$$\text{and } j \in \{1, 2, \ldots, n_2\}.$$

The equilibrium target controls of unallied ς nations are $(u_k^{(3)1**}, u_k^{(3)2**}, \ldots, u_k^{(3)n_2**})$ which is the solution of the linear equation system

$$\frac{p_k^{(3)\varsigma}}{2b_k^{(3)\varsigma(\varsigma)} + 2c_k^{(3)\varsigma}} = u_k^{(3)\varsigma} - \sum_{\substack{\ell \in S^3 \\ \ell \neq \varsigma}} \frac{\varpi_k^{(3)\varsigma(\ell)}}{2b_k^{(3)\varsigma(\varsigma)} + 2c_k^{(3)\varsigma}} u_k^{(3)\ell},$$

$$\text{for } \varsigma \in \{1, 2, \ldots, n_3\}. \tag{6.18}$$

Proof. Follow the proof Proposition 6.2. □

The agreed-upon sharing of gains proportional to the relative size of the nations' non-cooperative payoffs in Coalition 2 can be expressed as follows:

$$\xi^{(2)j}(k, x_k, y_k; \underline{\bar{u}}_{k-})$$
$$= \frac{W^{(2)j}(k, x, y; \underline{\bar{u}}_{k-})}{\sum_{\ell \in S^2} W^{(2)\ell}(k, x, y; \underline{\bar{u}}_{k-}^{**})} W^{(2)}(k, x, y; \underline{\bar{u}}_{k-}), j \in \{1, 2, \ldots, n_2\}. \tag{6.19}$$

Note that $\xi^{(2)j}(k, x_k, y_k; \underline{\bar{u}}_{k-})$ satisfies individual rationality and group optimality.

Invoking Theorem 2.2, the agreed-upon imputation $\xi^{(2)j}(k, x_k, y_k; \underline{\bar{u}}_{k-})$ in (6.19) can be realized by a payment

$$\beta_k^{(2)j}(x_k^{**}, y_k^{**}; \underline{\bar{u}}_{k-}^{**}) = \delta^{1-k}[\xi^{(2)j}(k, x_k^{**}, y_k^{**}; \underline{\bar{u}}_{k-}^{**})$$
$$- \xi^{(2)j}(k+1, x_{k+1}^{**}, y_{k+1}^{**}; \underline{\bar{u}}_{(k+1)-}^{**})]$$

given to nation $j \in S^2$ at period $k \in \{1, 2, \ldots, T\}$.

In a similar manner, we can perform the above analysis for Coalition 1.

Remark 6.1.

(i) The payoffs of nations forming a coalition would be higher than those when they act individually because $\xi^{(2)j}(k, x_k, y_k; \underline{\bar{u}}_{k-}) \geq W^{(2)j}(k, x, y, ; \underline{\bar{u}}_{k-})$.

(ii) The security level of nations belonging to coalition would be higher that when they act individually because

$$\bar{u}_t^{(2)j**} \geq \bar{u}_t^{(2)j***}, \text{ for } B_t^{(2)} = B_t^{(2)j}.$$

(iii) The income level of nations belonging to coalition would be higher that when they act individually because $u_k^{(2)j**}$ from the system

$$\frac{p_k^{(2)j}}{2b_k^{(2)j(j)} + 2c_k^{(2)j}} = u_k^{(2)j} - \sum_{\substack{\ell \in S^2 \\ \ell \neq j}} \frac{\varpi_k^{(2)j(\ell)} + \varpi_k^{(2)\ell(j)}}{2b_k^{(2)j(j)} + 2c_k^{(2)j}} u_k^{(2)\ell},$$

for $j \in \{1, 2, \ldots, n_2\}$, would be higher $u_k^{(2)j***}$ from the system

$$\frac{p_k^{(2)j}}{2b_k^{(2)j(j)} + 2c_k^{(2)j}} = u_k^{(2)j} - \sum_{\substack{\ell \in S^2 \\ \ell \neq j}} \frac{\varpi_k^{(2)j(\ell)}}{2b_k^{(2)j(j)} + 2c_k^{(2)j}} u_k^{(2)\ell},$$

for $j \in \{1, 2, \ldots, n_2\}$.

Remark 6.1 shows that nations with similar interests have incentive to form a coalition.

6.3. *Grand coalition cooperation and efficient outcomes*

A grand coalition cooperation of the coalitional blocs is studied. The Pareto optimal target controls can be obtained by solving the joint maximization problem as follows:

$$\max_{\substack{u_t^{(1)i}, \bar{u}_t^{(1)i}, i \in S^1 \\ u_t^{(2)j}, \bar{u}_t^{(2)j}, j \in S^2 \\ u_t^{(3)\varsigma}, \varsigma \in S^3 \\ t \in \{1,2,\ldots,T\}}} \left\{ \sum_{i \in S^1} \sum_{t=1}^{T} \left[\left(p_t^{(1)i} - b_t^{(1)i(i)} u_t^{(1)i} + \sum_{\substack{\ell \in S^1 \\ \ell \neq i}} \varpi_t^{(1)i(\ell)} u_t^{(1)\ell} \right) u_t^{(1)i} \right.\right.$$

$$\left. - c_t^{(1)i} (u_t^{(1)i})^2 - w_t^{(1)i} (\bar{u}_t^{(1)i})^2 + h_t^{(1)i} x_t - \hat{h}_t^{(1)i} y_t \right] \delta^{t-1}$$

$$+ \sum_{j \in S^2} \sum_{t=1}^{T} \left[\left(p_t^{(2)j} - b_t^{(2)j(j)} u_t^{(2)j} + \sum_{\substack{\ell \in S^2 \\ \ell \neq j}} \varpi_t^{(2)j(\ell)} u_t^{(2)\ell} \right) u_t^{(2)j} \right.$$

$$-c_t^{(2)j}(u_t^{(2)j})^2 - w_t^{(2)j}(\bar{u}_t^{(2)j})^2 + h_t^{(2)j} y_t - \hat{h}_t^{(2)j} x_t \Bigg] \delta^{t-1}$$

$$+ \sum_{\varsigma \in S^3} \sum_{t=1}^{T} \left[\left(p_t^{(3)\varsigma} - b_t^{(3)\varsigma(\varsigma)} u_t^{(3)\varsigma} + \sum_{\substack{\ell \in S^3 \\ \ell \neq \varsigma}} \varpi_t^{(3)\varsigma(\ell)} u_t^{(3)\varsigma} \right) \right.$$

$$\left. \times u_t^{(3)\varsigma} - c_t^{(3)\varsigma}(u_t^{(3)\varsigma})^2 - h_t^{(3)\varsigma} x_t - \hat{h}_t^{(3)\varsigma} y_t \right] \delta^{t-1}$$

$$+ \left[\sum_{i \in S^1} (h_{T+1}^{(1)i} x_{T+1} - \hat{h}_{T+1}^{(1)i} y_{T+1} + Q_{T+1}^{(1)i}) \right.$$

$$+ \sum_{j \in S^2} (h_{T+1}^{(2)j} y_{T+1} - \hat{h}_{T+1}^{(2)j} x_{T+1} + Q_{T+1}^{(2)j})$$

$$\left. + \sum_{\varsigma \in S^3} (Q_{T+1}^{(3)\varsigma} - h_{T+1}^{(3)\varsigma} x_{T+1} - \hat{h}_{T+1}^{(3)\varsigma} y_{T+1}) \right] \delta^T \Bigg\}, \qquad (6.20)$$

subject to dynamics (6.1)–(6.2).

Solving the grand coalition cooperation problem in (6.20) yields the payoff of the grand coalition and the Pareto optimal target controls. We use $\underline{u}_k^* = (\underline{u}_k^{(1)*}, \underline{u}_k^{(2)*}, \underline{u}_k^{(3)*})$ and $\underline{\bar{u}}_k^* = (\underline{\bar{u}}_k^{(1)*}, \underline{\bar{u}}_k^{(2)*})$ to denote the set of Pareto optimal target controls under the grand coalition in period k. We use $W(k, x, y; \underline{\bar{u}}_{k-}^*)$ to denote the maximized joint payoff of all nations at period k. The grand coalition cooperative payoff and the optimal target controlsis are obtained in the following propositions.

Proposition 6.5. *The grand coalition cooperative payoff at period $k \in \{1, 2, \ldots, T\}$ can be obtained as follows:*

$$W(k, x, y; \underline{\bar{u}}_{k-}^*) = (A_k x_k + B_k y_k + C_k)\delta^{k-1}, \qquad (6.21)$$

where

$$A_k = \sum_{i\in S^1} h_k^{(1)i} - \sum_{j\in S^2} \hat{h}_k^{(2)j} - \sum_{j\in S^3} h_k^{(3)\varsigma} + A_{k+1}(1-\lambda_k^{(1)})\delta,$$

$$B_k = \sum_{i\in S^2} h_k^{(2)j} - \sum_{i\in S^1} \hat{h}_k^{(1)i} - \sum_{\varsigma\in S^3} \hat{h}_k^{(3)\varsigma} + B_{k+1}(1-\lambda_k^{(2)})\delta, \text{ and}$$

$C_k^{(1)}$ *is a term which does not involve* (x,y) *but has previously executed controls* $(\bar{u}_{k-}^{(1)}, \bar{u}_{k-}^{(2)})$, *with*

$$A_{T+1} = \sum_{i\in S^1} h_{T+1}^{(1)i} - \sum_{j\in S^2} h_{T+1}^{(2)j} - \sum_{\varsigma\in S^3} h_{T+1}^{(3)\varsigma},$$

$$B_{T+1} = \sum_{j\in S^2} h_{T+1}^{(2)j} - \sum_{i\in S^1} \hat{h}_{T+1}^{(1)i} - \sum_{\varsigma\in S^3} \hat{h}_{T+1}^{(3)\varsigma},$$

$$C_{T+1} = \sum_{i\in S^1} Q_{T+1}^{(1)i} + \sum_{j\in S^2} Q_{T+1}^{(2)j} + \sum_{\varsigma\in S^3} Q_{T+1}^{(3)\varsigma}.$$

Proof. See Appendix D in the Mathematical Appendices. □

The set of Pareto optimal target controls under the grand coalition is obtained in the following proposition.

Proposition 6.6. *The set of Pareto optimal target controls under the grand coalition is* $(u_k^{(1)1*}, u_k^{(1)2*}, \ldots, u_k^{(1)n_1*})$ *which is the solution of the linear equation system*

$$\frac{p_k^{(1)i}}{2b_k^{(1)i(i)} + 2c_k^{(1)i}} = u_k^{(1)i} - \sum_{\substack{\ell\in S^1\\ \ell\neq i}} \frac{\varpi_k^{(1)i(\ell)} + \varpi_k^{(1)\ell(i)}}{2b_k^{(1)i(i)} + 2c_k^{(1)i}} u_k^{(1)\ell},$$

for $i \in \{1,2,\ldots,n_1\}$, *and*

$$\begin{aligned}
\bar{u}_T^{(1)i*} &= \frac{A_{T+1}\gamma_T^{(1)i}\delta}{2w_T^{(1)i}},\\
\bar{u}_k^{(1)i*} &= \frac{A_{k+1}\gamma_k^{(1)i}\delta + A_{k+2}\sigma_k^{(1)i}\delta^2}{2w_k^{(1)i}}, \text{ for } k \in \{1,2,\ldots,T-1\}\\
&\quad \text{and } i \in \{1,2,\ldots,n_1\}.
\end{aligned} \tag{6.22}$$

$(u_k^{(2)1*}, u_k^{(2)2*}, \ldots, u_k^{(2)n_2*})$ *which is the solution of the linear equation system*

$$\frac{p_k^{(2)j}}{2b_k^{(2)j(j)} + 2c_k^{(2)j}} = u_k^{(2)j} - \sum_{\substack{\ell \in S^2 \\ \ell \neq j}} \frac{\varpi_k^{(2)j(\ell)} + \varpi_k^{(2)\ell(j)}}{2b_k^{(2)j(j)} + 2c_k^{(2)j}} u_k^{(2)\ell},$$

for $j \in \{1, 2, \ldots, n_2\}$, *and*

$$\bar{u}_T^{(2)j*} = \frac{B_{T+1}\gamma_T^{(2)j}\delta}{2w_T^{(2)k}},$$

$$\bar{u}_k^{(2)j*} = \frac{B_{k+1}\gamma_k^{(2)j}\delta + B_{k+2}\sigma_k^{(2)j}\delta^2}{2w_k^{(2)j}}, \text{ for } k \in \{1, 2, \ldots, T-1\}$$

$$\text{and } j \in \{1, 2, \ldots, n_2\}. \tag{6.23}$$

$(u_k^{(3)1*}, u_k^{(3)2*}, \ldots, u_k^{(3)n_2*})$ *which is the solution of the linear equation system*

$$\frac{p_k^{(3)\varsigma}}{2b_k^{(3)\varsigma(\varsigma)} + 2c_k^{(3)\varsigma}} = u_k^{(3)\varsigma} - \sum_{\substack{\ell \in S^3 \\ \ell \neq \varsigma}} \frac{\varpi_k^{(3)\varsigma(\ell)}}{2b_k^{(3)\varsigma(\varsigma)} + 2c_k^{(3)\varsigma}} u_k^{(3)\ell},$$

$$\text{for } \varsigma \in \{1, 2, \ldots, n_3\}. \tag{6.24}$$

Proof. See Appendix D in the Mathematical Appendices. □

The Pareto optimal state dynamics becomes

$$x_{t+1} = x_t + \sum_{\ell=1}^{n_1} \gamma_t^{(1)\ell}\bar{u}_t^{(1)\ell*} + \sum_{\ell=1}^{n_1} \sigma_t^{(1)\ell}\bar{u}_{t-1}^{(1)\ell*} - \lambda_t^{(1)} x_t,$$

$$x_1 = x_1^0,$$

$$y_{t+1} = y_t + \sum_{\ell=1}^{n_1} \bar{u}_t^{(2)\ell*} + \sum_{\ell=1}^{n_2} \sigma_t^{(2)\ell}\bar{u}_{t-1}^{(2)\ell*} - \lambda_k^{(2)} x_k,$$

$$y_1 = y_1^0. \tag{6.25}$$

We use $\{(x_k^*, y_k^*)\}_{k=1}^{T+1}$ to denote the solution to the linear difference equation system (6.25).

The grand coalition would allocation the cooperative payoff to the coalitions and unallied nations proportional to the relative size of their coalitional equilibrium payoffs. Coalition 1 would receive a cooperative coalitional payoff

$$\begin{aligned}&\varphi^{(1)}(k, x^*, y^*; \underline{\bar{u}}_{k-}^*)\\&\quad = \frac{W^{(1)}(k, x^*, y^*; \underline{\bar{u}}_{k-}^*)}{\sum_{\ell=1}^{2} W^{(\ell)}(k, x^*, y^*; \underline{\bar{u}}_{k-}^*) + \sum_{\ell=1}^{n_3} W^{(3)\ell}(k, x^*, y^*; \underline{\bar{u}}_{k-}^*)}\\&\qquad \times W(k, x^*, y^*; \underline{\bar{u}}_{k-}^*).\end{aligned}$$

Coalition 2 would receive a cooperative coalitional payoff

$$\begin{aligned}&\varphi^{(2)}(k, x^*, y^*; \underline{\bar{u}}_{k-}^*)\\&\quad = \frac{W^{(2)}(k, x^*, y^*; \underline{\bar{u}}_{k-}^*)}{\sum_{\ell=1}^{2} W^{(\ell)}(k, x^*, y^*; \underline{\bar{u}}_{k-}^*) + \sum_{\ell=1}^{n_3} W^{(3)\ell}(k, x^*, y^*; \underline{\bar{u}}_{k-}^*)}\\&\qquad \times W(k, x^*, y^*; \underline{\bar{u}}_{k-}^*).\end{aligned}$$

Unallied nation $\varsigma \in S^3$ would receive a cooperative coalitional payoff

$$\begin{aligned}&\varphi^{(3)\varsigma}(k, x^*, y^*; \underline{\bar{u}}_{k-}^*)\\&\quad = \frac{W^{(3)\varsigma}(k, x^*, y^*; \underline{\bar{u}}_{k-}^*)}{\sum_{\ell=1}^{2} W^{(\ell)}(k, x^*, y^*; \underline{\bar{u}}_{k-}^*) + \sum\limits_{\ell=1}^{n_3} W^{(3)\ell}(k, x^*, y^*; \underline{\bar{u}}_{k-}^*)}\\&\qquad \times W(k, x^*, y^*; \underline{\bar{u}}_{k-}^*),\end{aligned}$$

for $\varsigma \in S^3$.

To ensure that the cooperative payoffs, $\varphi^{(1)}(k, x^*, y^*; \underline{\bar{u}}_{k-}^*)$, $\varphi^{(2)}(k, x^*, y^*; \underline{\bar{u}}_{k-}^*)$, and $\varphi^{(3)\varsigma}(k, x^*, y^*; \underline{\bar{u}}_{k-}^*)$, for $\varsigma \in S^3$, will be maintained in all period $k \in \{1, 2, \ldots, T\}$, a payment equaling

$$\begin{aligned}\hat{\beta}_k^{(\ell)}(x_k^*, y_k^*; \underline{\bar{u}}_{k-}^*) &= (\delta^k)^{-1}[\varphi^{(\ell)}(k, x_k^*, y_k^*; \underline{\bar{u}}_{k-}^*)\\&\quad - \varphi^{(\ell)}(k+1, x_{k+1}^*, y_{k+1}^*; \underline{\bar{u}}_{(k+1)-}^*)]\end{aligned}$$

is given to Coalition $\ell \in \{1, 2\}$ and

$$\hat{\beta}_k^{(3)\varsigma}(x_k^*, y_k^*; \underline{\bar{u}}_{k-}^*) = (\delta^k)^{-1}[\varphi^{(3)\varsigma}(k, x_k^*, y_k^*; \underline{\bar{u}}_{k-}^*) \\ - \varphi^{(3)\varsigma}(k+1, x_{k+1}^*, y_{k+1}^*; \underline{\bar{u}}_{(k+1)-}^*)]$$

is given to unallied nation $\varsigma \in S^3$.

Finally, nations within a coalition will share their gains according to the intra-coalition payoff distribution in Section 3.3.

Remark 6.2.

(i) The level of activities enhancing Coalition 1's defense level in the grand coalition

$$\bar{u}_k^{(1)i*} = \frac{A_{k+1}\gamma_k^{(1)i}\delta + A_{k+2}\sigma_k^{(1)i}\delta^2}{2w_k^{(1)i}}$$

is lower than the level of activities enhancing Coalition 1's defense level in the coalition equilibrium

$$\bar{u}_k^{(1)i**} = \frac{A_{k+1}^{(1)}\gamma_k^{(1)i}\delta + A_{k+2}^{(1)}\sigma_k^{(1)i}\delta^2}{2w_k^{(1)i}}$$

because grand coalition $A_k < A_k^{(1)}$.

Hence, Coalition 1's defense level is lower under the grand coalition.

Similarly, the level of activities enhancing Coalition 2's defense level in the grand coalition is lower than the level of activities enhancing Coalition 2's defense level in the coalition equilibrium. Coalition 2's defense level is lower under the grand coalition.

7. Chapter Notes

In many real-life scenarios, combinations of groups or nations with common interests form coalitions by treaty or agreement for mutual support and joint actions. Countries establish trading coalitions because they believe free trade benefits them and frequently such coalitions tend to be regional because it is more convenient to reach an agreement with nearby neighbors than with remote partners. Trade coalitions, currency blocs, and political and economic unions had been formed for more than a century. Studies of various kinds of blocs are not rare. Schott (1991) examined trading blocs and the world trading system. Kandogan (2008) obtained consistent estimates of the trade effects of regional blocs. Eichengreen and Irwin (1995), Frankel and Rose (1998), McDonald *et al.* (2011), and Wolf and Ritschl (2011) provided studies on currency areas and trade blocs. Yeung and Lui (2019) considered the political economy of coordinated coalition cooperation using China's Belt-Road initiative. This paper presents a dynamic game with the decomposition of player sets on coalitions. The common interests of players within a coalition are reflected through their common preference which is not shared by players outside the coalition, resources exclusively owned by players in the coalition, but state dynamics are affected by controls of all players (also outside the coalition). The paper by Petrosyan and Gromova (2014) first presents the basic structure of a cooperative differential game with coalitions as players. In this paper, the coalitional Nash bargaining solution which reflects the outcome of competition among coalitions is presented and this approach is different from the one represented in Petrosyan and Gromova (2014) where coalitions fully cooperate as in cooperative TU differential games. Intra-coalition payoff distribution is studied. To guarantee subgame consistency (dynamic stability) in a dynamic coalition cooperation scheme, the payoffs to players on the level of coalitions must satisfy individual rationality, and imputations inside the coalitions have to satisfy the property of subgame consistency. Analysis of dynamically consistent bi-level cooperation coalitional blocs can be found in Petrosyan and Yeung (2019). Petrosyan *et al.* (2024) and Yeung *et al.* (2021, 2023) studied discrete-time dynamic network games.

References

Eichengreen, B. & Irwin, D. (1995). Trade blocs, currency blocs, and the reorientation of trade in the 1930s. *Journal of International Economics*, 38, 1–24.

Frankel, J. A. & Rose, A. (1998). The endogeneity of the optimum currency area criteria. *The Economic Journal*, 108, 1009–1025.

Kandogan, Y. (2008). Consistent estimates of regional blocs' trade effects. *Review of International Economics*, 16(2), 301–314.

McDonald, F., Tuselmann, J. H., Voronkova, S., & Golesorkhi, S. (2011). The strategic development of subsidiaries in regional trade blocs. *Multinational Business Review*, 19(3), 256–271.

Petrosyan, L. A. & Gromova, E. V. (2014). Two-level cooperation in coalitional differential games. *Annals of Ekaterinburg Mathematical Institute*, 20(3), 193–203.

Petrosyan, L. A. & Yeung, D. W. K. (2019). Dynamically consistent bi-level cooperation of a dynamic game with coalitional blocs, in Petrosyan, L. A., V. V. Mazalov and N. A. Zenkevich (eds.), *Frontiers of Dynamic Games*, Birkhauser, Switzerland, 2019, pp. 209–230.

Petrosyan, L. A., Yeung, D. W. K., & Pankratova, Y. (2024). Characteristic functions in cooperative differential games on networks. *Journal of Dynamics and Games*. DOI: 10.3934/jdg.2023017.

Schott, J. J. (1991). Trading blocs and the world trading system. *World Economy*, 14(1), 1–17. https://doi.org/10.1111/j.1467-9701.1991.tb00748.x.

Wolf, N. & Ritschl, A. O. (2011). Endogeneity of currency areas and trade blocs: Evidence from a natural experiment. *KYKLOS*, 64(2), 291–312.

Yeung, D. W. K. & Lui, A. W. C. (2019). China's Belt-Road Initiative: The political economy of coordinated coalition cooperation. In Tony Yu and Dianna Kwan (eds.), *Contemporary Issues in International Political Economy*. Singapore: Palgrave Macmillan, pp. 197–226.

Yeung, D. W. K. & Petrosyan, L. A. (2022). *Durable-Strategies Dynamic Games: Theory, Solution Techniques and Applications*. Springer, Switzerland, 303 p.

Yeung, D. W. K., Petrosyan, L. A., & Zhang, Y. X. (2021). Trade with Technology spillover: A dynamic network game analysis. *International Game Theory Review*, 23(1), 205011 (1–31).

Yeung, D. W. K., Petrosyan, L. A., & Zhang, Y. X. (2023). A dynamic network game of the fintech industry. *Journal of the Operations Research Society of China*, https://doi.org/10.1007/s40305-022-00434-4.

Chapter 6

Random Termination: Mathematical Theory and Explication in International Joint Venture

> *"Everything comes to an end. A good bottle of wine, a summer's day, a long-running sitcom, one's life, and eventually our species. The question for many of us is not that everything will come to an end but when. And can we do anything vaguely useful until it does?"* — Jasper Forde

Durable strategies and random termination are frequently observed in real-life practical decisions involving strategic interactions. Durable strategies are controls that have influences lasting over a certain period of time. Revenue-generating investments, toxic waste disposal, durable goods, business contracts, taxes, regulations, coalition agreements, diffusion of knowledge, advertising, and investments to build up physical capital are vivid examples of durable strategies. The existence of at least some durable strategies in the decision-maker's strategy set is a rule rather than an exception in practical analysis. In addition, the terminal time of many international issues is not known with certainty. Commonly observed examples of horizontal uncertainties include the terms of offices of elected authorities, contract renewal, continuation of agreements subjected to periodic negotiations, trade wars, international sanctions, and the end of military conflicts. An empirically meaningful theory has to incorporate real-life phenomena. This chapter provides the underlying theory and techniques for dealing with problems in international economics in which there exist random termination and durable strategies. This chapter is organized as follows. In Section 1, we discuss the issue of termination of international institutions. Game formulation and mathematical solution theory are provided in Section 2. Analysis of dynamic cooperation under random termination is given in Section 3. An explication of international joint venture is presented in Section 4. A computational illustration is given in Sections 5 and 6. Chapter notes are provided in Section 7.

1. Termination of International Institutions

The termination of major international organizations has been quite common. For instance, African and Malagasy Coffee Organization (1960–2007), Central American Research Institute for Industry (1956–1998), Central Bureau of the International Map of the World (1909–1953), Commonwealth Air Transport Council (1945–1991), Council for Mutual Economic Aid (1949–1991), European Atomic Energy Community (1958–1992), Intergovernmental Bureau for Informatics (1974–1988), International Commission for the Decennial Revision of the International Lists of Diseases and Causes of Death (1900–1948), International Commission for the Protection of the Moselle against Pollution (1963–2004), International Office of Public Hygiene (1907–1946), International Technical Committee of Legal Experts on Air Questions (1926–1947), International Telecommunications Satellite Organization (1964–2001), Organization for European Economic Cooperation (1948–1961), Organization for the Management and Development of the Kagera River (1977–2004), Warsaw Treaty Organization (1955–1991), Western European Union (1955–2001), Arab Maghreb Union (1989–2007), International Allied Rhineland High Commission (1919–1934), International Institute of Agriculture (1905–1946), International Refugee Organization (1946–1952), and League of Nations (1919–1946).

Currently, international organizations are still under great pressure. The World Health Organization was heavily criticized over its handling of COVID. The World Trade Organization seemed incapable of mitigating an American-Chinese trade war. The United States quit UNESCO in 2018 and withdrew temporarily from Paris Agreement in 2017. The United Kingdom exited the EU in 2020. The ultimate way for states to show that international organizations have outlived their purpose is to disband them. This happens frequently: more than 30% of the international organizations created since 1815 have formally ceased to exist.

International governmental organizations are important subjects of contemporary international relations. As states face various global problems, although as the most important subjects of international law, they do not have enough power to face them on their own, therefore multilateral cooperation is more than necessary. This is the main reason why membership in international governmental organizations is a priority for many countries. However, there are cases where international organizations have faced the termination of the membership of states, which occurs in four major ways of terminating the membership of states from international organizations (i) the dissolution of the organization, (ii) the expulsion of the state from an international organization, (iii) the withdrawal of state from organization, and (iv) the disappearance of state (like in the case of the former Yugoslavia).

The termination of international joint ventures is also very frequent. Joint ventures between domestic companies in developing nations and foreign companies have become a popular means for both managements to satisfy their objectives. They offer, at least in principle, an opportunity for each partner to benefit significantly from the comparative advantages of the other. As attractive as joint ventures might seem, however, they frequently perform unsatisfactorily and are comparatively unstable. This seems to be true even when the partners are two companies from the same industrial nation; international joint ventures seem to be more vulnerable still. Many international joint ventures were rated by participants as having performed poorly. Joint venture relationships are often fragile and both difficult to negotiate and, once negotiated, to hold together.

In addition, international sanction termination is another issue that is hotly debated in the international political economy. With the proliferation of sanctions after the end of the Cold War, the termination of sanctions has become a ubiquitous phenomenon in international politics. Today, there are 15 ongoing sanctions regimes that focus on supporting political settlement of conflicts, nuclear nonproliferation, and counter-terrorism. Each regime is administered by a sanctions committee chaired by a non-permanent member of the Security Council. There are 11 monitoring groups, teams, and panels that support the work of 12 of the 15 sanctions committees.

There were terminations of UN, EU, US, as well as regional sanctions between 1990 and 2018 worldwide.

2. Game Formulation and Mathematical Solution Theory

In this section, we provide the mathematical foundation for analyzing an economic game system in which the game will terminate in the future. The probabilities of termination at different future periods are known.

2.1. *Characterization of the game*

Consider a $n-$nation dynamic game system with a horizon of $\hat{T}$ stages where $\hat{T}$ is a random variable with range $\{1, 2, \ldots, T\}$ and corresponding probabilities $\{\theta_1, \theta_2, \ldots, \theta_T\}$. Conditional upon the reaching of stage τ, the probability of the game would last up to stages $\tau, \tau+1, \ldots, T$ becomes, respectively,

$$\frac{\theta_\tau}{\sum_{\zeta=\tau}^{T}\theta_\zeta}, \frac{\theta_{\tau+1}}{\sum_{\zeta=\tau}^{T}\theta_\zeta}, \ldots, \frac{\theta_T}{\sum_{\zeta=\tau}^{T}\theta_\zeta}.$$

The constituent elements are as follows:

(*i*) *The set of non-durable target controls*

The set of non-durable target controls of all the n national governments is denoted by $\underline{u}_k$. It includes the non-durable target controls sets of nation 1, nation 2, to nation n. For notational convenience, we express $\underline{u}_k = (u_k^1, u_k^2, \ldots, u_k^{n_1})$ for $k \in \{1, 2, \ldots, T\}$, where u_k^i is the vector of non-durable targets controls of nation i in period k.

(*ii*) *The set of durable target controls*

The set of durable target controls of all the n national governments is denoted by $\underline{\bar{u}}_k$. It includes the durable target controls sets of nation 1, nation 2, to nation n. For notational convenience, we express $\underline{\bar{u}}_k = (\bar{u}_k^1, \bar{u}_k^2, \ldots, \bar{u}_k^{n_1})$ for $k \in \{1, 2, \ldots, T\}$, where $\bar{u}_k^i$ is the vector of durable targets controls of nation i in period k.

(*iii*) *The set of effective durable target controls previously executed*

The set of durable target controls of all nations executed before period k but still have effects is denoted it by $\underline{\bar{u}}_{k-}$. We express

$\underline{\bar{u}}_{k-} = (\bar{u}^1_{k-}, \bar{u}^2_{k-}, \ldots, \bar{u}^{n_1}_{k-})$, for $k \in \{1, 2, \ldots, T\}$, where $\bar{u}^i_{k-}$ is the vector of durable targets controls of nation i executed before period k but still has effects.

(iv) State variables and evolution dynamics

The vector of state variables of the economy at period k is denoted by $x_k \subset X$. The evolution dynamics of the vector of state variables are affected by $\underline{u}_k$, $\underline{\bar{u}}_k$, $\underline{\bar{u}}_{k-}$, and x_k. We denote the evolution dynamics by the vector difference equation:

$$x_{k+1} = f(x_k, \underline{u}_k, \underline{\bar{u}}_k; \underline{\bar{u}}_{k-}), \quad x_1 = x^0_1.$$

(v) Single-period gain of nation i in period k

The gain of nation i in period k depends on $(\underline{u}_k, \underline{\bar{u}}_k, \underline{\bar{u}}_{k-}, x_k)$ and is expressed as $\sum_{\varsigma=1}^{\Theta_i} w_k^{i(\varsigma)} g_k^{i(\varsigma)}(x_k, \underline{u}_k, \underline{\bar{u}}_k; \underline{\bar{u}}_{k-})$ if nation i has Θ_i objectives, where $g_k^{i(\varsigma)}(x_k, \underline{u}_k, \underline{\bar{u}}_k; \underline{\bar{u}}_{k-})$ is the ςth objective and $w_k^{i(\varsigma)}$ is the corresponding monetary weight. If $\Theta_i = 1$, the gain of nation i in period k reduces to $w_k^{i(1)} g_k^{i(1)}(x_k, \underline{u}_k, \underline{\bar{u}}_k; \underline{\bar{u}}_{k-})$.

(vi) Termination payoff

If the game ends after stage $\hat{T}$, player i will receive a terminal payment $q^i_{\hat{T}+1}(x_{\hat{T}+1}; \underline{\bar{u}}_{(\hat{T}+1)-})$ in stage $\hat{T}+1$, which can be zero, positive, or negative.

(vii) The payoffs of the nations

The expected payoff nation $i \in N$ can be expressed as follows:

$$E\left[\sum_{k=1}^{\hat{T}} \sum_{\varsigma=1}^{\Theta_i} w_k^{i(\varsigma)} g_k^{i(\varsigma)}(x_k, \underline{u}_k, \underline{\bar{u}}_k; \underline{\bar{u}}_{k-}) \delta_1^k \right.$$
$$\left. + q^i_{\hat{T}+1}(x_{T+1}; \underline{\bar{u}}_{(T+1)-}) \delta_1^{T+1}\right]$$
$$= \sum_{\hat{T}=1}^{T} \theta_{\hat{T}} \left[\sum_{k=1}^{\hat{T}} \sum_{\varsigma=1}^{\Theta_i} w_k^{i(\varsigma)} g_k^{i(\varsigma)}(x_k, \underline{u}_k, \underline{\bar{u}}_k; \underline{\bar{u}}_{k-}) \delta_1^k \right.$$
$$\left. + q^i_{\hat{T}+1}(x_{T+1}; \underline{\bar{u}}_{(T+1)-}) \delta_1^{T+1}\right].$$

Note that the influences of durable and non-durable strategies, previously executed durable strategies, and the current state of the nation's objectives and terminal payoffs are known realities. The influence on the evolution of the state by durable and non-durable strategies and previously executed durable strategies, and the current state are commonly observed. The relational contents exhibited in the system indeed reflect real-life phenomena in the global political economy.

2.2. *Mathematical solution theory*

The above scenario of a global political and economic arena involving n nations can be modeled as a random horizon durable-strategies dynamic game (see Yeung and Petrosyan, 2021) as follows:

$$\max_{u_k^i, \bar{u}_k^i k\in\{1,2,\ldots,T\}} \left\{ \sum_{\hat{T}=1}^{T} \theta_{\hat{T}} \left[\sum_{k=1}^{\hat{T}} \sum_{\varsigma=1}^{\Theta_i} w_k^{i(\varsigma)} g_k^{i(\varsigma)}(x_k, \underline{u}_k, \bar{u}_k; \underline{\bar{u}}_{k-}) \delta_1^k \right.\right.$$
$$\left.\left. + q_{\hat{T}+1}^i(x_{T+1}; \underline{\bar{u}}_{(T+1)-}) \delta_1^{T+1} \right] \right\}, \tag{2.1}$$

subject to

$$x_{k+1} = f_k(x_k, \underline{u}_k, \bar{u}_k; \underline{\bar{u}}_{k-}), \quad x_1 = x_1^0, \quad \text{for } i \in \{1, 2, \ldots, n\} \equiv N. \tag{2.2}$$

In addition, we assume that the functions $g_k^{i(\varsigma)}(x_k, \underline{u}_k, \bar{u}_k; \underline{\bar{u}}_{k-})$ and $f(x_k, \underline{u}_k, \bar{u}_k; \underline{\bar{u}}_{k-})$ are differentiable functions in $(x_k, \underline{u}_k, \bar{u}_k; \underline{\bar{u}}_{k-})$ and $\hat{q}_{T+1}^i(x_{T+1}; \underline{\bar{u}}_{(T+1)-})$ is differentiable in $(x_{T+1}; \underline{\bar{u}}_{(T+1)-})$.

Now consider the case when stage τ has arrived with the state being x_τ and the previously executed durable strategies $\underline{\bar{u}}_{\tau-}$. Then it becomes a game in which the payoff of player $i \in N$ is

$$E\left\{ \sum_{k=\tau}^{\hat{T}} \sum_{\varsigma=1}^{\Theta_i} w_k^{i(\varsigma)} g_k^{i(\varsigma)}(x_k, \underline{u}_k, \bar{u}_k; \underline{\bar{u}}_{k-}) \delta_1^k \right.$$
$$\left. + q_{\hat{T}+1}^i(x_{\hat{T}+1}; \underline{\bar{u}}_{(\hat{T}+1)-}) \delta_1^{\hat{T}+1} \right\}$$

$$= \sum_{\hat{T}=\tau}^{T} \frac{\theta_{\hat{T}}}{\sum_{\zeta=\tau}^{T} \theta_\zeta} \left[\sum_{k=\tau}^{\hat{T}} \sum_{\varsigma=1}^{\Theta_i} w_k^{i(\varsigma)} g_k^{i(\varsigma)}(x_k, \underline{u}_k, \bar{\underline{u}}_k; \bar{\underline{u}}_{k-}) \delta_1^k \right.$$

$$\left. + q_{\hat{T}+1}^i(x_{\hat{T}+1}; \bar{\underline{u}}_{(\hat{T}+1)-}) \delta_1^{\hat{T}+1} \right] \tag{2.3}$$

and the state dynamics are

$$x_{k+1} = f_k(x_k, \underline{u}_k, \bar{\underline{u}}_k : \bar{\underline{u}}_{k-}), \quad \text{for } k = \{\tau, \tau+1, \ldots, T\},\ x_\tau = x. \tag{2.4}$$

The solution of the durable-strategies dynamic game (2.1)–(2.2) can be characterized by a theorem from Yeung and Petrosyan (2021). In particular, a feedback Nash equilibrium of the game can be characterized by the following theorem.

Theorem 2.1. *Let* $\{\underline{u}_\tau^{**}, \bar{\underline{u}}_\tau^{**}, i \in N\}$ *denote the set of the players' feedback Nash equilibrium strategies and* $V^i(\tau, x; \bar{\underline{u}}_{\tau-}^{**})$ *denote the feedback Nash equilibrium payoff of player* i *in the non-cooperative game (2.1)–(2.2), then the function* $V^i(\tau, x; \bar{\underline{u}}_{\tau-}^{**})$ *satisfies the following system of recursive equations:*

$$V^i(T+1, x; \bar{\underline{u}}_{(T+1)-}^{**}) = q_{T+1}^i(x; \bar{\underline{u}}_{(T+1)-}^{**}) \delta_1^{T+1}, \tag{2.5}$$

$$V^i(T, x; \bar{\underline{u}}_{T-}^{**})$$

$$= \max_{u_T^i, \bar{u}_T^i} \left\{ \sum_{\varsigma=1}^{\Theta_i} w_T^{i(\varsigma)} g_T^{i(\varsigma)}(x, u_T^i, \bar{u}_T^i, \underline{u}_T^{(\neq i)**}, \bar{\underline{u}}_T^{(\neq i)**}; \bar{\underline{u}}_{T-}^{**}) \delta_1^T \right.$$

$$+ q_{T+1}^i [f_T(x, u_T^i, \bar{u}_T^i, \underline{u}_T^{(\neq i)**}, \bar{\underline{u}}_T^{(\neq i)**});$$

$$\left. u_T^i, \bar{u}_T^i, \underline{u}_T^{(\neq i)**}, \bar{\underline{u}}_T^{(\neq i)**}, \bar{\underline{u}}_{(T+1)-}^{**} \cap \bar{\underline{u}}_{T-}^{**}] \delta_1^{T+1} \right\}, \tag{2.6}$$

$$V^i(\tau, x; \underline{u}_{\tau-}^{**})$$

$$= \max_{u_\tau^i, \bar{u}_\tau^i} \left\{ \sum_{\varsigma=1}^{\Theta_i} w_\tau^{i(\varsigma)} g_\tau^{i(\varsigma)}(x, u_\tau^i, \bar{u}_\tau^i, \underline{u}_\tau^{(\neq i)**}, \bar{\underline{u}}_\tau^{(\neq i)**}; \bar{\underline{u}}_{\tau-}^{**}) \delta_1^\tau \right.$$

$$+\frac{\theta_\tau}{\sum_{\zeta=\tau}^{T}\theta_\zeta}\hat{q}_{\tau+1}[f_\tau(x,u_\tau^i,\bar{u}_\tau^i,\underline{u}_\tau^{(\neq i)**},\underline{\bar{u}}_\tau^{(\neq i)**};\underline{\bar{u}}_{\tau-}^{**});$$

$$u_\tau^i,\bar{u}_\tau^i,\underline{u}_\tau^{(\neq i)**},\underline{\bar{u}}_\tau^{(\neq i)**},\underline{\bar{u}}_{(\tau+1)-}^{**}\cap\underline{\bar{u}}_{\tau-}^{**}]\delta_1^{\tau+1}$$

$$+\frac{\sum_{\zeta=\tau+1}^{T}\theta_\zeta}{\sum_{\zeta=\tau}^{T}\theta_\zeta}V^i[\tau+1,f_\tau(x,u_\tau^i,\bar{u}_\tau^i,\underline{u}_\tau^{(\neq i)**},\underline{\bar{u}}_\tau^{(\neq i)**};\underline{\bar{u}}_{\tau-}^{**});$$

$$\left.u_\tau^i,\bar{u}_\tau^i\underline{u}_\tau^{(\neq i)**},\underline{\bar{u}}_\tau^{(\neq i)**},\underline{\bar{u}}_{(\tau+1)-}^{**}\cap\underline{\bar{u}}_{\tau-}^{**}]\delta_1^{\tau+1}\right\},$$

$$\textit{for } \tau\in\{1,2,\ldots,T-1\}, \tag{2.7}$$

where $\underline{u}_\tau^{(\neq i)**}=\underline{u}_\tau^{**}\backslash u_\tau^{i**}$ *and* $\underline{\bar{u}}_\tau^{(\neq i)**}=\underline{\bar{u}}_\tau^{**}\backslash\bar{u}_\tau^{i**}$.

Proof. See the proof of Theorem V of Mathematical Solution Mechanisms in Chapter 7. □

Substituting the set of feedback Nash equilibrium strategies $\underline{u}_k^{**}$ for $k\in\{1,2,\ldots,T\}$ from Theorem 2.1 into state dynamics (2.2) yields the game equilibrium dynamics

$$x_{k+1}=f(x_k,\underline{u}_k^{**},\underline{\bar{u}}_k^{**};\underline{\bar{u}}_{k-}^{**})\quad x_1=x_1^0. \tag{2.8}$$

Substituting the set of feedback Nash equilibrium strategies $(\underline{u}_k^{**},\underline{\bar{u}}_k^{**})$ for $k\in\{1,2,\ldots,T\}$ into the player i's payoff yields

$$V^i(\tau,x;\underline{\bar{u}}_{\tau-}^{**})=E\left\{\sum_{k=\tau}^{\hat{T}}\sum_{\varsigma=1}^{\Theta_i}w_k^{i(\varsigma)}g_k^{i(\varsigma)}(x_k,\underline{u}_k^{**},\underline{\bar{u}}_k^{**};\underline{\bar{u}}_{k-}^{**})\delta_1^k\right.$$
$$\left.+q_{\hat{T}+1}^i(x_{\hat{T}+1};\underline{\bar{u}}_{(\hat{T}+1)-}^{**})\delta_1^{\hat{T}+1}\right\}$$
$$=\sum_{\hat{T}=\tau}^{T}\frac{\theta_{\hat{T}}}{\sum_{\zeta=\tau}^{T}\theta_\zeta}\left[\sum_{k=\tau}^{\hat{T}}\sum_{\varsigma=1}^{\Theta_i}w_k^{i(\varsigma)}g_k^{i(\varsigma)}(x_k,\underline{u}_k^{**},\underline{\bar{u}}_k^{**};\underline{\bar{u}}_{k-}^{**})\delta_1^k\right.$$
$$\left.+q_{\hat{T}+1}^i(x_{\hat{T}+1};\underline{\bar{u}}_{(\hat{T}+1)-}^{**})\delta_1^{\hat{T}+1}\right],\quad\text{for } i\in N. \tag{2.9}$$

The value function $V^i(\tau,x;\underline{\bar{u}}_{\tau-}^{**})$ gives the expected game equilibrium payoff to player i from stage τ to the end of the game.

3. Dynamic Cooperation under Random Termination

Now we consider the case where the nations agree to cooperate and distribute the cooperative payoff among themselves according to an optimality principle. Two crucial properties that a cooperative scheme has to satisfy are group optimality and individual rationality. In addition to these crucial properties, the scheme has to satisfy the property of subgame consistency to guarantee dynamical stability. In this section, we first examine the properties of group optimality and individual rationality and then consider subgame-consistent cooperative solutions under random termination and durable strategies.

3.1. *Group optimality and individual rationality*

Group optimality requires the maximization of the players' expected joint payoff so that all gains are captured. To obtain group optimality, the involved nations have to solve the discrete-time dynamic optimization problem of maximizing their expected joint payoff

$$
\begin{aligned}
&E\left\{\sum_{j=1}^{n}\left[\sum_{k=1}^{\hat{T}}\sum_{\varsigma=1}^{\Theta_j} w_k^{j(\varsigma)} g_k^{j(\varsigma)}(x_k,\underline{u}_k,\bar{\underline{u}}_k;\bar{\underline{u}}_{k-})\delta_1^k\right.\right.\\
&\qquad\qquad\left.\left.+\,q_{\hat{T}+1}^{j}(x_{\hat{T}+1};\bar{\underline{u}}_{(\hat{T}+1)-})\delta_1^{\hat{T}+1}\right]\right\}\\
&=\sum_{\hat{T}=1}^{T}\theta_{\hat{T}}\left\{\sum_{j=1}^{n}\left[\sum_{k=1}^{\hat{T}}\sum_{\varsigma=1}^{\Theta_j} w_k^{j(\varsigma)} g_k^{j(\varsigma)}(x_k,\underline{u}_k,\bar{\underline{u}}_k;\bar{\underline{u}}_{k-})\delta_1^k\right.\right.\\
&\qquad\qquad\left.\left.+\,q_{\hat{T}+1}^{j}(x_{\hat{T}+1};\bar{\underline{u}}_{(\hat{T}+1)-})\delta_1^{\hat{T}+1}\right]\right\}
\end{aligned}
\tag{3.1}
$$

subject to dynamics

$$
x_{k+1}=f(x_k,\underline{u}_k,\bar{\underline{u}}_k;\bar{\underline{u}}_{k-}),\quad x_1=x_1^0. \tag{3.2}
$$

An optimal solution to the problem (3.1)–(3.2) can be characterized as follows.

Theorem 3.1. *Let the value function $W(\tau, x; \underline{\bar{u}}_{\tau-})$ denote the maximal value of the expected joint payoffs*

$$E\left\{\sum_{j=1}^{n}\left[\sum_{k=\tau}^{\hat{T}}\sum_{\varsigma=1}^{\Theta_j} w_k^{j(\varsigma)} g_k^{j(\varsigma)}(x_k, \underline{u}_k, \bar{u}_k; \underline{\bar{u}}_{k-})\delta_1^k \right.\right.$$
$$\left.\left. + q_{\hat{T}+1}^j(x_{\hat{T}+1}; \underline{\bar{u}}_{(\hat{T}+1)-})\delta_1^{\hat{T}+1}\right]\right\}$$
$$= \sum_{\hat{T}=\tau}^{T} \frac{\theta_{\hat{T}}}{\sum_{\zeta=\tau}^{T}\theta_\zeta}\left\{\sum_{j=1}^{n}\left[\sum_{k=\tau}^{\hat{T}}\sum_{\varsigma=1}^{\Theta_j} w_k^{j(\varsigma)} g_k^{j(\varsigma)}(x_k, \underline{u}_k, \bar{u}_k; \underline{\bar{u}}_{k-})\delta_1^k \right.\right.$$
$$\left.\left. + q_{\hat{T}+1}^j(x_{\hat{T}+1}; \underline{\bar{u}}_{(\hat{T}+1)-})\delta_1^{\hat{T}+1}\right]\right\},$$

for problem (3.1)–(3.2) starting at stage τ with state $x_\tau = x$ and previously executed controls $\underline{\bar{u}}_{\tau-}$. Then the function $W(\tau, x; \underline{\bar{u}}_{\tau-})$ satisfies the following system of recursive equations:

$$W(T+1, x; \underline{\bar{u}}_{(T+1)-}) = \sum_{j=1}^{j} q_{T+1}^j(x; \underline{\bar{u}}_{(T+1)-})\delta_1^{T+1}, \tag{3.3}$$

$$W(T, x; \underline{\bar{u}}_{T-})$$
$$= \max_{\underline{u}_T, \bar{u}_T}\left\{\sum_{j=1}^{n}\left[\sum_{\varsigma=1}^{\Theta_j} w_T^{j(\varsigma)} g_T^{j(\varsigma)}(x, \underline{u}_T, \bar{u}_T; \underline{\bar{u}}_{T-})\delta_1^T \right.\right.$$
$$\left.\left. + q_{T+1}^j[f_T(x, \underline{u}_T, \bar{u}_T; \underline{\bar{u}}_{T-}); \underline{u}_T, \bar{u}_T, \underline{\bar{u}}_{(T+1)-} \cap \underline{\bar{u}}_{T-}]\delta_1^{T+1}\right]\right\}, \tag{3.4}$$

$$W(\tau, x; \underline{\bar{u}}_{\tau-})$$
$$= \max_{\underline{u}_\tau, \underline{\bar{u}}_\tau} \left\{ \sum_{j=1}^{n} \left[\sum_{\varsigma=1}^{\Theta_j} w_\tau^{j(\varsigma)} g_\tau^{j(\varsigma)}(x, \underline{u}_\tau, \bar{u}_\tau; \underline{\bar{u}}_{\tau-}) \delta_1^\tau \right.\right.$$
$$\left. + \frac{\theta_\tau}{\sum_{\zeta=\tau}^{T} \theta_\zeta} \hat{q}_{\tau+1}^j (f_\tau(x, \underline{u}_\tau, \bar{u}_\tau; \underline{\bar{u}}_{\tau-}); \underline{u}_\tau, \bar{u}_\tau, \underline{\bar{u}}_{(\tau+1)-} \cap \underline{\bar{u}}_{\tau-}) \delta_1^{\tau+1} \right]$$
$$+ \frac{\sum_{\zeta=\tau+1}^{T} \theta_\zeta}{\sum_{\zeta=\tau}^{T} \theta_\zeta} W(\tau+1, f_\tau(x, \underline{u}_\tau, \bar{u}_\tau; \underline{\bar{u}}_{\tau-});$$
$$\left. \underline{u}_\tau, \bar{u}_\tau, \underline{\bar{u}}_{(\tau+1)-} \cap \underline{\bar{u}}_{\tau-}) \right\}, \quad \text{for } \tau \in \{1, 2, \ldots, T-1\}. \tag{3.5}$$

Proof. See the proof of Theorem IV of Mathematical Solution Mechanisms in Chapter 7. □

We use $(\underline{u}_\tau^*, \bar{u}_\tau^*)$, for $\tau \in \{1, 2, \ldots, T\}$, to denote the set of group optimal strategies derived in Theorem 5.1. Substituting the optimal strategies $(\underline{u}_\tau^*, \bar{u}_\tau^*)$ into the state dynamics (3.1), one can obtain the dynamics of the cooperative trajectory as follows:

$$x_{k+1} = f_k(x_k, \underline{u}_k^*, \bar{u}_k^*; \underline{\bar{u}}_{k-}^*), \quad x_1 = x_1^0. \tag{3.6}$$

We use $\{x_k^*\}_{k=1}^{T+1}$ to denote the solution generated by dynamics (3.6). Using the set of optimal strategies $(\underline{u}_\tau^*, \bar{u}_\tau^*)$, one can obtain the expected joint cooperative payoff as

$$W(\tau, x_\tau^*; \underline{\bar{u}}_{\tau-}^*)$$
$$= E\left\{ \sum_{j=1}^{n} \left[\sum_{k=\tau}^{\hat{T}} \sum_{\varsigma=1}^{\Theta_j} w_k^{j(\varsigma)} g_k^{j(\varsigma)}(x_k^*, \underline{u}_k^*, \bar{u}_k^*; \underline{\bar{u}}_{k-}^*) \delta_1^k \right.\right.$$
$$\left.\left. + q_{\hat{T}+1}^j (x_{\hat{T}+1}^*; \underline{\bar{u}}_{(\hat{T}+1)-}^*) \delta_1^{\hat{T}+1} \right] \right\}$$

$$= \sum_{\hat{T}=\tau}^{T} \frac{\theta_{\hat{T}}}{\sum_{\zeta=\tau}^{T} \theta_\zeta} \left\{ \sum_{j=1}^{n} \left[\sum_{k=\tau}^{\hat{T}} \sum_{\varsigma=1}^{\Theta_j} w_k^{j(\varsigma)} g_k^{j(\varsigma)}(x_k^*, \underline{u}_k^*, \underline{\bar{u}}_k^*; \underline{\bar{u}}_{k-}^*)\delta_1^k \right.\right.$$

$$\left.\left. + q_{\hat{T}+1}^j(x_{\hat{T}+1}^*; \underline{\bar{u}}_{(\hat{T}+1)-}^*)\delta_1^{\hat{T}+1} \right] \right\}, \quad (3.7)$$

for $\tau \in \{1, 2, \ldots, T\}$.

We then move to the property of individual rationality. The nations have to agree to an optimality principle in distributing the total cooperative payoff among themselves. For individual rationality to be upheld, the imputations under the agreed-upon optimality principle received by the players have to be no less than their expected non-cooperative payoff along the cooperative state trajectory. Let

$$\xi(\tau, x_\tau; \underline{\bar{u}}_{\tau-}) = [\xi^1(\tau, x_\tau; \underline{\bar{u}}_{\tau-}), \xi^2(\tau, x_\tau; \underline{\bar{u}}_{\tau-}), \ldots, \xi^n(\tau, x_\tau; \underline{\bar{u}}_{\tau-})]$$

denote the agreed-upon imputation vector guiding the distribution of the total cooperative payoff in stage $\tau \in \{1, 2, \ldots, T\}$. For individual rationality to be maintained throughout all the stages $\tau \in \{1, 2, \ldots, T\}$ along the optimal cooperative trajectory $\{x_k^*\}_{k=1}^{T+1}$, it is required that

$$\xi^i(\tau, x_\tau^*; \underline{\bar{u}}_{\tau-}^*) \geq V^i(\tau, x_\tau^*; \underline{\bar{u}}_{\tau-}^*), \quad \text{for } i \in N \text{ and } \tau \in \{1, 2, \ldots, T\},$$

along the cooperative trajectory $\{x_k^*\}_{k=1}^{T}$ for given previously executed strategies $\underline{\bar{u}}_{\tau-}^*$.

Finally, to satisfy group optimality, the imputation vector has to satisfy

$$W(\tau, x_\tau^*; \underline{\bar{u}}_{\tau-}^*) = \sum_{j=1}^{n} \xi^j(\tau, x_\tau^*; \underline{\bar{u}}_{\tau-}^*), \quad \text{for } \tau \in \{1, 2, \ldots, T\}.$$

3.2. *Subgame-consistent solutions and payment mechanism*

To guarantee dynamical stability in a dynamic cooperation scheme, the solution has to satisfy the property of subgame consistency. A cooperative solution is subgame-consistent if an extension of the solution policy to a subgame starting at a later time with a state along the optimal cooperative trajectory would remain optimal. In particular, subgame consistency ensures that as the game proceeds players are guided by the same optimality principle at each stage of the game and hence do not possess incentives to deviate from the previously adopted optimal behavior. Therefore, for subgame consistency to be satisfied, the agreed-upon imputation

$$\xi(\tau, x_\tau^*; \underline{\bar{u}}_{\tau-}^*) = [\xi^1(\tau, x_\tau^*; \underline{\bar{u}}_{\tau-}^*), \xi^2(\tau, x_\tau^*; \underline{\bar{u}}_{\tau-}^*), \ldots, \xi^n(\tau, x_\tau^*; \underline{\bar{u}}_{\tau-}^*)],$$
$$\tau \in \{1, 2, \ldots, T\}, \tag{3.8}$$

has to be maintained along the cooperative trajectory $\{x_\tau^*\}_{\tau=1}^T$ for given previously executed strategies $\underline{\bar{u}}_{\tau-}^*$.

Crucial to the analysis is the formulation of a payment mechanism so that the imputation in (3.8) can be realized as the game proceeds. Following the analysis of Yeung and Petrosyan (2021), we formulate a discrete-time random-horizon Payoff Distribution Procedure (PDP) so that the agreed-upon imputations (3.8) can be realized. Let $\beta_k^i(x_k^*; \underline{\bar{u}}_{k-}^*)$ denote the payment that player i will received at stage k under the cooperative agreement.

The payment scheme involving $\beta_k^i(x_k^*; \underline{\bar{u}}_{k-}^*)$ constitutes a PDP in the sense that along the cooperative trajectory $\{x_k^*\}_{k=1}^T$ with previously executed durable strategies $\underline{\bar{u}}_{k-}^*$ the imputation to player i over the stages from k to T can be expressed as follows:

$$\begin{aligned}
&\xi^i(\tau, x_\tau^*; \underline{\bar{u}}_{\tau-}^*) \\
&\quad = E\left[\sum_{k=\tau}^{\hat{T}} \beta_k^i(x_k^*; \underline{\bar{u}}_{k-}^*)\delta_1^k + q_{\hat{T}+1}^i(x_{\hat{T}+1}^*; \underline{\bar{u}}_{(\hat{T}+1)-}^*)\delta_1^{\hat{T}+1}\right] \\
&\quad = \sum_{\hat{T}=\tau}^{T} \frac{\theta_{\hat{T}}}{\sum_{\zeta=\tau}^{T}\theta_\zeta}\left[\sum_{k=\tau}^{\hat{T}} \beta_k^i(x_k^*; \underline{\bar{u}}_{k-}^*)\delta_1^k + q_{\hat{T}+1}^i(x_{\hat{T}+1}^*; \underline{\bar{u}}_{(\hat{T}+1)}^*)\delta_1^{\hat{T}+1}\right],
\end{aligned} \tag{3.9}$$

for $i \in N$ and $\tau \in \{1, 2, \ldots, T\}$.

A theorem characterizing a computational formula for $\beta_\tau^i(x_\tau^*; \underline{\bar{u}}_{\tau-}^*)$, for $\tau \in \{1, 2, \ldots, T\}$ and $i \in N$, which yields (3.9) is provided in the following.

Theorem 3.2. *A payment equaling*

$$\beta_\tau^i(x_\tau^*;\underline{\bar{u}}_{\tau-}^*)$$
$$= (\delta_1^\tau)^{-1}\left[\xi^i(\tau, x_\tau^*;\underline{u}_{\tau-}^*) - \frac{\theta_\tau}{\sum_{\zeta=\tau}^T \theta_\zeta} q_{\tau+1}^i(x_{\tau+1}^*;\underline{\bar{u}}_{(\tau+1)-}^*)\delta_1^{\tau+1}\right.$$
$$\left. - \frac{\sum_{\zeta=\tau+1}^T \theta_\zeta}{\sum_{\zeta=\tau}^T \theta_\zeta}\xi^i(\tau+1,\ x_{\tau+1}^*;\underline{\bar{u}}_{(\tau+1)-}^*)\right] \tag{3.10}$$

given to player $i \in N$ at stage $\tau \in \{1, 2, \ldots, T\}$ would lead to the realization of the imputation vector $\xi(\tau, x_\tau^*;\underline{\bar{u}}_{\tau-}^*)$ in (3.8).

Proof. See the proof of Theorem VI of Mathematical Solution Mechanisms in Chapter 7. □

Note that the payoff distribution procedure $\beta_\tau^i(x_\tau^*;\underline{\bar{u}}_{\tau-}^*)$ in Theorem 3.2 would give rise to the agreed-upon imputation vector in (3.8). Therefore, the property of subgame consistency is satisfied.

When all players are using the cooperative strategies, the payoff that player i will directly receive at stage $\tau \in \{1, 2, \ldots, T\}$ is $g_\tau^i(x^*, \underline{u}^*, \underline{\bar{u}}_\tau^*;\underline{\bar{u}}_{\tau-}^*)$. However, according to the agreed-upon imputation, player i is to receive $\beta_\tau^i(x_\tau^*;\underline{\bar{u}}_{\tau-}^*)$ at stage τ. Therefore, a side payment equaling

$$\beta_\tau^i(x_\tau^*;\underline{u}_{\tau-}^*) - g_\tau^i(x^*, \underline{u}^*, \underline{\bar{u}}_\tau^*;\underline{\bar{u}}_{\tau-}^*)$$

will be given to player $i \in N$ in stage $\tau \in \{1, 2, \ldots, T\}$.

4. An Explication in International Joint Venture

An international joint venture (IJV) is formed when two businesses based on two or more nations establish a partnership. A company that wants to explore international trade without taking full responsibility for cross-border business operations has the opportunity to create a joint venture with a foreign partner. International investors entering into a joint venture minimize the risk associated with a direct acquisition of a business. In the development of international business, performing due diligence in a foreign country and a partner limits the risks associated with such a business transaction. IJVs help companies to form strategic alliances and gain a competitive advantage through access to their partner's resources, markets, technology, capital, and people. International joint ventures are a practical means of transferring scientific knowledge, technology, and management skills from multinational experience to local companies. Such knowledge transfer can contribute to improving the productivity of local companies.

Examples of large and successful international joint ventures are not rare, for instance, Omega Navigation Enterprises Inc., IHS Holding and MTN Nigeria, Telstra and Telkom Indonesia, Chery Jaguar Land Rover Automotive Company, IndianOil Skytanking Limited, Starbucks, and Tata Global and Beverages.

However, though international joint ventures can be very profitable, the occurrence of dissolution of joint ventures is not uncommon.

Verizon and Vodafone is an example of a venture that was good while it lasted. Verizon Communications teamed up with European wireless network giant Vodafone in 2000. They had a huge impact on the North American market, but the JV ended in 2014 when Verizon offered to buy off its share for $130 billion. Even though the venture was highly profitable, at that time it suited Vodafone to sell its share and use the income for solving the issues within its own company.

Sony Ericsson was another example. Ericsson was a Swedish technology expert specializing in mobile networks. Sony and Ericsson decided to start a successful joint venture in 2001 with a 50-50 share in profits. The merger was so successful that hardly anyone remembered these were two different companies. However, the alliance ended in 2011 with Sony buying off Ericsson's share for over one billion dollars because it wanted to improve the overall offer to the consumers.

The major advantages of international joint ventures include the following:

(i) IJV provides faster and less costly access to foreign markets than can be achieved by purchasing an existing company or setting up a new venture in the foreign market.
(ii) IJV provides access to local brands, know-how, distribution channels, consumer market, knowledge of the local marketplace, local production facilities, or government contracts.

(iii) IJV is a practical vehicle for knowledge transfer and sharing of complementary intellectual property assets between the companies.
(iv) IJV lowers the costs and risks associated with doing business in a foreign market, especially for small to mid-size businesses.
(v) IJV is less risky than the acquisition of a company in a foreign market through more flexibility when local conditions change or if the market and relationship become unattractive.
(vi) IJV can be sold to the local partner when conditions prove undesirable.
(vii) IJV can be used to reduce risks associated with changes in the political, legal, and regulatory environment in the local market.

International joint ventures are formed with certain defined objectives and are not necessarily intended to function as long-term partnerships. Following are some of the common reasons for dissolving an IJV:

(i) The time period that was agreed upon for the joint venture to operate has been completed, and the parties agree that there is no further benefit to be gained from continuing the venture.
(ii) The individual objectives of each party are no longer aligned with the common objectives of the IJV partnership.
(iii) Legal or financial issues have arisen with one or both of the parties that make continuing the IJV no longer viable.
(iv) The parties discover that the benefits they had hoped to obtain from the IJV have not materialized and are not likely to be realized if the IJV were continued.
(v) Changes in market conditions and policies would make the IJV no longer likely to be profitable for either or both parties.

In international joint ventures, the most commonly used nondurable strategies $\underline{u}_k$ include various industrial productions, natural resource extraction, manpower application, and agricultural production. Durable strategies $\bar{u}_k$ include innovation and technology learning, knowledge-based technique innovation, capital investment, and management operations.

The state vector x_k includes financial assets, technology level, environmental state, number of joint venture plants, and level of capital stocks.

The longest expected lifespan of the IJV is T stages, the actual time of termination of the venture is a random variable $\hat{T}$ stage with range $\{1, 2, \ldots, T\}$ and corresponding probabilities are $\{\theta_1, \theta_2, \ldots, \theta_T\}$. Conditional upon the reaching of stage τ, the probability of the game that would last up to stages $\tau, \tau+1, \ldots, T$ becomes, respectively,

$$\frac{\theta_\tau}{\sum_{\zeta=\tau}^{T} \theta_\zeta}, \frac{\theta_{\tau+1}}{\sum_{\zeta=\tau}^{T} \theta_\zeta}, \ldots, \frac{\theta_T}{\sum_{\zeta=\tau}^{T} \theta_\zeta}.$$

Major objectives of international joint venture — like (i) profits, (ii) technology development, (iii) market sizes, (iv) financial integration, and (v) business growth — would appear in the vector of objectives of party i:

$$\begin{pmatrix} g_k^{i(1)}(x_k, \underline{u}_k, \bar{u}_k; \bar{\underline{u}}_{k-}) \\ g_k^{i(2)}(x_k, \underline{u}_k, \bar{u}_k; \bar{\underline{u}}_{k-}) \\ \vdots \\ g_k^{i(\Theta_{i_j})}(x_k, \underline{u}_k, \bar{u}_k; \bar{\underline{u}}_{k-}) \end{pmatrix} \quad \text{for } i \in N,$$

where N is the set of parties in the international joint venture.

If the venture terminates after stage $\hat{T}$, party i will receive a terminal payment $q^i_{\hat{T}+1}(x_{\hat{T}+1}; \bar{\underline{u}}_{(\hat{T}+1)-})$ in stage $\hat{T}+1$, which reflects the values of settlement between the venture parties.

The expected payoff of the joint venture is modeled as

$$\begin{aligned} E&\left\{\sum_{j=1}^{n}\left[\sum_{k=1}^{\hat{T}}\sum_{\varsigma=1}^{\Theta_j} w_k^{j(\varsigma)} g_k^{j(\varsigma)}(x_k, \underline{u}_k, \bar{u}_k; \bar{\underline{u}}_{k-})\delta_1^k \right.\right. \\ &\left.\left. + q^j_{\hat{T}+1}(x_{\hat{T}+1}; \bar{\underline{u}}_{(\hat{T}+1)-})\delta_1^{\hat{T}+1}\right]\right\} \\ &= \sum_{\hat{T}=1}^{T}\theta_{\hat{T}}\left\{\sum_{j=1}^{n}\left[\sum_{k=1}^{\hat{T}}\sum_{\varsigma=1}^{\Theta_j} w_k^{j(\varsigma)} g_k^{j(\varsigma)}(x_k, \underline{u}_k, \bar{u}_k; \bar{\underline{u}}_{k-})\delta_1^k \right.\right. \\ &\left.\left. + q^j_{\hat{T}+1}(x_{\hat{T}+1}; \bar{\underline{u}}_{(\hat{T}+1)-})\delta_1^{\hat{T}+1}\right]\right\}. \end{aligned} \tag{4.1}$$

The state dynamics is

$$x_{k+1} = f(x_k, \underline{u}_k, \bar{u}_k; \underline{\bar{u}}_{k-}), \quad x_1 = x_1^0. \tag{4.2}$$

4.1. *Maximization of joint venture profits*

The expected venture profit can be obtained by maximizing the expected profit (4.1) subject to state dynamics (4.2). Applying Theorem 3.1, an optimal solution to the problem (4.1)–(4.2) can be characterized with the set of optimal joint venture strategies $(\underline{u}_\tau^*, \bar{u}_\tau^*)$, for $\tau \in \{1, 2, \ldots, T\}$. Substituting the optimal strategies $(\underline{u}_\tau^*, \bar{u}_\tau^*)$ into the state dynamics (3.1), one can obtain the dynamics of the cooperative trajectory as follows:

$$x_{k+1} = f_k(x_k, \underline{u}_k^*, \bar{u}_k^*; \underline{\bar{u}}_{k-}^*), \quad x_1 = x_1^0. \tag{4.3}$$

We use $\{x_k^*\}_{k=1}^{T+1}$ to denote the solution generated by dynamics (4.3). Using the set of optimal strategies $(\underline{u}_\tau^*, \bar{u}_\tau^*)$, one can obtain the expected joint cooperative payoff as

$$\begin{aligned} &W(\tau, x_\tau^*; \underline{\bar{u}}_{\tau-}^*) \\ &= E\left\{\sum_{j=1}^{n}\left[\sum_{k=\tau}^{\hat{T}}\sum_{\varsigma=1}^{\Theta_j} w_k^{j(\varsigma)} g_k^{j(\varsigma)}(x_k^*, \underline{u}_k^*, \bar{u}_k^*; \underline{\bar{u}}_{k-}^*)\delta_1^k \right.\right. \\ &\qquad \left.\left. + q_{\hat{T}+1}^j(x_{\hat{T}+1}^*; \underline{\bar{u}}_{(\hat{T}+1)-}^*)\delta_1^{\hat{T}+1}\right]\right\} \\ &= \sum_{\hat{T}=\tau}^{T} \frac{\theta_{\hat{T}}}{\sum_{\zeta=\tau}^{T}\theta_\zeta}\left\{\sum_{j=1}^{n}\left[\sum_{k=\tau}^{\hat{T}}\sum_{\varsigma=1}^{\Theta_j} w_k^{j(\varsigma)} g_k^{j(\varsigma)}(x_k^*, \underline{u}_k^*, \bar{u}_k^*; \underline{\bar{u}}_{k-}^*)\delta_1^k \right.\right. \\ &\qquad \left.\left. + q_{\hat{T}+1}^j(x_{\hat{T}+1}^*; \underline{\bar{u}}_{(\hat{T}+1)-}^*)\delta_1^{\hat{T}+1}\right]\right\}, \quad \text{for } \tau \in \{1, 2, \ldots, T\}. \end{aligned} \tag{4.4}$$

4.2. *Subgame-consistent sharing of venture payoff*

Let $\xi(\tau, x_\tau; \underline{\bar{u}}_{\tau-}) = [\xi^1(\tau, x_\tau; \underline{\bar{u}}_{\tau-}), \xi^2(\tau, x_\tau; \underline{\bar{u}}_{\tau-}), \ldots, \xi^n(\tau, x_\tau; \underline{\bar{u}}_{\tau-})]$ denote the agreed-upon sharing of venture in stage $\tau \in \{1, 2, \ldots, T\}$.

For individual rationality to be maintained throughout all the stages $\tau \in \{1, 2, \ldots, T\}$ along the optimal cooperative trajectory $\{x_k^*\}_{k=1}^{T+1}$, it is required that

$$\xi^i(\tau, x_\tau^*; \underline{\bar{u}}_{\tau-}^*) \geq V^i(\tau, x_\tau^*; \underline{\bar{u}}_{\tau-}^*), \quad \text{for } i \in N \text{ and } \tau \in \{1, 2, \ldots, T\},$$

along the cooperative trajectory $\{x_k^*\}_{k=1}^{T}$ for given previously executed strategies $\underline{\bar{u}}_{\tau-}^*$, where $V^i(\tau, x_\tau; \underline{\bar{u}}_{\tau-})$ is the payoff to party i before forming the joint venture.

The payoff of party $i \in N$ before establishing the joint venture can be obtained from Theorem 2.1.

Finally, to satisfy group optimality, the imputation vector has to satisfy

$$W(\tau, x_\tau^*; \underline{\bar{u}}_{\tau-}^*) = \sum_{j=1}^{n} \xi^j(\tau, x_\tau^*; \underline{\bar{u}}_{\tau-}^*), \quad \text{for } \tau \in \{1, 2, \ldots, T\}.$$

Crucial to the sustainability of the joint venture, the following condition has to be maintained:

$$\begin{aligned} \xi(\tau, x_\tau^*; \underline{\bar{u}}_{\tau-}^*) &= [\xi^1(\tau, x_\tau; \underline{\bar{u}}_{\tau-}), \xi^2(\tau, x_\tau^*; \underline{\bar{u}}_{\tau-}^*), \ldots, \xi^n(\tau, x_\tau^*; \underline{\bar{u}}_{\tau-}^*)], \\ &\quad \tau \in \{1, 2, \ldots, T\}. \end{aligned} \tag{4.5}$$

Let $\beta_k^i(x_k^*; \underline{\bar{u}}_{k-}^*)$ denote the payment that player i will receive at stage k under the cooperative agreement. The payment scheme involving $\beta_k^i(x_k^*; \underline{\bar{u}}_{k-}^*)$ constitutes a PDP in the sense that along the cooperative trajectory $\{x_k^*\}_{k=1}^{T}$ with previously executed durable strategies $\underline{\bar{u}}_{k-}^*$, the imputation to player i over the stages from k to T can be expressed as follows:

$$\begin{aligned} &\xi^i(\tau, x_\tau^*; \underline{\bar{u}}_{\tau-}^*) \\ &= E\left[\sum_{k=\tau}^{\hat{T}} \beta_k^i(x_k^*; \underline{\bar{u}}_{k-}^*)\delta_1^k + q_{\hat{T}+1}^i(x_{\hat{T}+1}^*; \underline{\bar{u}}_{(\hat{T}+1)-}^*)\delta_1^{\hat{T}+1}\right] \\ &= \sum_{\hat{T}=\tau}^{T} \frac{\theta_{\hat{T}}}{\sum_{\zeta=\tau}^{T}\theta_\zeta}\left[\sum_{k=\tau}^{\hat{T}} \beta_k^i(x_k^*; \underline{\bar{u}}_{k-}^*)\delta_1^k + q_{\hat{T}+1}^i(x_{\hat{T}+1}^*; \underline{\bar{u}}_{(\hat{T}+1)}^*)\delta_1^{\hat{T}+1}\right], \end{aligned} \tag{4.6}$$

for $i \in N$ and $\tau \in \{1, 2, \ldots, T\}$.

Invoking Theorem 3.2, a theorem characterizing a computational formula for $\beta_\tau^i(x_\tau^*; \underline{\bar{u}}_{\tau-}^*)$, for $\tau \in \{1, 2, \ldots, T\}$ and $i \in N$, can be obtained as follows:

$$\beta_\tau^i(x_\tau^*; \underline{\bar{u}}_{\tau-}^*)$$

$$= (\delta_1^\tau)^{-1} \Bigg[\xi^i(\tau, x_\tau^*; \underline{\bar{u}}_{\tau-}^*) - \frac{\theta_\tau}{\sum_{\zeta=\tau}^{T} \theta_\zeta} q_{\tau+1}^i(x_{\tau+1}^*; \underline{\bar{u}}_{(\tau+1)-}^*) \delta_1^{\tau+1}$$

$$- \frac{\sum_{\zeta=\tau+1}^{T} \theta_\zeta}{\sum_{\zeta=\tau}^{T} \theta_\zeta} \xi^i(\tau+1,\ x_{\tau+1}^*; \underline{\bar{u}}_{(\tau+1)-}^*) \Bigg]. \tag{4.7}$$

When all parties are using the cooperative strategies, the payoff that party i will directly receive at stage $\tau \in \{1, 2, \ldots, T\}$ is $g_\tau^i(x^*, \underline{u}^*, \underline{\bar{u}}_\tau^*; \underline{\bar{u}}_{\tau-}^*)$. However, according to the agreed-upon imputation, player i is to receive $\beta_\tau^i(x_\tau^*; \underline{\bar{u}}_{\tau-}^*)$ at stage τ. Therefore, a side payment equaling

$$\beta_\tau^i(x_\tau^*; \underline{u}_{\tau-}^*) - g_\tau^i(x^*, \underline{u}_\tau^*, \underline{\bar{u}}_\tau^*; \underline{\bar{u}}_{\tau-}^*)$$

will be given to player $i \in N$ in stage $\tau \in \{1, 2, \ldots, T\}$.

5. A Computational Illustration: Part I

In this section, we present a computational illustration with an example in Yeng and Petrosyan (2021). It involves two parties with interactive investments with technology spillover to demonstrate the computation of the non-cooperative equilibrium in a dynamic game under random horizon with durable strategies.

5.1. *Game formulation*

Consider the case of two economic agents (regions or firms) that are given a certain lease to trade with each. The lease for trade has to be renewed after each stage (year or decade) for up to a maximum of 10 stages. At stage 1, it is known that the probabilities that the lease will last for $1, 2, 3, \ldots, 10$ stages with respective probabilities $\theta_1, \theta_2, \theta_3, \ldots, \theta_{10}$. Conditional upon the reaching of stage $\tau > 1$, the probabilities of the game that would last up to stages $\tau, \tau+1$, to 10 are

$$\frac{\theta_\tau}{\sum_{\zeta=\tau}^{10}\theta_\zeta}, \frac{\theta_{\tau+1}}{\sum_{\zeta=\tau}^{10}\theta_\zeta}, \frac{\theta_{\tau+2}}{\sum_{\zeta=\tau}^{10}\theta_\zeta}, \ldots, \frac{\theta_{10}}{\sum_{\zeta=\tau}^{10}\theta_\zeta}.$$

Each agent has a private production capital $x_k^i \in X^i \subset R$, for $i \in \{1,2\}$. Investment in private capital takes three stages to complete its transformation into the agent's capital stock, and we use u_k^{Ii} to denote the investment in the private capital by agent i in stage k. The increment in the private capital in stage k brought about by agent i's investment u_k^{Ii} in stage k is $\varepsilon_k^i u_k^{Ii}$. There is technology spillover through trade so that the private capital of its trading partner can positively affect the increment of the agent's private capital. The spillover of agent $\gamma_k^{(i)j}$'s private capital to the increment of agent i's private capital stock is $\gamma_k^{(i)j} x_k^j$. The terms ε_k^i and $\gamma_k^{(i)j}$ are positive parameters.

The accumulation process of the private capital stock of agent i is governed by the dynamical equation

$$\begin{aligned} x_{k+1}^i &= x_k^i + \varepsilon_k^i u_k^{Ii} - \lambda_k^i x_k^i + \gamma_k^{(i)j} x_k^j, \\ x_1^i &= x_1^{i(0)}, \;\; i,j \in \{1,2\} \quad \text{and } i \neq j, \end{aligned} \tag{5.1}$$

where λ_k^i is the depreciation rate of capital x_k^i.

The cost of investment in physical capital is $c_k^{Ii}(u_k^{Ii})^2$. Agent i uses a non-durable input u_k^{IIi} along with its private capital x_k^i to produce an output. The net operating profit of agent i's output in stage k is $R_k^i(u_k^{IIi}x_k^i)^{1/2} - c_k^{IIi}u_k^{IIi}$, where $(u_k^{IIi}x_k^i)^{1/2}$ is the production function of the output and the unit cost of the non-durable input u_k^{IIi} is c_k^{IIi}. The terms R_k^i, c_k^{Ii}, and c_k^{IIi} are positive parameters.

Each agent has a revenue-generating investment which yields an income stream over four stages. We use $\bar{u}_k^{(4)i}$ to denote agent i's revenue-generating investment in stage k. The income generated by agent i's stage k generating investment is $p_t^{|k|i}\bar{u}_k^{(4)i}$ in stage $t \in \{k, k+1, k+3, k+4\}$. The cost of the revenue-generating investment $u_k^{(4)i}$ is $\varphi_k^i(\bar{u}_k^{(4)i})^2$, with φ_k^i being a positive parameter.

In addition, each agent has a project investment $\bar{u}_k^{(3)i}$ which would yield a profit $\chi_k^i(\bar{u}_k^{(3)i})^{1/2}$ in stage k and the cost of investment will be paid by three installments in stages k, $k+1$, and $k+2$. The payment in stage t is $w_t^{|k|i}\bar{u}_t^{(3)i}$, for $t \in \{k, k+1, k+2\}$.

If the agent is still in operation in stage k, agent i's income net of all costs in stage k is

$$R_k^i(u_k^{IIi}x_k^i)^{\frac{1}{2}} - c_k^{IIi}u_k^{IIi} - c_k^{Ii}(u_k^{Ii})^2 - \varphi_k^i(\bar{u}_k^{(4)i})^2 + \sum_{t=k-3}^{k} p_k^{|t|i}\bar{u}_t^{(4)i}$$

$$+\chi_k^i(\bar{u}_k^{(3)i})^{\frac{1}{2}} - \sum_{t=k-2}^{k} w_k^{|t|i}\bar{u}_t^{(3)i}. \tag{5.2}$$

If the agents remain in operation until stage 10, agent $i \in \{1, 2\}$ will receive a terminal payoff in stage 11 equaling

$$q^i(11, x; \underline{\bar{u}}_{11-}^{(3)}, \underline{\bar{u}}_{11-}^{(4)})$$
$$= \left(Q_{11}^{(i)i}x_{11}^i + \sum_{t=8}^{10}\bar{p}_{11}^{|t|i}\bar{u}_t^{(4)i} - \sum_{t=9}^{10}\bar{w}_{11}^{|t|i}\bar{u}_t^{(3)i} + \varpi_{11}^i\right).$$

If the game terminates after stage τ, agent $i \in \{1, 2\}$ will receive a terminal payoff in stage $\tau+1$ equaling

$$q^i(\tau+1, x; \underline{\bar{u}}_{(\tau+1)-}^{(3)}, \underline{\bar{u}}_{(\tau+1)-}^{(4)})$$
$$= \left(Q_{\tau+1}^{(i)i}x_{\tau+1}^i + \sum_{t=\tau-2}^{\tau}\bar{p}_{\tau+1}^{|t|i}\bar{u}_t^{(4)i}\right) - \sum_{t=\tau-1}^{\tau}\bar{w}_{\tau+1}^{|t|i}\bar{u}_t^{(3)i} + \varpi_{\tau+1}^i).$$

In particular, $Q_{\tau+1}^{(i)i}x_{\tau+1}^i$ is the value (or salvage price) of agent i private capital $x_{\tau+1}^i$ in stage $\tau+1$ if the agent terminates its operation after stage τ. The term $\sum_{t=\tau-2}^{\tau}\bar{p}_{\tau+1}^{|t|i}\bar{u}_t^{(4)i}$ is the realizable value of the returns for previously procured revenue-generating investments at stage $\tau+1$ if the agent terminates its operation after stage τ. Some or all of the terms $\bar{p}_{\tau+1}^{(t)i}$ can be zero.

The term $\sum_{t=\tau-2}^{\tau}\bar{w}_{\tau+1}^{|t|i}u_t^{(3)i}$ is the payment of the debt obligations from previously started project investments at stage $\tau+1$ if the agent terminates its operation after stage τ. Moreover, the term $\sum_{t=\tau-2}^{\tau}\bar{w}_{\tau+1}^{|t|i}u_t^{(3)i}$ may involve a penalty for terminating the financing plan prematurely.

The expected payoff of agent i to be maximized is

$$\sum_{\hat{T}=1}^{10} \theta_{\hat{T}} \left\{ \sum_{k=1}^{\hat{T}} \left[R_k^i (u_k^{IIi} x_k^i)^{\frac{1}{2}} - c_k^{IIi} u_k^{IIi} - c_k^{Ii} (u_k^{Ii})^2 \right.\right.$$

$$\left. - \varphi_k^i (\bar{u}_k^{(4)i})^2 + \sum_{t=k-3}^{k} p_k^{|t|i} \bar{u}_t^{(4)i} + \chi_k^i (\bar{u}_k^{(3)i})^{\frac{1}{2}} - \sum_{t=k-2}^{k} w_k^{|t|i} \bar{u}_\tau^{(3)i} \right] \delta^{k-1}$$

$$\left. + \left[Q_{\hat{T}+1}^{(i)i} x^i + \sum_{t=\hat{T}-2}^{\hat{T}} \bar{p}_{\hat{T}+1}^{|t|i} \bar{u}_t^{(4)i} - \sum_{t=\hat{T}-1}^{\hat{T}} \bar{w}_{\hat{T}+1}^{|t|i} \bar{u}_t^{(3)i} + \varpi_{\hat{T}+1}^i \right] \delta^{\hat{T}} \right\}, \tag{5.3}$$

where δ is the discount factor, and the controls executed before the start of the operation in stage 1, that is, $(\underline{\bar{u}}_{1-}^{(3)}, \underline{\bar{u}}_{1-}^{(4)})$, are zeros.

5.2. *Game equilibrium outcome*

Invoking Theorem 3.1, a feedback Nash equilibrium of the game (5.1)–(5.3) can be characterized as follows.

Corollary 5.1. *Let* $\{\underline{u}_\tau^{I(1)**}, \underline{u}_\tau^{II(1)**}, \underline{\bar{u}}_\tau^{(3)**}, \underline{\bar{u}}_\tau^{(4)**}\}$ *denote the set of the agents' feedback Nash equilibrium strategies and* $V^i(\tau, x; \underline{\bar{u}}_{\tau-}^{(3)**}, \underline{\bar{u}}_{\tau-}^{(4)**})$ *denote the feedback Nash equilibrium payoff of agent i in the non-cooperative game* (5.1)–(5.2), *then the function* $V^i(\tau, x; \underline{\bar{u}}_{\tau-}^{(3)**}, \underline{\bar{u}}_{\tau-}^{(4)**})$ *satisfies the following system of recursive equations:*

$$V^i(11, x; \underline{\bar{u}}_{11-}^{(3)**}, \underline{\bar{u}}_{11-}^{(4)**})$$
$$= \left(Q_{11}^{(i)i} x_{11}^i + \sum_{t=11-3}^{10} \bar{p}_{11}^{|t|i} \bar{u}_t^{(4)i**} - \sum_{t=11-2}^{10} \bar{w}_{11}^{|t|i} \bar{u}_t^{(3)i**} + \varpi_{11}^i \right) \delta^{10}, \tag{5.4}$$

$$V^i(10, x; \underline{\bar{u}}_{10-}^{(3)**}, \underline{\bar{u}}_{10-}^{(4)**})$$
$$= \max_{\substack{u_{10}^{Ii}, u_{10}^{IIi} \\ \bar{u}_{10}^{(3)i}, \bar{u}_{10}^{(4)i}}} \left\{ \left[R_{10}^i (u_{10}^{IIi} x^i)^{\frac{1}{2}} - c_{10}^{IIi} u_{10}^{IIi} - c_{10}^{Ii} (u_{10}^{Ii})^2 - \varphi_{10}^i (\bar{u}_{10}^{(4)i})^2 \right.\right.$$

$$+ p_{10}^{|10|i}\bar{u}_{10}^{(4)i} + \sum_{t=10-3}^{9} p_{10}^{|t|i}\bar{u}_t^{(4)i**} + \chi_k^i(\bar{u}_{10}^{(3)i})^{\frac{1}{2}} - w_{10}^{|t|i}\bar{u}_{10}^{(3)i}$$

$$- \sum_{t=10-2}^{9} w_{10}^{|t|i}\bar{u}_t^{(3)i**}\Bigg]\delta^{k-1} + \Bigg[Q_{11}^{(i)i}(x^i + \varepsilon_{10}^i u_{10}^{Ii} - \lambda_{10}^i x^i$$

$$+ \gamma_{10}^{(i)j}x^j) + \bar{p}_{11}^{|10|i}\bar{u}_{10}^{(4)i} + \sum_{t=11-3}^{9} \bar{p}_{11}^{|t|i}\bar{u}_t^{(4)i**} + \bar{w}_{11}^{|10|i}\bar{u}_{10}^{(4)i}$$

$$+ \sum_{t=11-2}^{9} w_{11}^{|t|i}\bar{u}_t^{(4)i**} + \varpi_{11}^i\Bigg]\delta^{10}\Bigg\}, \tag{5.5}$$

$$V^i(\tau, x; \underline{\bar{u}}_{\tau-}^{(3)**}, \underline{\bar{u}}_{\tau-}^{(4)**})$$

$$= \max_{\substack{u_\tau^{Ii}, u_\tau^{IIi} \\ \bar{u}_\tau^{(3)i}, \bar{u}_\tau^{(4)i}}} \Bigg\{\Bigg[R_\tau^i(u_\tau^{IIi}x^i)^{\frac{1}{2}} - c_\tau^{IIi}u_\tau^{IIi} - c_\tau^{Ii}(u_\tau^{Ii})^2 - \varphi_\tau^i(\bar{u}_\tau^{(4)i})^2$$

$$+ p_\tau^{|\tau|i}\bar{u}_\tau^{(4)i} + \sum_{t=\tau-3}^{\tau} p_\tau^{|t|i}u_t^{(4)i**} + \chi_\tau^i(\bar{u}_\tau^{(4)i})^{\frac{1}{2}} - w_\tau^{|\tau|i}\bar{u}_\tau^{(3)i}$$

$$- \sum_{t=\tau-2}^{\tau} w_\tau^{|t|i}u_t^{(3)i**}\Bigg]\delta^{\tau-1} + \frac{\theta_\tau}{\sum_{\zeta=\tau}^{T}\theta_\zeta}\Bigg[Q_{\tau+1}^{(i)i}x_{\tau+1}^{i***} + \bar{p}_{\tau+1}^{(\tau)i}\bar{u}_\tau^{(4)i}$$

$$+ \sum_{t=\tau+1-3}^{\tau-1} \bar{p}_{\tau+1}^{(t)i}\bar{u}_t^{(4)i**} - \bar{w}_{\tau+1}^{(\tau)i}\bar{u}_\tau^{(3)i} + \sum_{t=\tau+1-2}^{\tau-1} \bar{w}_{\tau+1}^{(t)i}\bar{u}_t^{(3)i**}$$

$$+ \varpi_{\tau+1}^i\Bigg]\delta^\tau + \frac{\sum_{\zeta=\tau+1}^{T}\theta_\zeta}{\sum_{\zeta=\tau}^{T}\theta_\zeta}V^i\Bigg[\tau+1, x_{\tau+1}^{i***}, x_{\tau+1}^{j***}; \bar{u}_\tau^{(3)i}, \underline{\bar{u}}_{(\tau+1)-}^{(3)i**}$$

$$\cap \underline{\bar{u}}_{\tau-}^{(3)i**}, \underline{\bar{u}}_{(\tau+1)-}^{(3)j**}, \bar{u}_\tau^{(4)i}, \underline{\bar{u}}_{(\tau+1)-}^{(4)i**} \cap \underline{\bar{u}}_{\tau-}^{(4)i**}, \underline{\bar{u}}_{(\tau+1)-}^{(4)j**}\Bigg]\Bigg\},$$

$$\text{for } \tau \in \{1, 2, \ldots, 9\}, \tag{5.6}$$

where $x_{\tau+1}^{i***} = x^i + \varepsilon_\tau^i u_\tau^{Ii} - \lambda_\tau^i x^i + \gamma_\tau^{(i)j}x^j$,

$x_{\tau+1}^{j***} = x^j + \varepsilon_\tau^j u_\tau^{Ij**} - \lambda_\tau^j x^j + \gamma_\tau^{(j)i}x^i$, for $i, j \in \{1, 2\}$ and $i \neq j$.

Performing the indicated maximization operator in (5.5)–(5.6) of Corollary 5.1, we obtain a feedback Nash equilibrium solution of the game (5.1)–(5.2) as follows.

Proposition 5.1. *System (5.4)–(5.5) admits a solution with the game equilibrium payoff of agent i being*

$$V^i(\tau, x; \underline{\bar{u}}_{\tau-}^{(4)**}) = (A_\tau^{(i)i}x^i + A_\tau^{(i)j}x^j + C_\tau^i)\delta^{\tau-1}, \quad \text{for } \tau \in \{1, 2, \ldots, 11\},$$

where

$$\begin{aligned}
A_{11}^{(i)i} &= Q_{11}^{(i)i}, A_{11}^{(i)j} = 0, C_{11}^i = \sum_{t=11-3}^{10} \bar{p}_{11}^{(t)i}\bar{u}_t^{(4)i**} \\
&\quad - \sum_{t=11-2}^{10} \bar{w}_{11}^{(t)i}\bar{u}_t^{(3)i**} + \varpi_{11}^i, \\
A_{10}^{(i)i} &= \frac{(R_{10}^i)^2}{4c_{10}^{IIi}} + \delta A_{11}^{(i)i}(1 - \lambda_{10}^i), A_{10}^{(i)j} = \delta A_{11}^{(i)i}\gamma_{10}^{(i)j}; \\
A_\tau^{(i)i} &= \frac{(R_\tau^i)^2}{4c_\tau^{IIi}} + \frac{\theta_\tau}{\sum_{\zeta=\tau}^{10}\theta_\zeta}\delta Q_{\tau+1}^{(i)i}(1 - \lambda_\tau^i) \\
&\quad + \frac{\sum_{\zeta=\tau+1}^{10}\theta_\zeta}{\sum_{\zeta=\tau}^{10}\theta_\zeta}(A_{\tau+1}^{(i)i}(1 - \lambda^i) + A_{\tau+1}^{(i)j}\gamma_\tau^{(j)i})\delta, \\
A_\tau^{(i)j} &= \frac{\theta_\tau}{\sum_{\zeta=\tau}^{10}\theta_\zeta}(\delta Q_{\tau+1}^{(i)i}\gamma_\tau^{(i)j}) + \frac{\sum_{\zeta=\tau+1}^{10}\theta_\zeta}{\sum_{\zeta=\tau}^{10}\theta_\zeta} \\
&\quad \times [A_{\tau+1}^{(i)i}\gamma_\tau^{(i)j} + A_{\tau+1}^{(i)j}(1 - \lambda_\tau^j)]\delta,
\end{aligned} \tag{5.7}$$

for $\tau \in \{9, 8, \ldots, 1\}$, $i, j \in \{1, 2\}$ and $i \neq j$, and C_τ^i is an expression containing previously executed durable controls $(\underline{\bar{u}}_{\tau-}^{(3)**}, \underline{\bar{u}}_{\tau-}^{(4)**})$, for $\tau \in \{1, 2, \ldots, 10\}$.

Proof. See Appendix E in the Mathematical Appendices. □

The values of $A_\tau^{(i)i}$ and $A_\tau^{(i)j}$, for $\tau \in \{1, 2, \ldots, 10\}$ and $i, j \in \{1, 2\}$ and $i \neq j$, in Proposition 5.1 can be solved backwardly from $\tau = 10$ to $\tau = 1$. Using Corollary 5.1 and Proposition 5.1, the game equilibrium strategies of the agent $i \in \{1,2\}$ can be obtained as follows (see derivation details in Appendix B):

$$u_\tau^{Ii**} = \frac{A_{\tau+1}^{(i)i}\varepsilon_\tau^i\delta}{2c_\tau^{Ii}}, \quad u_\tau^{IIi**} = \frac{(R_\tau^i)^2}{4(c_\tau^{IIi})^2}x^i, \quad \text{for } \tau \in \{1,2,\ldots,10\},$$

$$\bar{u}_{10}^{(3)i**} = \frac{(\chi_{10}^i)^2}{4(w_{10}^{|10|i} + \delta\bar{w}_{11}^{|10|i})^2},$$

$$\bar{u}_9^{(3)i**} = (\chi_9^i)^2 \div 4\left[w_9^{|9|i} + \frac{\theta_9}{\sum_{\zeta=9}^{10}\theta_\zeta}w_9^{|9|i}\delta + \frac{\sum_{\zeta=10}^{10}\theta_\zeta}{\sum_{\zeta=9}^{10}\theta_\zeta}(w_{10}^{|9|i}\delta + \bar{w}_{11}^{|9|i}\delta^2)\right]^2,$$

$$\bar{u}^{(3)i**} = (\chi_\tau^i)^2 \div 4\left[w_\tau^{|\tau|i} + \frac{\theta_\tau}{\sum_{\zeta=\tau}^{10}\theta_\zeta}w_{\tau+1}^{|\tau|i}\delta + \frac{\sum_{\zeta=\tau+1}^{10}\theta_\zeta}{\sum_{\zeta=\tau}^{10}\theta_\zeta}(w_{\tau+1}^{|\tau|i}\delta + w_{\tau+2}^{|\tau|i}\delta^2)\right]^2, \quad \text{for } \tau \in \{1,2,\ldots,8\},$$

$$\bar{u}_\tau^{(4)i**} = \frac{1}{2\varphi_\tau^i}\left[p_\tau^{|\tau|i} + \frac{\theta_\tau}{\sum_{\zeta=\tau}^{10}\theta_\zeta}\bar{p}_{\tau+1}^{|\tau|i}\delta + \frac{\sum_{\zeta=\tau+1}^{10}\theta_\zeta}{\sum_{\zeta=\tau}^{10}\theta_\zeta}(p_{\tau+1}^{|\tau|i}\delta + p_{\tau+2}^{|\tau|i}\delta^2 + p_{\tau+3}^{|\tau|i}\delta^2)\right],$$

$$\text{for } \tau \in \{1,2,\ldots,10\} \text{ with } p_{11}^{|\tau|i} = \bar{p}_{11}^{|\tau|i},$$
$$\text{for } \tau \in \{8,9,10\}, \quad \text{and } p_t^{|\tau|i} = 0, \quad \text{for } t > 11. \tag{5.8}$$

With the values of $A_\tau^{(i)i}$, for $\tau \in \{1,2,\ldots,10\}$ and $i \in \{1,2\}$, the game equilibrium capital investment strategies $u_\tau^{(3)i**}$ and $u_\tau^{(3)j**}$ in (5.8) can be obtained explicitly. Substituting the game equilibrium capital investment strategies into the accumulation processes of the private capital stock of agents in (5.1) yields

$$x_{k+1}^i = x_k^i + \varepsilon_k^i\frac{A_{\tau+1}^{(i)i}\varepsilon_\tau^i\delta}{2c_\tau^{Ii}} - \lambda_k^i x_k^i + \gamma_k^{(i)j}x_k^j,$$
$$x_1^i = x_1^{i(0)}, \quad \text{for } i \in \{1,2\} \quad \text{and } j \neq i. \tag{5.9}$$

Equation (5.9) is a system of linear difference equations whose solution can be solved readily with standard techniques. We use $\{x_t^{1**}, x_t^{2**}\}_{t=1}^{T+1}$ to denote the game equilibrium trajectory of private capitals.

Novel results in dynamic game analysis are identified in the following remarks.

Remark 5.1. The terminal values of the agents' previously executed durable strategies may be low or equal to zeros in premature termination of the game. For instance, the termination return of the income-generating investment $(\bar{p}_{\tau+1}^{|t|i})$ may be substantially below the regular return $(p_{\tau+1}^{|t|i})$ or even zero. Loss of the values of previously executed durable strategies in a premature termination is a new outcome in dynamic game analysis.

Remark 5.2. In formulating durable investment strategies, the agents have to take into consideration the durable effects and the random termination effects.

In particular, in determining the durable income-earning investment $\bar{u}_\tau^{(4)i**}$ in (5.8), the agent i has to consider

(i) the probability that the game would terminate in the next stage, that is, $\theta_\tau / \sum_{\zeta=\tau}^{10} \theta_\zeta$, and the probability that the game would continue onward, that is, $\sum_{\zeta=\tau+1}^{10} \theta_\zeta / \sum_{\zeta=\tau}^{10} \theta_\zeta$ and
(ii) the future return of the investment if the game would terminate in the next stage, that is, $\bar{p}_{\tau+1}^{|\tau|i}\delta$, and the future returns of the investment if the game would continue onward, that is, $(\sum_{t=\tau}^{\tau+2} p_{t+1}^{|\tau|i} \delta^{t+1-\tau})$.

Taking into consideration of the durable effects and the random termination effects in formulating strategies is a new outcome in dynamic game analysis.

Remark 5.3. The presence of random horizon has put a significant discounting effect on the benefits of durable strategies. Consider agent i's revenue generating investment

$$\bar{u}_{\tau}^{(4)i**} = \frac{1}{2\varphi_{\tau}^{i}} \left[p_{\tau}^{|\tau|i} + \frac{\theta_{\tau}}{\sum_{\zeta=\tau}^{10} \theta_{\zeta}} \bar{p}_{\tau+1}^{|\tau|i} \delta + \frac{\sum_{\zeta=\tau+1}^{10} \theta_{\zeta}}{\sum_{\zeta=\tau}^{10} \theta_{\zeta}} \right.$$

$$\left. \times \left(p_{\tau+1}^{|\tau|i} \delta + p_{\tau+2}^{|\tau|i} \delta^{2} + p_{\tau+3}^{|\tau|i} \delta^{2} \right) \right].$$

If the terminal payoff $\bar{p}_{\tau+1}^{|\tau|i}$ in stage $\tau+1$ is negligible, the benefit of agent i's revenue-generating investment is discounted by the factor $\sum_{\zeta=\tau+1}^{10} \theta_{\zeta} / \sum_{\zeta=\tau}^{10} \theta_{\zeta}$.

Moreover, there may exist cases where there are no redemption values for previously executed durable capital investment strategies when operations stop prematurely. This brings about the situation where durable investments are subject to the risk of early termination.

Remark 5.4. The analysis establishes a new dynamic theory of the agent $V^{i}(\tau, x; \underline{\bar{u}}_{\tau-}^{(3)**}, \underline{\bar{u}}_{\tau-}^{(4)**})$ which shows that the agent's future investment returns, debt obligations, and private capital are determinants of the value of the agent. In addition, the stochastic elements affecting the lifespan of the agent are fully captured in the value of the agent.

The outcomes identified in Remarks 5.1, 5.2, 5.3, and 5.4 are outcomes that are novel in dynamic game analysis. However, these outcomes are not uncommon in real-life operations. Random horizon durable strategies dynamic games indeed enrich game theory in analyzing decision-making.

6. A Computational Illustration: Part II

In this section, we consider the case where the agents in Section 5 form a joint venture and present a random horizon cooperative durable strategies dynamic game of investments to demonstrate the computation of the cooperative solution.

6.1. *Game formulation and group payoff maximization*

Consider the game (5.1)–(5.2). Now, the agents agree to cooperate, maximize their expected joint profit, and adopt an optimality principle that allows the agents' cooperative payoffs proportional to the sizes of their non-cooperative payoffs. To maximize the expected joint payoff, the agents have to solve the discrete-time dynamic programming problem of maximizing

$$\sum_{i=1}^{2}\left\{\sum_{\hat{T}=1}^{10}\theta_{\hat{T}}\left\{\sum_{k=1}^{\hat{T}}\left[R_k^i(u_k^{IIi}x_k^i)^{\frac{1}{2}} - c_k^{IIi}u_k^{IIi} - c_k^{Ii}(u_k^{Ii})^2 - \varphi_k^i(\bar{u}_k^{(4)i})^2\right.\right.\right.$$
$$\left. + \sum_{t=k-3}^{k} p_k^{|t|i}\bar{u}_t^{(4)i} + \chi_k^i(\bar{u}_k^{(3)i})^{\frac{1}{2}} - \sum_{t=k-2}^{k} w_k^{|t|i}\bar{u}_\tau^{(3)i}\right]\delta^{k-1}$$
$$\left.\left. + \left(Q_{\hat{T}+1}^{(i)i}x^i + \sum_{t=\hat{T}-2}^{\hat{T}} \bar{p}_{\hat{T}+1}^{|t|i}\bar{u}_t^{(4)i} - \sum_{t=\hat{T}-1}^{\hat{T}} \bar{w}_{\hat{T}+1}^{|t|i}\bar{u}_t^{(3)i} + \varpi_{\hat{T}+1}^i\right)\delta^{\hat{T}}\right\}\right\}, \tag{6.1}$$

subject to

$$\begin{aligned} x_{k+1}^i &= x_k^i + \varepsilon_k^i u_k^{Ii} - \lambda_k^i x_k^i + \gamma_k^{(i)j} x_k^j, \\ x_1^i &= x_1^{i(0)}, \ i, j \in \{1,2\} \quad \text{and } i \neq j. \end{aligned} \tag{6.2}$$

Invoking Theorem 5.1, the expected joint payoff of the cooperative game (6.1)–(6.2) can be characterized as follows.

Corollary 6.1. *Let the value function $W(\tau, x; \underline{\bar{u}}_{\tau-}^{(3)}, \underline{\bar{u}}_{\tau-}^{(4)})$ denote the maximal value of the expected joint payoffs of the game (6.1)–(6.2) starting at stage $\tau \in \{1, 2, \ldots, 10\}$ with state $x_\tau = x$ and previously executed controls $(\underline{\bar{u}}_{\tau-}^{(3)}, \underline{\bar{u}}_{\tau-}^{(4)})$. Then the function $W(\tau, x; \underline{\bar{u}}_{\tau-}^{(3)}, \underline{\bar{u}}_{\tau-}^{(4)})$ satisfies the following system of recursive equations:*

$$W(11, x; \underline{\bar{u}}_{11-}^{(3)}, \underline{\bar{u}}_{11-}^{(4)})$$
$$= \sum_{i=1}^{2}\left[Q_{11}^{(i)i}x_{11}^i + \sum_{t=11-3}^{10} \bar{p}_{11}^{|t|i}\bar{u}_t^{(4)i**} - \sum_{t=11-2}^{10} \bar{w}_{11}^{|t|i}\bar{u}_t^{(3)i**} + \varpi_{11}^i\right]\delta^{10}, \tag{6.3}$$

$$
\begin{aligned}
&W(10,x;\underline{\bar{u}}^{(3)}_{10-},\underline{\bar{u}}^{(4)}_{10-})\\
&\quad=\max_{\substack{u^{Ii}_{\tau},u^{IIi}_{10},\bar{u}^{(3)i}_{10}\\ \bar{u}^{(4)i}_{10},i\in\{1,2\}}}\left\{\sum_{i=1}^{2}\left\{\left[R^i_{10}(u^{IIi}_{10}x^i)^{\frac{1}{2}}-c^{IIi}_{10}u^{IIi}_{10}-c^{Ii}_{10}(u^{Ii}_{10})^2\right.\right.\right.\\
&\qquad-\varphi^i_{10}(\bar{u}^{(4)i}_{10})^2+p^{|10|i}_{10}\bar{u}^{(4)i}_{10}+\sum_{t=10-3}^{9}p^{|t|i}_{10}\bar{u}^{(4)i}_t+\chi^i_k(\bar{u}^{(3)i}_{10})^{\frac{1}{2}}\\
&\qquad\left.-w^{|t|i}_{10}\bar{u}^{(3)i}_{10}-\sum_{t=10-2}^{9}w^{|t|i}_{10}\bar{u}^{(3)i}_t\right]\delta^{k-1}+\left[Q^{(i)i}_{11}(x^i+\varepsilon^i_{10}u^{Ii}_{10}\right.\\
&\qquad-\lambda^i_{10}x^i+\gamma^{(i)j}_{10}x^j)+\bar{p}^{|10|i}_{11}\bar{u}^{(4)i}_{10}+\sum_{t=11-3}^{9}\bar{p}^{|t|i}_{11}\bar{u}^{(4)i}_t+\bar{w}^{|10|i}_{11}\bar{u}^{(4)i}_{10}\\
&\qquad\left.\left.\left.+\sum_{t=11-2}^{9}w^{|t|i}_{11}\bar{u}^{(4)i}_t+\varpi^i_{11}\right]\delta^{10}\right\}\right\},
\end{aligned}
\tag{6.4}
$$

$$
\begin{aligned}
&W(\tau,x;\underline{\bar{u}}^{(3)}_{\tau-},\underline{\bar{u}}^{(4)}_{\tau-})\\
&\quad=\max_{\substack{u^{Ii}_{\tau},u^{IIi}_{10},\bar{u}^{(3)i}_{10}\\ \bar{u}^{(4)i}_{10},i\in\{1,2\}}}\left\{\sum_{i=1}^{2}\left\{\left[R^i_{\tau}(u^{IIi}_{\tau}x^i)^{\frac{1}{2}}-c^{IIi}_{\tau}u^{IIi}_{\tau}-c^{Ii}_{\tau}(u^{Ii}_{\tau})^2\right.\right.\right.\\
&\qquad-\varphi^i_{\tau}(\bar{u}^{(4)i}_{\tau})^2+p^{|\tau|i}_{\tau}\bar{u}^{(4)i}_{\tau}+\sum_{t=\tau-3}^{\tau}p^{|t|i}_{\tau}u^{(4)i}_t+\chi^i_{\tau}(\bar{u}^{(4)i}_{\tau})^{\frac{1}{2}}\\
&\qquad\left.-w^{|\tau|i}_{\tau}\bar{u}^{(3)i}_{\tau}-\sum_{t=\tau-2}^{\tau}w^{|t|i}_{\tau}\bar{u}^{(3)i}_t\right]\delta^{\tau-1}\\
&\qquad+\frac{\theta_\tau}{\sum_{\zeta=\tau}^{T}\theta_\zeta}\left[Q^{(i)i}_{\tau+1}x^i_{\tau+1}+\bar{p}^{(\tau)i}_{\tau+1}\bar{u}^{(4)i}_{\tau}\right.\\
&\qquad+\sum_{t=\tau+1-3}^{\tau-1}\bar{p}^{(t)i}_{\tau+1}\bar{u}^{(4)i}_t-\bar{w}^{(\tau)i}_{\tau+1}\bar{u}^{(3)i}_{\tau}
\end{aligned}
$$

$$+ \sum_{t=\tau+1-2}^{\tau-1} \bar{w}_{\tau+1}^{(t)i} \bar{u}_t^{(3)i} + \varpi_{\tau+1}^i \Bigg] \delta^\tau + \frac{\sum_{\zeta=\tau+1}^T \theta_\zeta}{\sum_{\zeta=\tau}^T \theta_\zeta} W \Bigg[\tau+1, x_{\tau+1};$$

$$u_\tau^{(3)}, \underline{\bar{u}}_{(\tau+1)-}^{(3)} \cap \underline{\bar{u}}_{\tau-}^{(3)}, u_\tau^{(4)}, \underline{\bar{u}}_{(\tau+1)-}^{(4)} \cap \underline{\bar{u}}_{\tau-}^{(4)}] \Bigg\} \Bigg\},$$

$$\text{for } \tau \in \{1, 2, \ldots, 9\}, \tag{6.5}$$

where $x_{\tau+1}$ *is a vector containing two elements*

$$x^1 + \varepsilon_\tau^1 u_\tau^{I1} - \lambda_\tau^1 x^1 + \gamma_\tau^{(1)2} x^2 \quad \text{and}$$

$$x^2 + \varepsilon_\tau^2 u_\tau^{I2} - \lambda_\tau^2 x^2 + \gamma_\tau^{(2)1} x^1.$$

Performing the indicated maximization operator in (6.4)–(6.5) of Corollary 6.1, we obtain an optimal cooperative solution of the game (6.1)–(6.2) as follows.

Proposition 6.1. *System* (6.3)–(6.5) *admits an optimal cooperative solution with the expected joint payoff being*

$$W(\tau, x; \underline{\bar{u}}_{\tau-}^{(3)}, \underline{\bar{u}}_{\tau-}^{(4)}) = (A_\tau^1 x^1 + A_\tau^2 x^2 + C_\tau)\delta^{\tau-1},$$

$$\text{for } \tau \in \{1, 2, \ldots, 11\}, \tag{6.6}$$

where

$$A_{11}^1 = Q_{T+1}^{(1)1}, A_{11}^2 = Q_{T+1}^{(2)2},$$

$$C_{11} = \sum_{i=1}^2 \left(\sum_{t=11-3}^{10} \bar{p}_{11}^{(t)i} \bar{u}_t^{(4)i} - \sum_{t=11-2}^{10} \bar{w}_{11}^{(t)i} \bar{u}_t^{(3)i} + \varpi_{11}^i \right);$$

$$A_{10}^1 = \sum_{i=1}^2 \frac{(R_{10}^i)^2}{4c_{10}^{IIi}} + \delta A_{11}^1 (1 - \lambda_{10}^1) + \delta A_{11}^2 \gamma_{10}^{(2)1},$$

$$A_{10}^2 = \sum_{i=1}^2 \frac{(R_{10}^i)^2}{4c_{10}^i} + \delta A_{11}^2 (1 - \lambda_{10}^2) + \delta A_{11}^1 \gamma_{10}^{(1)2};$$

$$A_\tau^1 = \frac{(R_\tau^1)^2}{4c_\tau^{II1}} + \frac{\theta_\tau}{\sum_{\zeta=\tau}^{10} \theta_\zeta} (Q_{\tau+1}^{(1)1}(1 - \lambda_\tau^1) + Q_{\tau+1}^{(2)2} \gamma_\tau^{(2)1})\delta$$

$$+ \frac{\sum_{\zeta=\tau+1}^{10} \theta_\zeta}{\sum_{\zeta=\tau}^{10} \theta_\zeta} (A_{\tau+1}^1 \ (1 - \lambda_\tau^1) + A_{\tau+1}^2 \gamma_\tau^{(2)1})\delta,$$

$$A_\tau^2 = \frac{(R_\tau^2)^2}{4c_\tau^{II2}} + \frac{\theta_\tau}{\sum_{\zeta=\tau}^{10}\theta_\zeta}(Q_{\tau+1}^{(2)2}(1-\lambda_\tau^2) + Q_{\tau+1}^{(1)1}\gamma_\tau^{(1)2})\delta$$
$$+ \frac{\sum_{\zeta=\tau+1}^{10}\theta_\zeta}{\sum_{\zeta=\tau}^{10}\theta_\zeta}(A_{\tau+1}^2(1-\lambda_\tau^2) + A_{\tau+1}^1\gamma_\tau^{(1)2})\delta,$$
$$\text{for } \tau \in \{1,2,\ldots,9\}. \tag{6.7}$$

In addition, C_τ is an expression containing previously executed controls $(\underline{\bar{u}}_{\tau-}^{(3)}, \underline{\bar{u}}_{\tau-}^{(4)})$,$\tau \in \{1,2,\ldots,10\}$.

Proof. See Appendix F in the Mathematical Appendices. □

The values of A_τ^1 and A_τ^2, for $\tau \in \{1,2,\ldots,10\}$, in Proposition 6.1 can be solved backwardly from $\tau = 10$ to $\tau = 1$. In addition, using Corollary 6.1 and Proposition 6.1, the optimal cooperative strategies can be obtained as follows (see derivation details in Appendix D):

$$u_\tau^{(1)i*} = \frac{(R_\tau^i)^2}{4(c_\tau^i)}x^i, \quad \text{for } i \in \{1,2\} \quad \text{and } \tau \in \{1,2,\ldots,10\};$$

$$u_{10}^{(3)i*} = \frac{A_{11}^i\varepsilon_{10}^{(10)i} + \hat{v}_{11}^{(10)i}}{2w_{10}^i}\delta,$$

$$u_9^{(3)i*} = \frac{1}{2w_9^i}\left[\frac{\theta_9}{\sum_{\zeta=9}^{10}\theta_\zeta}\left(Q_{10}^{(i)i}\varepsilon_9^{(9)i}\delta + \hat{v}_{10}^{(9)i}\delta\right)\right.$$
$$\left.+\frac{\sum_{\zeta=10}^{10}\theta_\zeta}{\sum_{\zeta=9}^{10}\theta_\zeta}(A_{10}^{(i)i}\varepsilon_9^{(9)i}\delta + A_{11}^i\varepsilon_{10}^{(9)i}\delta^2 + \hat{v}_{11}^{(9)i}\delta^2)\right],$$

$$u_8^{(3)i*} = \frac{1}{2w_8^i}\left[\frac{\theta_8}{\sum_{\zeta=8}^{10}\theta_\zeta}(Q_9^{(i)i}\varepsilon_8^{(8)i}\delta + \hat{v}_9^{(8)i}\delta)\right.$$
$$\left.+\frac{\sum_{\zeta=9}^{10}\theta_\zeta}{\sum_{\zeta=8}^{10}\theta_\zeta}(A_9^{(i)i}\varepsilon_8^{(8)i}\delta + A_{10}^i\varepsilon_9^{(8)i}\delta^2 + A_{11}^i\varepsilon_{10}^{(8)i}\delta^3 + \hat{v}_{11}^{(8)i}\delta^3)\right],$$

$$u_\tau^{(3)i*} = \frac{1}{2w_\tau^i}\left[\frac{\theta_\tau}{\sum_{\zeta=\tau}^{10}\theta_\zeta}(Q_{\tau+1}^{(i)i}\varepsilon_\tau^{(\tau)i}\delta + \hat{v}_{\tau+1}^{(\tau)i}\delta)\right.$$
$$\left.+\frac{\sum_{\zeta=\tau+1}^{10}\theta_\zeta}{\sum_{\zeta=\tau}^{10}\theta_\zeta}\left(\sum_{t=\tau}^{\tau+2}A_{t+1}^i\varepsilon_t^{(\tau)i}\delta^{t+1-\tau}\right)\right],$$
$$\text{for } i \in \{1,2\} \quad \text{and } \tau \in \{1,2,\ldots,7\},$$

$$u_{10}^{(4)i*} = \frac{p_{10}^{(10)i} + \delta\bar{p}_{11}^{(10)i}}{2\phi_{10}^i},$$

$$u_9^{(4)i*} = \frac{1}{2\varphi_9^i}\left[p_9^{(9)i} + \frac{\theta_9}{\sum_{\zeta=9}^{10}\theta_\zeta}\bar{p}_{10}^{(9)i}\delta + \frac{\sum_{\zeta=10}^{10}\theta_\zeta}{\sum_{\zeta=9}^{10}\theta_\zeta}(p_{10}^{(9)i}\delta + \bar{p}_{11}^{(9)i}\delta^2)\right],$$

$$u_8^{(4)i*} = \frac{1}{2\varphi_8^i}\left[p_8^{(8)i} + \frac{\theta_8}{\sum_{\zeta=8}^{10}\theta_\zeta}\bar{p}_9^{(8)i}\delta + \frac{\sum_{\zeta=9}^{10}\theta_\zeta}{\sum_{\zeta=8}^{10}\theta_\zeta}\right.$$
$$\left.\times (p_9^{(8)i}\delta + \bar{p}_{10}^{(8)i}\delta^2 + \bar{p}_{11}^{(8)i}\delta^3)\right],$$

$$u_\tau^{(4)i*} = \frac{1}{2\varphi_\tau^i}\left[p_\tau^{(\tau)i} + \frac{\theta_\tau}{\sum_{\zeta=\tau}^{10}\theta_\zeta}\bar{p}_{\tau+1}^{(\tau)i}\delta\right.$$
$$\left.+\frac{\sum_{\zeta=\tau+1}^{10}\theta_\zeta}{\sum_{\zeta=\tau}^{10}\theta_\zeta}\left(\sum_{t=\tau}^{\tau+2}p_{t+1}^{(\tau)i}\delta^{t+1-\tau}\right)\right],$$
$$\text{for } i \in \{1,2\} \quad \text{and} \quad \tau \in \{1,2,\ldots,7\}. \tag{6.8}$$

Comparing Proposition 6.1 with Proposition 5.1, one can obtain that $A_\tau^1 > A_\tau^{(1)1}$ and $A_\tau^2 > A_\tau^{(2)2}$, for $\tau \in \{1,2,\ldots,11\}$. Therefore, private capital investments under cooperation in (6.8) are greater than private capital investments under non-cooperation in (5.8). Substituting the optimal cooperative capital investment strategies into the accumulation processes of the private capital stock of agents in (6.2) yields

$$x_{k+1}^1 = x_k^1 + \varepsilon_k^{(k)1}u_k^{(3)1*} + \sum_{t=k-2}^{k-1}\varepsilon_k^{(t)1}u_t^{(3)1*}$$
$$-\lambda_k^1 x_k^1 + \gamma_k^{(1)2}x_k^2, \quad x_1^1 = x_1^{1(0)},$$

$$x_{k+1}^2 = x_k^2 + \varepsilon_k^{(k)2} u_k^{(3)2*} + \sum_{t=k-2}^{k-1} \varepsilon_k^{(t)2} u_t^{(3)2*}$$
$$- \lambda_k^2 x_k^2 + \gamma_k^{(2)1} x_k^1, x_1^2 = x_1^{2(0)}. \quad (6.9)$$

In initial stage 1, $u_{\tau-}^{(3)1}$ and $u_{\tau-}^{(3)2}$ are zeros. Given $x_1^1 = x_1^{1(0)}$ and $x_1^2 = x_1^{2(0)}$, the values of x_2^1 and x_2^2 can be obtained. With x_2^1 and x_2^2, the values of x_3^1 and x_3^2 can be obtained. Performing the analysis repeatedly, the values of x_t^1 and x_t^2 for $t \in \{4, 5, \ldots, 11\}$ can be obtained. We use $\{x_t^{1*}, x_t^{2*}\}_{t=1}^{T+1}$ to denote the optimal cooperative trajectory of private capitals. With private capital investments under cooperation being greater than private capital investments under non-cooperation, we have $x_t^{1*} > x_t^{1**}$ and $x_t^{2*} > x_t^{2**}$, and outputs produced using private capitals will be higher under cooperation.

6.2. *Subgame-consistent payoff distribution*

Let $\xi(\tau, x; \underline{u}_{\tau-}^{(3)}, \underline{u}_{\tau-}^{(4)}) = [\xi^1(\tau, x; \underline{u}_{\tau-}^{(3)}, \underline{u}_{\tau-}^{(4)}), \xi^2(\tau, x; \underline{u}_{\tau-}^{(3)}, \underline{u}_{\tau-}^{(4)})]$ denote the agreed-upon imputation vector guiding the distribution of the total cooperative payoff in stage $\tau \in \{1, 2, \ldots, 10\}$. Since the agents agree to distribute the cooperative payoff proportional to their non-cooperative payoffs, the imputation to agent i is

$$\xi^i(\tau, x^*; \underline{u}_{\tau-}^{(3)*}, \underline{u}_{\tau-}^{(4)*})$$
$$= \frac{V^i(\tau, x^*; \underline{u}_{\tau-}^{(3)*}, \underline{u}_{\tau-}^{(4)*})}{\sum_{\ell=1}^2 V^\ell(\tau, x^*; \underline{u}_{\tau-}^{(3)*}, \underline{u}_{\tau-}^{(4)*})} W(\tau, x^*; \underline{u}_{\tau-}^{(3)*}, \underline{u}_{\tau-}^{(4)*})$$
$$= \frac{A_\tau^{(i)i} x_\tau^{i*} + A_\tau^{(i)j} x_\tau^{j*} + C_\tau^i}{\sum_{\ell=1}^2 (A_\tau^{(\ell)\ell} x_\tau^{\ell*} + A_\tau^{(\ell)j} x_\tau^{j*} + C_\tau^\ell)} (A_\tau^1 x_\tau^{1*} + A_\tau^2 x_\tau^{2*} + C_\tau)\delta^{\tau-1}, \quad (6.10)$$

for $i \in \{1, 2\}$ at stage $\tau \in \{1, 2, \ldots, 10\}$.

Invoking Theorem 5.2, a payment equaling

$$\beta_\tau^i(x_\tau^*; \underline{u}_{\tau-}^{(3)*}, \underline{u}_{\tau-}^{(4)*})$$
$$= \delta^{1-\tau} \left[\xi^i(\tau, x_\tau^*; \underline{u}_{\tau-}^{(3)*}, \underline{u}_{\tau-}^{(4)*}) - \frac{\theta_\tau}{\sum_{\zeta=\tau}^T \theta_\zeta} \right.$$

$$
\begin{aligned}
&\times q^i_{\tau+1}(x^*_{\tau+1}; \underline{u}^{(3)*}_{(\tau+1)-}, \underline{u}^{(4)*}_{(\tau+1)-})\delta^\tau \\
&\left. - \frac{\sum_{\zeta=\tau+1}^T \theta_\zeta}{\sum_{\zeta=\tau}^T \theta_\zeta} \xi^i(\tau+1, x^*_{\tau+1}; \underline{u}^{(3)*}_{(\tau+1)-}, \underline{u}^{(4)*}_{(\tau+1)-}) \right] \\
= {} & \frac{A^{(i)i}_\tau x^{i*}_\tau + A^{(i)j}_\tau x^{j*}_\tau + C^i_\tau}{\sum_{\ell=1}^2 (A^{(\ell)\ell}_\tau x^{\ell*}_\tau + A^{(\ell)j}_\tau x^{j*}_\tau + C^\ell_\tau)} (A^1_\tau x^{1*}_\tau + A^2_\tau x^{2*}_\tau + C_\tau) \\
&- \frac{\theta_\tau}{\sum_{\zeta=\tau}^T \theta_\zeta} \left[Q^{(i)i}_{\tau+1} x^{i*}_{\tau+1} + \bar{v}^{(\tau)i}_{\tau+1} u^{(3)i*}_\tau \right. \\
&\qquad + \sum_{t=\tau-2}^{\tau-1} \bar{v}^{(t)i}_{\tau+1} u^{(3)i*}_t + \bar{p}^{(\tau)i}_{\tau+1} u^{(4)i*}_\tau \\
&\qquad \left. + \sum_{t=\tau+1-3}^{\tau-1} \bar{p}^{(t)i}_{\tau+1} u^{(4)i*}_t + \varpi^i_{\tau+1} \right] \delta - \frac{\sum_{\zeta=\tau+1}^T \theta_\zeta}{\sum_{\zeta=\tau}^T \theta_\zeta} \\
&\times \left(\frac{A^{(i)i}_{\tau+1} x^{i*}_{\tau+1} + A^{(i)j}_{\tau+1} x^{j*}_{\tau+1} + C^i_{\tau+1}}{\sum_{l=1}^2 (A^{(l)l}_{\tau+1} x^{l*}_{\tau+1} + A^{(l)j}_{\tau+1} x^{j*}_{\tau+1} + C^l_{\tau+1})} \right. \\
&\qquad \left. \times (A^1_{\tau+1} x^{1*}_{\tau+1} + A^2_{\tau+1} x^{2*}_{\tau+1} + C_{\tau+1})\delta \right) \qquad (6.11)
\end{aligned}
$$

given to agent $i \in \{1, 2\}$ at stage $\tau \in \{1, 2, \ldots, 10\}$ would lead to the realization of the imputation vector $\xi(\tau, x^*; \underline{u}^{(3)*}_{\tau-}, \underline{u}^{(4)*}_{\tau-})$ in (6.10).

Note that previously executed durable strategies $(\underline{u}^{(3)*}_{\tau-}, \underline{u}^{(4)*}_{\tau-})$ are embedded in $\beta^i_\tau(x^*_\tau; \underline{u}^{(3)*}_{\tau-}, \underline{u}^{(4)*}_{\tau-})$. Under cooperation, all agents would use the cooperative strategies and the single-stage payment that agent i will directly receive at stage τ along the cooperative trajectory $\{x^*_\tau\}_{\tau=1}^{T}$ with previously executed durable strategies $(\underline{u}^{(3)*}_{\tau-}, \underline{u}^{(4)*}_{\tau-})$ become

$$
\begin{aligned}
&\pi^i_\tau(x^*_\tau; \underline{u}^{(3)*}_{\tau-}, \underline{u}^{(4)*}_{\tau-}) \\
&\quad = R^i_\tau(u^{(1)i*}_\tau x^{i*}_\tau)^{\frac{1}{2}} - c^i_\tau u^{(1)i*}_\tau - w^i_\tau(u^{(3)i*}_\tau)^2 - \phi^i_\tau(u^{(4)i*}_\tau)^2 \\
&\qquad + p^{(k)i}_k u^{(4)i}_k + \sum_{t=\tau-3}^{\tau-1} p^{(t)i}_\tau u^{(4)i*}_t.
\end{aligned}
$$

However, according to the agreed upon imputation, agent i will receive $\beta_\tau^i(x_\tau^*; \underline{u}_{\tau-}^{(3)*}, \underline{u}_{\tau-}^{(4)*})$ at stage τ. Therefore, a side payment equaling

$$\beta_\tau^i(x_\tau^*; \underline{u}_{\tau-}^{(3)*}, \underline{u}_{\tau-}^{(4)*}) - \pi_\tau^i(x_\tau^*; \underline{u}_{\tau-}^{(3)*}, \underline{u}_{\tau-}^{(4)*}) \tag{6.12}$$

has to be given to agent $i \in \{1,2\}$ in stage $\tau \in \{1,2,\ldots,10\}$ to yield the cooperative imputation $\xi(\tau, x^*; \underline{u}_{\tau-}^{(3)*}, \underline{u}_{\tau-}^{(4)*})$ in (6.10).

A couple of novel results in cooperative dynamic game analysis are identified in the following remarks.

Remark 6.1. The positive spillover effect from agent i's capital investment to agent j's private capital stock could be substantial and cooperation would enhance the effects of the agents' durable capital investments to the expected joint payoff. In formulating the durable capital investment strategies $u_\tau^{(3)i}$ in (6.8), the agents have to consider

(i) the potential cooperative gains

$$\left(\sum_{t=\tau}^{\tau+2} A_{t+1}^i \varepsilon_t^{(\tau)i} \delta^{t+1-\tau}\right),$$

(ii) the probability that the game would terminate in the next stage, that is, $\theta_\tau / \sum_{\zeta=\tau}^{10} \theta_\zeta$, and the probability that the game would continue onward, that is, $\sum_{\zeta=\tau+1}^{10} \theta_\zeta / \sum_{\zeta=\tau}^{10} \theta_\zeta$, and

(iii) the future return of capital investments if the game would terminate in the next stage, that is, $\sum_{i=1}^{2} (Q_{\tau+1}^{(i)i} \varepsilon_\tau^{(\tau)i} \delta + \bar{v}_{\tau+1}^{(\tau)i} \delta)$.

Remark 6.2. Under cooperation, the maximization of the expected joint payoff would take into consideration the terminal payoffs of the agents jointly if the game stops prematurely after stage τ, that is,

$$\left(Q_{\tau+1}^{(i)i} x_{\tau+1}^i + \bar{v}_{\tau+1}^{(\tau)i} u_\tau^{(3)i} + \sum_{t=\tau-2}^{\tau-1} \bar{v}_{\tau+1}^{(t)i} u_t^{(3)i} + \bar{p}_{\tau+1}^{(\tau)i} u_\tau^{(4)i}\right.$$

$$\left. + \sum_{t=\tau+1-3}^{\tau-1} \bar{p}_{\tau+1}^{(t)i} u_t^{(4)i} + \varpi_{\tau+1}^i\right),$$

for $i \in \{1,2\}$ and $\tau \in \{1,2,\ldots,9\}$.

The outcomes identified in Remarks 6.1 and 6.2 reflect outcomes which are novel in cooperative dynamic game analysis.

7. Chapter Notes

Durable strategies and random horizon are frequently observed in real-life practical decision involving strategic interactions. Durable strategies are controls that have influences lasting over a certain period of time. Revenue generating investments, toxic waste disposal, durable goods, business contracts, taxes, regulations, coalition agreements, diffusion of knowledge, advertising, and investments to build up physical capital are vivid examples of durable strategies. In many cases, both the decision-maker's payoff and the evolution of the state dynamics can be subjected to effects arising from durable strategies. For instance, durable goods and revenue-generating investments affect the decision-maker's payoffs for more than one stage. Tariffs have delayed impacts on the economy's growth dynamics. Conversion of investments into physical capital stock involves a certain number of stages, and the corresponding costs may be paid by installment over a certain period of time. Advertising is well known for having promotion lags under which reputation takes time to establish (see Mela *et al.*, 1997; Tellis, 2006; Ruffino, 2008). Sarkar and Zhang (2015) and Agliardi and Koussisc (2013) showed the time-to-build characteristics in investment and capital formation. Tsuruga and Shota (2019) and Sarkar and Zhang (2013) studied implementation lags which generate lagged effects on fiscal policy and investment decisions. Rigidities is another source of delay in policy implementation (see Christiano *et al.*, 2015; Shibata *et al.*, 2019). The existence of at least some durable strategies in the decision-maker's strategy set is a rule rather than an exception in practical analysis.

In addition, the terminal time of many decision problems is not known with certainty. Commonly observed examples of horizontal uncertainties include the terms of offices of elected authorities, contract renewal, continuation of agreements subjected to periodic negotiations, and the life span of business firms. A classic case is that business firms are normally hypothesized to have an indefinitely long life expectancy, but in reality, the average life span of today's multinational Fortune 500-size corporation is less than a hundred years.

Petrosyan and Murzov (1966) first developed the Isaacs–Bellman equations under random horizon for zero-sum differential games. Petrosyan and Shevkoplyas (2003) gave the first analysis of dynamically consistent solutions for cooperative differential with random duration. Shevkoplyas (2011) considered Shapley value in cooperative differential games with random horizon. Yeung and Petrosyan (2011) first developed discrete-time dynamic games with random horizons. Yeung and Petrosyan (2012) presented subgame-consistent solution for cooperative stochastic dynamic games with random horizons. Yeung and Petrosyan (2014) studied subgame-consistent cooperative solutions for randomly furcating stochastic dynamic games with uncertain horizons.

The reality of random horizon on top of durable strategies posts an interesting challenge to dynamic game theory analysis. In the presence of durable strategies and uncertainty in planning horizon, significant modification of the dynamic optimization techniques is required to accommodate these phenomena. Most of the existing works on lagged control optimization involved lag in the state dynamics (see Bellman, 1957; Arthur, 1977; Burdet and Sethi, 1976; Hartl and Sethi, 1984; Brandt-Pollmann *et al.*, 2008; Huschto *et al.*, 2011; Sethi and Mcguire, 1977; Winkler *et al.*, 2003; Yeung and Petrosyan, 2019). Bokov (2011) considered continuous-time optimal control with control lags in the payoff and time lags in the state dynamics. Petrosyan and Yeung (2020) considered discrete-time dynamic optimization with durable controls affecting both the payoffs and dynamics. Dynamic optimization techniques with random horizon in continuous time can be found in Shevkoplyas (2011). Dynamic optimization techniques with random horizon in discrete time were developed by Yeung and Petrosyan (2011, 2012, 2014).

Finally, the analysis in this chapter provides a comprehensive theory with solution techniques to a general class of dynamic games with random horizons and durable strategies. It can be reduced to the Petrosyan and Yeung (2020) dynamic games with durable control, the Yeung and Petrosyan (2011) random horizon dynamic games, and standard dynamic games.

References

Agliardi, E. & Koussisc, N. (2013). Optimal capital structure and the impact of time-to-build. *Finance Research Letters*, 10, 124–130.

Arthur, W. B. (1977). Control of linear processes with distributed lags using dynamic programming from first principles. *Journal of Optimization Theory and Applications*, 23, 429–443.

Bellman, R. (1957). Terminal control, time lags, and dynamic programming. *Proceedings of the National Academy of Sciences of the United States of America*, 43, 927–930.

Bokov, G. V. (2011). Pontryagin's maximum principle of optimal control problems with time-delay. *Journal of Mathematical Sciences*, 172, 623–634.

Brandt-Pollmann, U., Winkler, R., Sager, S., Moslener, U., & Schlöder, J. (2008). Numerical solution of optimal control problems with constant control delays. *Computational Economics*, 31, 181–206.

Burdet, C. A. & Sethi, S. P. (1976). On the maximum principle for a class of discrete dynamical systems with lags. *Journal of Optimization Theory and Applications*, 19, 445–454.

Christiano, L. J., Eichenbaum, M., & Evans, C. L. (2005). Nominal rigidities and the dynamic effects of a shock to monetary policy. *Journal of Political Economy*, 113, 1–45.

Hartl, R. F. & Sethi, S. P. (1984). Optimal control of a class of systems with continuous lags: Dynamic programming approach and economic interpretations. *Journal of Optimization Theory and Applications*, 43(1), 73–88.

Huschto, T., Feichtinger, G., Hartl, R. F., Kort, P. M., & Sager, S. (2011). Numerical solution of a conspicuous consumption model with constant control delay. *Automatica*, 47, 1868–1877.

Mela, C. F., Gupta, S., & Lehmann, D. R. (1997). The long-term impact of promotion and advertising on consumer brand choice. *Journal of Marketing Research*, 34, 248–261.

Petrosyan, L. A. & Murzov, N. V. (1966). Game theoretic problems in mechanics. *Litovskiy Matematicheskiy Sbornik*, 6, 423–433.

Petrosyan, L. A. & Shevkoplyas, E. V. (2003). Cooperative solutions for games with random duration. *Game Theory and Applications*, IX, 125–139.

Petrosyan, L. A. & Yeung, D. W. K. (2020). Cooperative dynamic games with durable controls: Theory and application. *Dynamic Games and Applications*. DOI: 10.1007/s13235-019-00336-w.

Ruffino, C. C. (2008). Lagged effects of TV advertising on sales of an intermittently advertised product. *Business and Economic Review*, 18, 1–12.

Sarkar, S. & Zhang, C. (2015). Investment policy with time-to-build. *Journal of Banking & Finance*, 55, 142–156.

Sarkar, S. & Zhang, H. (2013). Implementation lag and the investment decision. *Economics Letters*, 119, 136–140.

Sethi, S. P. & Mcguire, T. W. (1977). Optimal skill mix: An application of the maximum principle for systems with retarded controls. *Journal of Optimization Theory and Applications*, 23, 245–275.

Shevkoplyas, E. V. (2011). The Shapley value in cooperative differential games with random duration. *Advances in Dynamic Games*, 11, 359–373.

Shibata, A., Shintani, M., & Tsurugac, T. (2019). Current account dynamics under information rigidity and imperfect capital mobility. *Journal of International Money and Finance*, 92, 153–167.

Tellis, G. J. (2006). Optimal data interval for estimating advertising response. *Marketing Science*, 25, 217–229.

Tsuruga, T. & Shota, W. (2019). Money-financed fiscal stimulus: The effects of implementation lag. *Journal of Economic Dynamics and Control*, 104, 132–151.

Winkler, R., Brandt-Pollmann, U., Moslener, U., & Schlöder, J. (2003). Time-lags in capital accumulation. In Ahr, D., Fahrion, R., Oswald, M., & Reinelt, G. (eds.), *Operations Research Proceedings*, pp. 451–458.

Yeung, D. W. K. & Petrosyan, L. A. (2011). Subgame consistent cooperative solution of dynamic games with random horizon. *Journal of Optimization Theory and Applications*, 150, 78–97.

Yeung, D. W. K. & Petrosyan, L. A. (2012). Subgame consistent solution for cooperative stochastic dynamic games with random horizon. *International Game Theory Review*, 14(2), 1250012.

Yeung, D. W. K. & Petrosyan, L. A. (2014). Subgame consistent cooperative solutions for randomly furcating stochastic dynamic games with uncertain horizon. *International Game Theory Review*, 16(2), 1440012.01–1440012.29.

Yeung, D. W. K. & Petrosyan, L. A. (2019). Cooperative dynamic games with control lags. *Dynamic Games and Applications*, 9, 550–567.

Yeung, D. W. K. & Petrosyan, L. A. (2021). Generalized dynamic games with durable strategies under uncertain planning horizon. *Journal of Computational and Applied Mathematics*, 395, 113595. https://doi.org/10.1016/j.cam.2021.113595.

Chapter 7

Mathematical Solution Mechanisms

"One cannot escape the feeling that these mathematical formulas have an independent existence and an intelligence of their own, that they are wiser than we are, wiser even than their discoverers." — Heinrich Hertz

In this chapter, we present the solution mechanisms for (i) durable-strategies dynamic optimization, (ii) durable-strategies dynamic games, (iii) dynamically consistent solutions and payment mechanism, (iv) random horizon durable-strategies dynamic optimization, (v) random horizon durable-strategies dynamic games, and (vi) subgame-consistent payment mechanism under random termination.

1. Durable-Strategies Dynamic Optimization (Theorem I)

We first consider the optimization technique for durable strategies problems. A durable-strategies dynamic game can be expressed as follows:

$$\max_{\substack{u_k, \bar{u}_k \\ k \in \{1,2,\ldots,T\}}} \left\{ \sum_{k=1}^{T} \sum_{\varsigma=1}^{\Theta} w_k^{(\varsigma)} \sum_{\varsigma=1}^{\Theta} w_k^{(\varsigma)} g_k^{(\varsigma)}(x_k, u_k, \bar{u}_k; \bar{u}_{k-}) \delta_1^k \right.$$
$$\left. + q_{T+1}(x_{T+1};\ \bar{u}_{(T+1)-}) \delta_1^{T+1} \right\} \tag{1.1}$$

subject to

$$x_{k+1} = f(x_k, u_k, \bar{u}_k; \bar{u}_{k-}), \quad x_1 = x_1^0. \tag{1.2}$$

Theorem I. *Let $V(k, x; \bar{u}_{k-})$ be the maximal value of the payoff*

$$\sum_{t=k}^{T}\sum_{\varsigma=1}^{\Theta} w_t^{(\varsigma)} g_t^{(\varsigma)}(x_t, u_t, \bar{u}_k; \bar{u}_{t-})\delta_1^t + q_{T+1}(x_{T+1}; \bar{u}_{(T+1)-})\delta_1^{T+1}$$

for problem (1.1)–(1.2) *starting at period k with state $x_k = x$ and previously executed controls $\bar{u}_{k-}$, then the function $V(k, x; \bar{u}_{k-})$ satisfies the following system of recursive equations:*

$$V(T+1, x; \bar{u}_{(T+1)-}) = q_{T+1}(x_{T+1}; \bar{u}_{(T+1)-})\delta_1^{T+1}, \tag{1.3}$$

$$\begin{aligned} V(k, x; \bar{u}_{k-}) &= \max_{u_k, \bar{u}_k}\left\{\sum_{\varsigma=1}^{\Theta} w_k^{(\varsigma)} g_k^{(\varsigma)}(x, u_k, \bar{u}_k; \bar{u}_{k-})\delta_1^k \right. \\ &\qquad \left. + V[k+1, f_k(x, u_k, \bar{u}_k; \bar{u}_{k-}); \bar{u}_{(k+1)-}]\right\} \\ &= \max_{u_k, \bar{u}_k}\left\{\sum_{\varsigma=1}^{\Theta} w_k^{(\varsigma)} g_k^{(\varsigma)}(x, u_k, \bar{u}_k; \bar{u}_{k-})\delta_1^k \right. \\ &\qquad \left. + V[k+1, f_k(x, u_k, \bar{u}_k; \bar{u}_{k-}); \bar{u}_k, \bar{u}_{(k+1)-} \cap \bar{u}_{k-}]\right\}, \\ &\text{for } k \in \{1, 2, \ldots, T\}. \end{aligned} \tag{1.4}$$

Proof. To prove Theorem I, we adopt the technique of backward induction. Consider first the last operational period T, invoking Theorem I, we have

$$\begin{aligned} V(T, x; \bar{u}_{T-}) = \max_{u_T, \bar{u}_T}&\left\{\sum_{\varsigma=1}^{\Theta} w_T^{(\varsigma)} g_T^{(\varsigma)}(x, u_T, \bar{u}_T; \bar{u}_{T-})\delta_1^T \right. \\ &\left. + q_{T+1}[f_T(x, u_T, \bar{u}_T; \bar{u}_{T-}); \bar{u}_{(T+1)-}]\delta_1^{T+1}\right\} \end{aligned}$$

$$= \max_{u_T, \bar{u}_T} \left\{ \sum_{\varsigma=1}^{\Theta} w_T^{(\varsigma)} g_T^{(\varsigma)}(x, u_T, \bar{u}_T; \bar{u}_{T-}) \delta_1^T \right.$$

$$\left. + q_{T+1}[f_T(x, u_T, \bar{u}_T; \bar{u}_{T-}); \bar{u}_{(T+1)-} \cap \bar{u}_{T-}] \delta_1^{T+1} \right\}. \tag{1.5}$$

The maximization operator in period T involves u_T and $\bar{u}_T$ only. The current state x and the previously executed controls $\bar{u}_{T-}$ appear in the period T maximization problem as given parameters. If the first-order conditions of the maximization problem in (1.5) satisfy the implicit function theorem, one can obtain the optimal controls u_T and $\bar{u}_T$ as functions of x and $\bar{u}_{T-}$. Substituting these optimal controls into the function on the RHS of (1.5) yields the function $V(T, x; \bar{u}_{T-})$, which satisfies the optimal conditions of a maximum for given x and $\bar{u}_{T-}$.

Consider then the second last operational period $T - 1$, using $V(T, x; \bar{u}_{T-})$ derived from (1.5) and invoking Theorem I, we obtain

$$V(T-1, x; \bar{u}_{(T-1)-})$$

$$= \max_{u_{T-1}, \bar{u}_{T-1}} \left\{ \sum_{\zeta=1}^{\Theta} w_{T-1}^{(\varsigma)} g_{T-1}^{(\varsigma)}(x, u_{T-1}, \bar{u}_{T-1}; \bar{u}_{(T-1)-}) \delta_1^{T-1} \right.$$

$$\left. + V[T, f_{T-1}(x, u_{T-1}, \bar{u}_{T-1}; \bar{u}_{(T-1)-}); \bar{u}_{T-}] \right\}$$

$$= \max_{u_{T-1}, \bar{u}_{T-1}} \left\{ \sum_{\zeta=1}^{\Theta} w_{T-1}^{i(\zeta)} g_{T-1}^{i(\zeta)}(x, u_{T-1}, \bar{u}_{T-1}; \bar{u}_{(T-1)-}) \delta_1^{T-1} \right.$$

$$\left. + V\left[T, f_{T-1}(x, u_{T-1}, \bar{u}_{T-1}; \bar{u}_{(T-1)-}); \bar{u}_{T-1}, \bar{u}_{T-} \cap \bar{u}_{(T-1)-}\right] \right\}. \tag{1.6}$$

The maximization operator in period $T - 1$ involves u_{T-1} and $\bar{u}_{T-1}$. The current state x and the previously executed controls

$\bar{u}_{(T-1)-}$ appear in the period $T-1$ maximization problem as given parameters. If the first-order conditions of the maximization problem in (1.6) satisfy the implicit function theorem, one can obtain the optimal controls u_{T-1} and $\bar{u}_{T-1}$ as functions of x and previously determined controls $\bar{u}_{(T-1)-}$. Substituting these optimal controls into the function on the RHS of (1.6) yields the function $V(T-1, x; \bar{u}_{(T-1)-})$.

Now consider period $k \in \{T-2, T-3, \ldots, 2, 1\}$, invoking Theorem I, we have

$$\begin{aligned} V(k, x; \bar{u}_{k-}) &= \max_{u_k, \bar{u}_k} \left\{ \sum_{\varsigma=1}^{\Theta} w_k^{(\varsigma)} g_k^{(\varsigma)}(x, u_k, \bar{u}_k; \bar{u}_{k-}) \delta_1^k \right. \\ &\qquad \left. + V\left[k+1, f_k(x, u_k, \bar{u}_k; \bar{u}_{k-}); \bar{u}_{(k+1)-}\right] \right\} \\ &= \max_{u_k, \bar{u}_k} \left\{ \sum_{\varsigma=1}^{\Theta} w_k^{(\varsigma)} g_k^{(\varsigma)}(x, u_k, \bar{u}_k; \bar{u}_{k-}) \delta_1^k \right. \\ &\qquad \left. + V\left[k+1, f_k(x, u_k, \bar{u}_k; \bar{u}_{k-}); \bar{u}_k, \bar{u}_{(k+1)-} \cap \bar{u}_{k-}\right] \right\}. \end{aligned} \tag{1.7}$$

The maximization operator involves u_k and $\bar{u}_k$. Again, the current state x and the previously executed controls $\bar{u}_{k-}$ appear in the period k optimization problem. If the first-order conditions of the maximization problem in (1.7) satisfy the implicit function theorem, one can obtain the optimal controls u_k and $\bar{u}_k$ as functions of x and $\bar{u}_{k-}$. Substituting these optimal controls into the function on the RHS of (1.7) yields the function $(k, x; \bar{u}_{k-})$. □

Theorem I was established by Petrosyan and Yeung (2020) and Yeung and Petrosyan (2022). It is worth noting that both the current state x_k and previously executed controls $\bar{u}_{k-}$ appear as given in the period k optimization problem. While the state variables x_k have transition equations governing their transition from one period to another, the previously executed controls $\bar{u}_{k-}$ have no such equations of motion.

2. Solution of Durable-Strategies Dynamic Games (Theorem II)

A durable-strategies dynamic game can be expressed as follows:

$$\max_{\substack{u_k^i, \bar{u}_k^i \\ k \in \{1,2,\ldots T\}}} \left\{ \sum_{k=1}^{T} \sum_{\varsigma=1}^{\Theta_i} w_k^{i(\varsigma)} g_k^{i(\varsigma)} \left(x_k, \underline{u}_k, \underline{\bar{u}}_k; \underline{\bar{u}}_{k-}\right) \delta_1^k \right.$$

$$\left. + q_{T+1}^i \left(x_{T+1}; \underline{\bar{u}}_{(T+1)-}\right) \delta_1^{T+1} \right\} \tag{2.1}$$

subject to

$$x_{k+1} = f(x_k, \underline{u}_k, \underline{\bar{u}}_k;\ \underline{\bar{u}}_{k-}),\ x_1 = x_1^0, \quad \text{for } i \in \{1, 2, \ldots, n\} \equiv N. \tag{2.2}$$

In particular, n is the number of players, T is the operation horizon which can be large or even approaches infinity, x_k is a vector of the states of the economy at time period k, u_k^i is the vector of non-durable targets controls of player i in period k, $\underline{u}_k = (u_k^1, u_k^2, \ldots, u_k^n)$ is the vector of non-durable targets controls of all players, $\bar{u}_k^i$ is the vector of durable target controls of player i in period k, $\underline{\bar{u}}_k = (\bar{u}_k^1, \bar{u}_k^2, \ldots, \bar{u}_k^n)$ is the concatenation of the vectors of durable target controls of all players, $\bar{u}_{k-}^i$ is the vector of the player i's durable target controls executed before period k but still in effect in period k, $\underline{\bar{u}}_{k-} = (\bar{u}_{k-}^1, \bar{u}_{k-}^2, \ldots, \bar{u}_{k-}^n)$ is the concatenation of all players' vectors of durable target controls executed before period k but still in effect in period k, $\sum_{\varsigma=1}^{\Theta_i} w_k^{i(\varsigma)} g_k^{i(\varsigma)}(x_k, \underline{u}_k, \underline{\bar{u}}_k; \underline{\bar{u}}_{k-})$ is the objective of player i, $q_{T+1}^i(x_{T+1}; \underline{\bar{u}}_{(T+1)-})$ is the terminal objective of player i at period $T+1$, $x_{k+1} = f(x_k, \underline{u}_k, \underline{\bar{u}}_k; \underline{\bar{u}}_{k-})$ is the evolution dynamics of the vector of states of the economy, and δ_1^k is the discount factor from period 1 to period k.

The controls executed before the start of the operation in period 1, that is, $\underline{\bar{u}}_{1-}$, are known and some or all of them can be zeros. The $w_k^{i(\varsigma)} g_k^{i(\varsigma)}(x_k, \underline{u}_k, \underline{\bar{u}}_k; \underline{\bar{u}}_{k-})$, $f_k^i(x_k, \underline{u}_k, \underline{\bar{u}}_k; \underline{\bar{u}}_{k-})$, and $q_{T+1}^i(x_{T+1}; \underline{\bar{u}}_{(T+1)-})$ are continuously differentiable functions.

The information set of every player includes the knowledge in

(i) all the possible moves by himself and other players, that is, u_k^i and $\bar{u}_k^i$, for $k \in \{1, 2, \ldots, T\}$ and $i \in N$,
(ii) the set of controls which are executed before period k by all players but still in effect in period k, that is, $\underline{\bar{u}}_{k-} = (\bar{u}_{k-}^1, \bar{u}_{k-}^2, \ldots, \bar{u}_{k-}^n)$, for $k \in \{1, 2, \ldots, T\}$,
(iii) the payoff functions of all players, that is, $\sum_{k=1}^{T} \sum_{\varsigma=1}^{\Theta_i} w_k^{i(\varsigma)} g_k^{i(\varsigma)}(x_k, \underline{u}_k, \underline{\bar{u}}_k; \underline{\bar{u}}_{k-})\delta_1^k + q_{T+1}^i(x_{T+1}; \underline{\bar{u}}_{(T+1)-})\delta_1^{T+1}$, for $i \in N$, and
(iv) the state dynamics $x_{k+1} = f_k(x_k, \underline{u}_k, \underline{\bar{u}}_k; \underline{\bar{u}}_{k-})$ and the values of present and past states $(x_k, x_{k-1}, \ldots, x_1)$.

The non-cooperative payoffs of the players in a Nash equilibrium of the dynamic game (2.1)–(2.2) can be characterized by the following theorem.

Theorem II. *Let $(\underline{u}_k^{**}, \underline{\bar{u}}_k^{**})$ be the set of feedback Nash equilibrium strategies and $V^i(k, x; \underline{\bar{u}}_{k-}^{**})$ be the feedback Nash equilibrium payoff of player i at period k in the non-cooperative dynamic game (2.1)–(2.2), then the function $V^i(k, x; \underline{\bar{u}}_{k-}^{**})$ satisfies the following recursive equations:*

$$V^i(T+1, x; \underline{u}_{(T+1)-}^{**}) = q_{T+1}^i(x; \underline{\bar{u}}_{(T+1)-}^{**})\delta_1^{T+1}, \tag{2.3}$$

$$\begin{aligned} V^i(k, x; \underline{u}_{k-}^{**}) = \max_{u_k^i, \bar{u}_k^i} \Bigg\{ & \sum_{\varsigma=1}^{\Theta_i} w_k^{i(\varsigma)} g_k^{i(\varsigma)} \left(x, u_k^i, \bar{u}_k^i, \underline{u}_k^{**(\neq i)}, \underline{\bar{u}}_k^{**(\neq i)}; \underline{\bar{u}}_{k-}^{**}\right) \delta_1^k \\ & + V^i \Bigg[k+1, f_k \left(x, u_k^i, \bar{u}_k^i, \underline{u}_k^{**(\neq i)}, \underline{\bar{u}}_k^{**(\neq i)}; \underline{\bar{u}}_{k-}^{**}\right); \\ & \bar{u}_k^i, \underline{\bar{u}}_k^{**(\neq i)}, \underline{\bar{u}}_{(k+1)-}^{**} \cap \underline{\bar{u}}_{k-}^{**} \Bigg] \Bigg\}, \end{aligned} \tag{2.4}$$

for $k \in \{1, 2, \ldots, T\}$ and $i \in N$,
where $\underline{u}_k^{(\neq i)**} = (u_k^{1**}, u_k^{2**}, \ldots, u_k^{i-1**}, u_k^{i+1**}, \ldots, u_k^{n**})$ and

$$\underline{\bar{u}}_k^{(\neq i)**} = (\bar{u}_k^{1**}, \bar{u}_k^{2**}, \ldots, \bar{u}_k^{i-1**}, \bar{u}_k^{i+1**}, \ldots, \bar{u}_k^{n**}).$$

Proof. Conditions (2.3)–(2.4) show that $V^i(k, x; \underline{\bar{u}}_{k-}^{**})$ is the maximized payoff of player $i \in N$ according to Theorem I given the game equilibrium strategies of the other $n-1$ players. Hence a Nash equilibrium results. □

System (2.3)–(2.4) can be regarded as the Hamilton–Jacobi–Bellman equations for durable-strategies dynamic games. Worth noting is that this class of games cannot be handled by the standard approach of dynamic programming. Theorem II was established by Petrosyan and Yeung (2020) and Yeung and Petrosyan (2019).

3. Dynamically Consistent Solutions and Payment Mechanism (Theorem III)

For dynamical/subgame consistency to be satisfied, the players' agreed-upon optimality principle has to be effective along the cooperative state trajectory x_k^* contingent upon $\underline{\bar{u}}_{k-}^{**}$.

Let

$$\xi(1, x_1^0; \underline{\bar{u}}_{1-}) = [\xi^1(1, x_1^0; \underline{\bar{u}}_{1-}), \xi^2(1, x_1^0; \underline{\bar{u}}_{1-}), \ldots, \xi^n(1, x_1^0; \underline{\bar{u}}_{1-})], \tag{3.1}$$

denote the agreed-upon distribution of cooperative payoffs among the players. In particular, player i would receive $\xi^i(1, x_1^0; \underline{u}_{1-}^*)$ over period 1 to period T, for $i \in N$. For the agreed-upon imputation principle in (1.8) to be maintained at all periods $k \in \{1, 2, \ldots, T\}$ along the cooperative trajectory $\{x_k^*\}_{k=1}^{T+1}$, the cooperative payoff given to player i has to satisfy the following condition:

$$\xi(k, x_k^*; \underline{\bar{u}}_{k-}^*) = [\xi^1(k, x_k^*; \underline{\bar{u}}_{k-}^*), \xi^2(k, x_k^*; \underline{\bar{u}}_{k-}^*), \ldots, \xi^n(k, x_k^*; \underline{\bar{u}}_{k-}^*)],$$
$$\text{for } k \in \{1, 2, \ldots, T\}. \tag{3.2}$$

To satisfy group optimality along the cooperative trajectory $\{x_k^*\}_{k=1}^{T+1}$, it is required that

$$W(k, x^*; \underline{\bar{u}}_{k-}^*) = \sum_{j=1}^{n} \xi^j(k, x_k^*; \underline{\bar{u}}_{k-}^*), \quad \text{for } k \in \{1, 2, \ldots, T\}. \tag{3.3}$$

For individual rationality to be maintained throughout the game, it is required that

$$\xi^i(k, x_k^*; \underline{\bar{u}}_{k-}^*) \geq V^i(k, x_k^*; \underline{\bar{u}}_{k-}^{**}), \quad \text{for } i \in N \quad \text{and} \quad k \in \{1, 2, \ldots, T\}. \tag{3.4}$$

Crucial to the analysis is the derivation of an Imputation Distribution Procedure (IDP) leading to the realization of the agreed imputations in (2.1). To do this, we follow Yeung and Petrosyan (2010, 2016a, 2019) and use $\beta_k^i(x_k^*;\underline{\bar{u}}_{k-}^*)$ to denote the payment that player i receives in period k under the cooperative agreement along the cooperative trajectory $\{x_k^*\}_{k=1}^T$ with durable strategies executive before but still in effect being $\underline{\bar{u}}_{k-}^*$. The payment scheme involving $\beta_k^i(x_k^*;\underline{\bar{u}}_{k-}^*)$ constitutes an IDP in the sense that the payoff to player i over the periods from k to $T+1$ satisfies the following condition:

$$\xi^i\left(k, x_k^*;\underline{\bar{u}}_{k-}^*\right) = \beta_k^i\left(x_k^*;\underline{\bar{u}}_{k-}^*\right)\delta_1^k + \left[\sum_{\zeta=k+1}^{T}\beta_\zeta^i\left(x_\zeta^*;\underline{\bar{u}}_{\zeta-}^*\right)\delta_1^\zeta + q_{T+1}^i\left(x_{T+1}^*;\underline{\bar{u}}_{(T+1)-}^*\right)\delta_1^{T+1}\right], \tag{3.5}$$

for $i \in N$ and $k \in \{1, 2, \ldots, T\}$.

A theorem for the derivation of $\beta_k^i(x_k^*;\underline{\bar{u}}_{k-}^*)$, for $k \in \{1, 2, \ldots, T\}$ and $i \in N$, that satisfies (2.4) is provided as follows.

Theorem III. *The agreed-upon imputation $\xi(k, x_k^*;\underline{\bar{u}}_{k-}^*)$, for $k \in \{1, 2, \ldots, T\}$ along the cooperative trajectory $\{x_k^*\}_{k=1}^T$, can be realized by a payment equaling*

$$\beta_k^i(x_k^*;\underline{\bar{u}}_{k-}^*) = (\delta_1^k)^{-1}[\xi^i(k, x_k^*;\underline{\bar{u}}_{k-}^*) - \xi^i(k+1, f_k(x_k^*, \underline{u}_k^*, \underline{\bar{u}}_k^*;\underline{\bar{u}}_{k-}^*);\underline{\bar{u}}_{(k+1)-}^*)] \tag{3.6}$$

given to player $i \in N$ at period $k \in \{1, 2, \ldots, T\}$.

Proof. Using (3.5), one can obtain

$$\xi^i\left(k+1, x_{k+1}^*;\underline{\bar{u}}_{k-}^*\right) = B_{k+1}^i(x_{k+1}^*;\underline{\bar{u}}_{k-}^*)\delta_1^{k+1} + \left[\sum_{\zeta=k+2}^{T}B_\zeta^i\left(x_\zeta^*;\underline{\bar{u}}_{\zeta-}^*\right)\delta_1^\zeta + q_{T+1}^i\left(x_{T+1}^*;\underline{\bar{u}}_{(T+1)-}^*\right)\delta_1^{T+1}\right]. \tag{3.7}$$

Upon substituting (3.7) into (3.5) yields

$$\xi^i(k, x_k^*;\underline{\bar{u}}_{k-}^*) = \beta_k^i(x_k^*;\underline{\bar{u}}_{k-}^*)\delta_1^k + \xi^i(k+1, x_{k+1}^*;\underline{\bar{u}}_{(k+1)-}^*),$$

which can be expressed as

$$\xi^i(k, x_k^*; \underline{\bar{u}}_{k-}^*) = \beta_k^i(x_k^*; \underline{\bar{u}}_{k-}^*)\delta_1^k + \xi^i(k+1, f_k(x_k^*, \underline{u}_k^*, \underline{\bar{u}}_k^*; \underline{\bar{u}}_{k-}^*); \underline{u}_{(k+1)-}^*). \tag{3.8}$$

From (3.8), one can obtain Theorem III. □

The payment scheme in Theorem III gives rise to the realization of the imputation guided by the agreed-upon optimality principle and constitutes a dynamically consistent payment scheme. More specifically, the payment of $\beta_k^i(x_k^*; \underline{\bar{u}}_{k-}^*)$ allotted to player $i \in N$ in period $k \in \{1, 2, \ldots, T\}$ will establish a cooperative plan that matches with the agreed-upon imputation to every player along the cooperative path:

$$\pi_k^i(x_k^*; \underline{\bar{u}}_{k-}^*) = \beta_k^i(x_k^*; \underline{\bar{u}}_{k-}^*) - g_k^i(x_k^*, \underline{u}_k^*, \underline{\bar{u}}_k^*; \underline{\bar{u}}_{k-}^*), \tag{3.9}$$

for $k \in \{1, 2, \ldots, T\}$, has to be given to player $i \in N$ to yield the cooperative imputation $\xi^i(k, x_k^*; \underline{\bar{u}}_{k-}^*)$. Theorem III was developed in Yeung and Petrosyan (2022).

4. Random Horizon Dynamic Optimization under Durable Strategies (Theorem IV)

We develop a novel dynamic programming technique for solving inter-temporal problems with random horizon and durable strategies which will serve as the foundation for solving the game problem. Consider a durable strategies dynamic optimization problem with a random horizon of $\widehat{T}$ stages, where $\widehat{T}$ is a random variable with range $\{1, 2, \ldots, T\}$ and corresponding probabilities $\{\theta_1, \theta_2, \ldots, \theta_T\}$. Conditional upon the reaching of stage τ, the probability of the game that would last up to stages $\tau, \tau+1, \ldots, T$ becomes, respectively,

$$\frac{\theta_\tau}{\sum_{\zeta=\tau}^T \theta_\zeta}, \frac{\theta_{\tau+1}}{\sum_{\zeta=\tau}^T \theta_\zeta}, \ldots, \frac{\theta_T}{\sum_{\zeta=\tau}^T \theta_\zeta}.$$

The state dynamics is

$$x_{k+1} = f_k(x_k, u_k; \bar{u}_{k-}), \quad x_1 = x_1^0. \tag{4.1}$$

The expected payoff to be maximized becomes

$$E\left[\sum_{k=1}^{\widehat{T}}\sum_{\varsigma=1}^{\Theta} w_k^{(\varsigma)} g_k^{(\varsigma)}\left(x_k, u_k, \bar{u}_k; \bar{u}_{k-}\right)\delta_1^k\right.$$

$$\left. + q_{\widehat{T}+1}\left(x_{\widehat{T}+1}; \bar{u}_{(\widehat{T}+1)-}\right)\delta_1^{\widehat{T}+1}\right]$$

$$= \sum_{\widehat{T}}^{T}\theta_{\widehat{T}}\left[\sum_{k=1}^{\widehat{T}}\sum_{\varsigma=1}^{\Theta} w_k^{(\varsigma)} g_k^{(\varsigma)}(x_k, u_k, \bar{u}_k; \bar{u}_{k-})\delta_1^k\right.$$

$$\left. + q^i_{\widehat{T}+1}\left(x_{\widehat{T}+1}; \bar{u}_{(\widehat{T}+1)-}\right)\delta_1^{\widehat{T}+1}\right], \tag{4.2}$$

where δ_1^k is the discount factor from stage 1 to stage k.

Now consider the case when stage τ has arrived with the state being x_τ and the previously executed durable strategies $u_{\tau-}$. The problem can be formulated as the maximization of the payoff:

$$E\left[\sum_{k=\tau}^{\widehat{T}}\sum_{\varsigma=1}^{\Theta} w_k^{(\varsigma)} g_k^{(\varsigma)}\left(x_k, u_k, \bar{u}_k; \bar{u}_{k-}\right)\delta_1^k + q_{\widehat{T}+1}\left(x_{\widehat{T}+1}; \bar{u}_{(\widehat{T}+1)-}\right)\delta_1^{\widehat{T}+1}\right]$$

$$= \sum_{\widehat{T}=\tau}^{T}\frac{\theta_{\widehat{T}}}{\sum_{\zeta=\tau}^{T}\theta_\zeta}\left[\sum_{k=\tau}^{\widehat{T}}\sum_{\varsigma=1}^{\Theta} w_k^{(\varsigma)} g_k^{(\varsigma)}\left(x_k, u_k, \bar{u}_k; \bar{u}_{k-}\right)\delta_1^k\right.$$

$$\left. + q_{\widehat{T}+1}\left(x_{\widehat{T}+1}; \bar{u}_{(\widehat{T}+1)-}\right)\delta_1^{\widehat{T}+1}\right], \tag{4.3}$$

subject to the dynamics

$$x_{k+1} = f_k(x_k, u_k), \quad \text{for } k \in \{\tau, \tau+1, \ldots, T\}. \tag{4.4}$$

A theorem characterizing an optimal solution to the random-horizon problem (4.1)–(4.2) is provided in the following.

Theorem IV. *Let the value function $V(\tau, x; \bar{u}_{\tau-})$ denote the maximal value of the expected payoffs*

$$E\left[\sum_{k=\tau}^{\widehat{T}}\sum_{\varsigma=1}^{\Theta} w_k^{(\varsigma)} g_k^{(\varsigma)}\left(x_k, u_k, \bar{u}_k; \bar{u}_{k-}\right)\delta_1^k + q_{\widehat{T}+1}\left(x_{\widehat{T}+1}; \bar{u}_{(\widehat{T}+1)-}\right)\delta_1^{\widehat{T}+1}\right]$$

$$= \sum_{\widehat{T}=\tau}^{T}\frac{\theta_{\widehat{T}}}{\sum_{\varsigma=\tau}^{T}\theta_\zeta}\left[\sum_{k=\tau}^{\widehat{T}}\sum_{\varsigma=1}^{\Theta} w_k^{(\varsigma)} g_k^{(\varsigma)}\left(x_k, u_k, \bar{u}_k; \bar{u}_{k-}\right)\delta_1^k + q_{\widehat{T}+1}\left(x_{\widehat{T}+1}; \bar{u}_{(\widehat{T}+1)-}\right)\delta_1^{\widehat{T}+1}\right],$$

for problem (4.3)–(4.4) starting at stage τ with state $x_\tau = x$ and previously executed durable controls $\bar{u}_{\tau-}$. Then the function $V(\tau, x; \bar{u}_{\tau-})$ satisfies the following system of recursive equations:

$$V(T+1, x; \bar{u}_{(T+1)-}) = q_{T+1}(x; \bar{u}_{(T+1)-})\delta_1^{T+1}, \tag{4.5}$$

$$V\left(T, x; \bar{u}_{T-}\right) = \max_{u_T, \bar{u}_T}\left\{\sum_{\varsigma=1}^{\Theta} w_T^{(\varsigma)} g_T^{(\varsigma)}\left(x, u_T, \bar{u}_T; \bar{u}_{T-}\right)\delta_1^T + q_{T+1}\left[f_T(x, u_T); \bar{u}_T, \bar{u}_{(T+1)-}\cap \bar{u}_{T-}\right]\delta_1^{T+1}\right\}, \tag{4.6}$$

$$V\left(\tau, x; \bar{u}_{\tau-}\right) = \max_{u_\tau, \bar{u}_\tau}\left\{\sum_{\varsigma=1}^{\Theta} w_\tau^{(\varsigma)} g_\tau^{(\varsigma)} g_\tau\left(x, u_\tau, \bar{u}_\tau; \bar{u}_{\tau-}\right)\delta_1^\tau + \frac{\theta_\tau}{\sum_{\varsigma=\tau}^{T}\theta_\zeta} q_{\tau+1}\left[f_\tau\left(x, u_\tau\right); \bar{u}_\tau, \bar{u}_{(\tau+1)-}\cap \bar{u}_{\tau-}\right]\delta_1^{\tau+1} + \frac{\sum_{\zeta=\tau+1}^{T}\theta_\zeta}{\sum_{\zeta=\tau}^{T}\theta_\zeta} V\left[\tau+1, f_\tau\left(x, u_\tau\right); \bar{u}_\tau, \bar{u}_{(\tau+1)-}\cap \bar{u}_{\tau-}\right]\right\}, \quad \text{for } \tau \in \{1, 2, \ldots, T-1\}. \tag{4.7}$$

Proof. According to (4.5), the value function at stage $T+1$ is

$$\begin{aligned} V(T+1,x;\bar{u}_{(T+1)-}) &= q_{T+1}(x_{T+1};\bar{u}_{(T+1)-})\delta_1^{T+1} \\ &= q_{T+1}(x_{T+1};\bar{u}_T,\bar{u}_{(T+1)-}\cap\bar{u}_{T-})\delta_1^{T+1}. \end{aligned} \tag{4.8}$$

We first consider the case when the last stage T has been reached. The optimization problem then becomes

$$\begin{aligned} V(T,x;u_{T-}) = \max_{u_T,\bar{u}_T} \Bigg\{ &\sum_{\varsigma=1}^{\Theta} w_T^{(\varsigma)} g_T^{(\varsigma)}\left(x,u_T,\bar{u}_T;\bar{u}_{T-}\right)\delta_1^T \\ &+ q_{T+1}\left(f_T\left(x,u_T\right);\bar{u}_T,\bar{u}_{(T+1)-}\cap\bar{u}_{T-}\right)\delta_1^{T+1}\Bigg\}. \end{aligned} \tag{4.9}$$

The maximization operator in stage T involves $(u_T, \bar{u}_T)$ only, and $\bar{u}_{(T+1)-} \cap \bar{u}_{T-}$ is a subset of $\bar{u}_{T-}$. The current state x and the previously executed durable controls $\bar{u}_{T-}$ appear in the stage T maximization problem as given parameters. If the first-order conditions of the maximization problem in (4.9) satisfy the implicit function theorem, one can obtain the optimal controls u_T as functions of x and $\bar{u}_{T-}$. Substituting these optimal controls into the function on the right-hand side of (4.9) yields the function $V(T, x; \bar{u}_{T-})$, which satisfies the optimal conditions of a maximum for given x and $\bar{u}_{T-}$.

Now consider the problem in stage $\tau \in \{T-1, T-2, \ldots, 1\}$, according to (2.3) one has to maximize the expected payoff

$$\begin{aligned} \sum_{\hat{T}=\tau}^{T} \frac{\theta_{\hat{T}}}{\sum_{\varsigma=\tau}^{T}\theta_\varsigma} \Bigg[&\sum_{k=\tau}^{\hat{T}} \sum_{\varsigma=1}^{\Theta} w_k^{(\varsigma)} g_k^{(\varsigma)} g_k\left(x_k,u_k,\bar{u}_k;\bar{u}_{k-}\right)\delta_1^k \\ &+ q_{\hat{T}+1}\left(x_{\hat{T}+1};\bar{u}_{(\hat{T}+1)-}\right)\delta_1^{k+1}\Bigg]. \end{aligned} \tag{4.10}$$

The expected payoff in (4.10) can be expressed as

$$\sum_{\varsigma=1}^{\Theta} w_{\tau}^{(\varsigma)} g_{\tau}^{(\varsigma)}\left(x, u_{\tau}, \bar{u}_{\tau}; \bar{u}_{\tau-}\right) \delta_1^{\tau} + \frac{\theta_{\tau}}{\sum_{\zeta=\tau}^{T} \theta_{\zeta}} q_{\tau+1}\left(x_{\tau+1}; \bar{u}_{(\tau+1)-}\right) \delta_1^{\tau+1}$$

$$+ \frac{\sum_{\widehat{T}=\tau+1}^{T} \theta_{\widehat{T}}}{\sum_{\zeta=\tau}^{T} \theta_{\zeta}} \left[\sum_{k=\tau+1}^{\widehat{T}} \sum_{\varsigma=1}^{\Theta} w_k^{(\varsigma)} g_k^{(\varsigma)}\left(x_k, u_k, \bar{u}_k; \bar{u}_{k-}\right) \delta_1^k \right.$$

$$\left. + q_{\widehat{T}+1}\left(x_{\widehat{T}+1}; \bar{u}_{(T+1)-}\right) \delta_1^{T+1} \right]$$

$$= \sum_{\varsigma=1}^{\Theta} w_{\tau}^{(\varsigma)} g_{\tau}^{(\varsigma)}\left(x, u_{\tau}, \bar{u}_{\tau}; \bar{u}_{\tau-}\right) \delta_1^{\tau} + \frac{\theta_{\tau}}{\sum_{\zeta=\tau}^{T} \theta_{\zeta}} q_{\tau+1}\left(x_{\tau+1}; \bar{u}_{(\tau+1)-}\right) \delta_1^{\tau+1}$$

$$+ \frac{\sum_{\zeta=\tau+1}^{T} \theta_{\zeta}}{\sum_{\zeta=\tau}^{T} \theta_{\zeta}} \left\{ \frac{\sum_{\widehat{T}=\tau+1}^{T} \theta_{\widehat{T}}}{\sum_{\zeta=\tau+1}^{T} \theta_{\zeta}} \left[\sum_{k=\tau+1}^{\widehat{T}} \sum_{\varsigma=1}^{\Theta} w_k^{(\varsigma)} g_k^{(\varsigma)}\left(x_k, u_k, \bar{u}_k; \bar{u}_{k-}\right) \delta_1^k \right.\right.$$

$$\left.\left. + q_{\widehat{T}+1}\left(x_{\widehat{T}+1}; \bar{u}_{(T+1)-}\right) \delta_1^{\widehat{T}+1} \right] \right\}. \tag{4.11}$$

Note that the term inside the curly brackets of (4.11) can be expressed as

$$\sum_{\widehat{T}=\tau+1}^{T} \frac{\theta_{\widehat{T}}}{\sum_{\varsigma=\tau+1}^{T} \theta_{\zeta}} \left[\sum_{k=\tau+1}^{\widehat{T}} \sum_{\varsigma=1}^{\Theta} w_k^{(\varsigma)} g_k^{(\varsigma)} g_k\left(x_k, u_k, \bar{u}_k; \bar{u}_{k-}\right) \delta_1^k \right.$$

$$\left. + q_{\widehat{T}+1}\left(x_{\widehat{T}+1}; \bar{u}_{(\widehat{T}+1)-}\right) \delta_1^{\widehat{T}+1} \right]$$

$$= E\left[\sum_{k=\tau+1}^{\widehat{T}} \sum_{\varsigma=1}^{\Theta} w_k^{(\varsigma)} g_k^{(\varsigma)}\left(x_k, u_k, \bar{u}_k; \bar{u}_{k-}\right) \delta_1^k \right.$$

$$\left. + q_{\widehat{T}+1}\left(x_{\widehat{T}+1}; \bar{u}_{(\widehat{T}+1)-}\right) \delta_1^{\widehat{T}+1} \right]. \tag{4.12}$$

Invoking Theorem IV, $V(\tau+1, x; \bar{u}_{(\tau+1)-})$ is the maximal value of

$$E\left[\sum_{k=\tau+1}^{\widehat{T}} \sum_{\varsigma=1}^{\Theta} w_k^{(\varsigma)} g_k^{(\varsigma)}\left(x_k, u_k, \bar{u}_k; \bar{u}_{k-}\right) \delta_1^k \right.$$
$$\left. + q_{\widehat{T}+1}\left(x_{\widehat{T}+1}; \bar{u}_{(\widehat{T}+1)-}\right) \delta_1^{\widehat{T}+1}\right].$$

Therefore, the expected payoff to be maximized in stage τ, as specified in (4.11), can be expressed as

$$\sum_{\varsigma=1}^{\Theta} w_\tau^{(\varsigma)} g_\tau^{(\varsigma)} g_\tau\left(x, u_\tau, \bar{u}_\tau; \bar{u}_{\tau-}\right) \delta_1^\tau$$
$$+ \frac{\theta_\tau}{\sum_{\zeta=\tau}^{T} \theta_\zeta} q_{\tau+1}\left[f_\tau(x, u_\tau), \bar{u}_\tau, \bar{u}_{(\tau+1)-} \cap \bar{u}_{\tau-}\right] \delta_1^{\tau+1}$$
$$+ \frac{\sum_{\zeta=\tau+1}^{T} \theta_\zeta}{\sum_{\zeta=\tau}^{T} \theta_\zeta} V\left[\tau+1, f_\tau\left(x, u_\tau\right); \bar{u}_\tau, \bar{u}_{(\tau+1)-} \cap \bar{u}_{\tau-}\right]. \quad (4.13)$$

The τth equation in (2.7) of Theorem IV indeed leads to

$$V(\tau, x; \bar{u}_{\tau-}) = \max_{u_\tau, \bar{u}_\tau}\left\{\sum_{\varsigma=1}^{\Theta} w_\tau^{(\varsigma)} g_\tau^{(\varsigma)}\left(x, u_\tau, \bar{u}_\tau; \bar{u}_{\tau-}\right) \delta_1^\tau\right.$$
$$+ \frac{\theta_\tau}{\sum_{\zeta=\tau}^{T} \theta_\zeta} q_{\tau+1}\left[f_\tau(x, u_\tau), \bar{u}_\tau, \bar{u}_{(\tau+1)-} \cap \bar{u}_{\tau-}\right] \delta_1^{\tau+1}$$
$$\left. + \frac{\sum_{\zeta=\tau+1}^{T} \theta_\zeta}{\sum_{\zeta=\tau}^{T} \theta_\zeta} V\left[\tau+1, f_\tau\left(x, u_\tau\right); \bar{u}_\tau, \bar{u}_{(\tau+1)-} \cap \bar{u}_{\tau-}\right]\right\},$$

for $\tau \in \{1, 2, \ldots, T-2\}$. (4.14)

The maximization operator involves $(u_\tau, \bar{u}_\tau)$. Again, the current state x and the previously executed controls $\bar{u}_{\tau-}$ appear in the stage τ optimization problem. If the first-order conditions of the maximization problem in (4.14) satisfy the implicit function theorem, one can obtain the optimal controls u_τ as functions of x and $\bar{u}_{\tau-}$. Substituting these optimal controls into the function on the right-hand side of (4.14) yields the function $V(\tau, x; \bar{u}_{\tau-})$.

Hence Theorem IV follows. □

Theorem IV was established by Yeung and Petrosyan (2021).

5. Solution for Random Horizon Dynamic Games with Durable Strategies (Theorem V)

Consider the $n-$person dynamic game with $\widehat{T}$ stages where $\widehat{T}$ is a random variable with range $\{1, 2, \ldots, T\}$ and corresponding probabilities $\theta_1, \theta_2, \ldots, \theta_T$. Conditional upon the reaching of stage τ, the probability of the game that would last up to stages $\tau, \tau+1, \ldots, T$ becomes, respectively,

$$\frac{\theta_\tau}{\sum_{\zeta=\tau}^{T} \theta_\zeta}, \frac{\theta_{\tau+1}}{\sum_{\zeta=\tau}^{T} \theta_\zeta}, \ldots, \frac{\theta_T}{\sum_{\zeta=\tau}^{T} \theta_\zeta}.$$

The state dynamics is

$$x_{k+1} = f_k(x_k, \underline{u}_k), \quad x_1 = x_1^0. \tag{5.1}$$

The expected payoff of player i is

$$E\left[\sum_{k=1}^{\widehat{T}} \sum_{\varsigma=1}^{\Theta_i} w_k^{(\varsigma)} g_k^{i(\varsigma)}(x_k, \underline{u}_k, \bar{\underline{u}}_k; \bar{\underline{u}}_{k-})\delta_1^k \right.$$
$$\left. + q_{\widehat{T}+1}^i(x_{\widehat{T}+1}; \bar{\underline{u}}_{(\widehat{T}+1)-})\delta_1^{\widehat{T}+1}\right]$$
$$= \sum_{\widehat{T}=1}^{T} \theta_{\widehat{T}} \left[\sum_{k=1}^{\widehat{T}} \sum_{\varsigma=1}^{\Theta_i} w_k^{i(\varsigma)} g_k^{i(\varsigma)}(x_k, \underline{u}_k, \bar{\underline{u}}_k; \bar{\underline{u}}_{k-})\delta_1^k \right.$$
$$\left. + q_{\widehat{T}+1}^i(x_{\widehat{T}+1}; \bar{\underline{u}}_{(\widehat{T}+1)-})\delta_1^{\widehat{T}+1}\right],$$

for $i \in N$, (5.2)

where δ_1^k is the discount factor from stage 1 to stage k. Theorem IV was established by Yeung and Petrosyan (2021).

Now consider the case when stage τ has arrived with the state being x_τ and the previously executed durable strategies $\underline{\bar{u}}_{\tau-}$. Then it becomes a game in which the payoff of player i is

$$E\left[\sum_{k=\tau}^{\widehat{T}}\sum_{\varsigma=1}^{\Theta_i} w_k^{i(\varsigma)} g_k^{i(\varsigma)}(x_k, \underline{u}_k, \bar{u}_k; \underline{\bar{u}}_{k-})\delta_1^k + q_{\widehat{T}+1}^i(x_{\widehat{T}+1}; \underline{\bar{u}}_{(\widehat{T}+1)-})\delta_1^{\widehat{T}+1}\right]$$

$$= \sum_{\widehat{T}=\tau}^{T} \frac{\theta_{\widehat{T}}}{\sum_{\varsigma=\tau}^{T}\theta_\varsigma}\left[\sum_{k=\tau}^{\widehat{T}}\sum_{\varsigma=1}^{\Theta_i} w_k^{i(\varsigma)} g_k^{i(\varsigma)}\left(x_k, \underline{u}_k, \bar{u}_k; \underline{\bar{u}}_{k-}\right)\delta_1^k + q_{\widehat{T}+1}^i\left(x_{\widehat{T}+1}; \underline{\bar{u}}_{(\widehat{T}+1)-}\right)\delta_1^{\widehat{T}+1}\right],$$

for $i \in N$, (5.3)

and the state dynamics are

$$x_{k+1} = f_k(x_k, \underline{u}_k), \quad \text{for } k = \{\tau, \tau+1, \ldots, T\},\ x_\tau = x. \tag{5.4}$$

A feedback Nash equilibrium of the game can be characterized by the following theorem.

Theorem V. *Let $\{\underline{u}_\tau^{i**}, \bar{u}_\tau^{i**}, i \in N\}$ denote the set of the players' feedback Nash equilibrium strategies and $V^i(\tau, x; \underline{\bar{u}}_{\tau-}^{**})$ denote the feedback Nash equilibrium payoff of player i in the non-cooperative game* (5.3)–(5.4), *then the function $V^i(\tau, x; \underline{\bar{u}}_{\tau-}^{**})$ satisfies the following system of recursive equations:*

$$V^i(T+1, x; \underline{\bar{u}}_{(T+1)-}^{**}) = q_{T+1}^i(x; \underline{\bar{u}}_{(T+1)-}^{**})\delta_1^{T+1}, \tag{5.5}$$

$$V^i\left(T, x; \underline{\bar{u}}_{T-}^{**}\right) = \max_{\underline{u}_T^i, \bar{u}_T^i}\left\{\sum_{\varsigma=1}^{\Theta_i} w_T^{i(\varsigma)} g_T^{i(\varsigma)}\left(x, \underline{u}_T^i, \bar{u}_T^i, \underline{u}_T^{(\neq i)**}, \bar{u}_T^{(\neq i)**}; \underline{\bar{u}}_{T-}^{**}\right)\delta_1^T + q_{T+1}^i\left[f_T\left(x, \underline{u}_T^i, \bar{u}_T^i, \underline{u}_T^{(\neq i)**}, \bar{u}_T^{(\neq i)**}\right); \underline{u}_T^i, \bar{u}_T^i, \underline{u}_T^{(\neq i)**}, \bar{u}_T^{(\neq i)**}, \underline{\bar{u}}_{(T+1)-}^{**} \cap \underline{\bar{u}}_{T-}^{**}\right]\delta_1^{T+1}\right\}, \tag{5.6}$$

$$V^i\left(\tau,x;\underline{u}^{**}_{\tau-}\right)$$
$$=\max_{u^i_\tau,\bar{u}^i_\tau}\left\{\sum_{\varsigma=1}^{\Theta_i}w^{i(\varsigma)}_\tau g^{i(\varsigma)}_\tau\left(x,u^i_\tau,\bar{u}^i_\tau,\underline{u}^{(\neq i)**}_\tau,\underline{\bar{u}}^{(\neq i)**}_\tau;\underline{\bar{u}}^{**}_{\tau-}\right)\delta^\tau_1\right.$$
$$+\frac{\theta_\tau}{\sum_{\zeta=\tau}^{T}\theta_\zeta}\hat{q}_{\tau+1}\left[f_\tau\left(x,u^i_\tau,\underline{u}^{(\neq i)**}_\tau\right);u^i_\tau,\bar{u}^i_\tau,\underline{u}^{(\neq i)**}_\tau,\underline{\bar{u}}^{(\neq i)**}_\tau,\right.$$
$$\left.\underline{\bar{u}}^{**}_{(\tau+1)-}\cap\underline{\bar{u}}^{**}_{\tau-}\right]\delta^{\tau+1}_1+\frac{\sum_{\zeta=\tau+1}^{T}\theta_\zeta}{\sum_{\zeta=\tau}^{T}\theta_\zeta}V^i\left[\tau+1,f_\tau\left(x,u^i_\tau,\underline{u}^{(\neq i)**}_\tau\right);\right.$$
$$\left.\left.u^i_\tau,\bar{u}^i_\tau\underline{u}^{(\neq i)**}_\tau,\underline{\bar{u}}^{(\neq i)**}_\tau,\underline{\bar{u}}^{**}_{(\tau+1)-}\cap\underline{\bar{u}}^{**}_{\tau-}\right]\right\},$$

for $\tau\in\{1,2,\ldots,T-1\}$, (5.7)

where $\underline{u}^{(\neq i)**}_\tau=\underline{u}^{**}_\tau\backslash u^{i**}_\tau$.

Proof. The conditions in (5.5)–(5.7) show that the optimal random horizon dynamic optimization result under durable strategies in Theorem IV holds for each player given other players' equilibrium strategies $\underline{u}^{**}_\tau$, for $\tau\in\{1,2,\ldots,T-1\}$. Hence the conditions of a Nash equilibrium are satisfied and Theorem V follows. □

Theorem V was established by Yeung and Petrosyan (2021).

6. Subgame-Consistent Payment Mechanism under Random Termination (Theorem VI)

Consider a random horizon cooperative dynamic game that maximizes the joint payoff

$$E\left\{\sum_{j=1}^{n}\left[\sum_{k=1}^{\widehat{T}}\sum_{\varsigma=1}^{\Theta_i}w^{j(\varsigma)}_k g^{j(\varsigma)}_k(x_k,\underline{u}_k,\underline{\bar{u}}_k;\underline{\bar{u}}_{k-})\delta^k_1\right.\right.$$
$$\left.\left.+q^j_{\widehat{T}+1}(x_{\widehat{T}+1};\underline{\bar{u}}_{(\widehat{T}+1)-})\delta^{\widehat{T}+1}_1\right]\right\}$$

$$= \sum_{\widehat{T}=1}^{T} \theta_{\widehat{T}} \left\{ \sum_{j=1}^{n} \left[\sum_{k=1}^{\widehat{T}} w_k^{j(\varsigma)} g_k^{j(\varsigma)} \left(x_k, \underline{u}_k, \bar{u}_k; \underline{\bar{u}}_{k-} \right) \delta_1^k \right. \right.$$

$$\left. \left. + q^j_{\widehat{T}+1}(x_{\widehat{T}+1}; \underline{\bar{u}}_{(\widehat{T}+1)-}) \delta_1^{\widehat{T}+1} \right] \right\}, \tag{6.1}$$

subject to dynamics

$$x_{k+1} = f(x_k, \underline{u}_k, \bar{u}_k; \underline{\bar{u}}_{k-}), \quad x_1 = x_1^0. \tag{6.2}$$

Let the agreed-upon payoff imputation under cooperation be denoted by

$$\xi(\tau, x_\tau; \underline{\bar{u}}_{\tau-}) = [\xi^1(\tau, x_\tau; \underline{\bar{u}}_{\tau-}), \xi^2(\tau, x_\tau; \underline{\bar{u}}_{\tau-}), \ldots, \xi^n(\tau, x_\tau; \underline{\bar{u}}_{\tau-})],$$
$$\tau \in \{1, 2, \ldots, T\}. \tag{6.3}$$

To satisfy subgame consistency, the imputation in (6.3) has to be maintained along the cooperative trajectory $\{x_\tau^*\}_{\tau=1}^T$ for given previously executed strategies $\underline{\bar{u}}_{\tau-}^*$ and state x_τ^*. Hence, the following condition has to hold:

$$\xi\left(\tau, x_\tau^*; \underline{\bar{u}}_{\tau-}^*\right) = [\xi^1\left(\tau, x_\tau^*; \underline{\bar{u}}_{\tau-}^*\right), \xi^2(\tau, x_\tau^*; \underline{\bar{u}}_{\tau-}^*), \ldots, \xi^n(\tau, x_\tau^*; \underline{\bar{u}}_{\tau-}^*)],$$
$$\tau \in \{1, 2, \ldots, T\}. \tag{6.4}$$

Crucial to the analysis is the formulation of a payment mechanism so that the imputation in (6.4) can be realized as the game proceeds. Following the analysis of Yeung and Petrosyan (2021), we formulate a discrete-time random-horizon Payoff Distribution Procedure (PDP) so that the agreed-upon imputations (6.4) can be realized. Let $\beta_k^i(x_k^*; \underline{\bar{u}}_{k-}^*)$ denote the payment that player i will received at stage k under the cooperative agreement.

The payment scheme involving $\beta_k^i(x_k^*; \underline{\bar{u}}_{k-}^*)$ constitutes a PDP in the sense that along the cooperative trajectory $\{x_k^*\}_{k=1}^T$ with previously executed durable strategies $\underline{\bar{u}}_{k-}^*$ the imputation to player i over the stages from k to T can be expressed as follows:

$$\begin{aligned}
&\xi^i\left(\tau, x_\tau^*; \underline{\bar{u}}_{\tau-}^*\right)\\
&= E\left[\sum_{k=\tau}^{\widehat{T}} \beta_k^i\left(x_k^*; \underline{\bar{u}}_{k-}^*\right)\delta_1^k + q_{\widehat{T}+1}^i\left(x_{\widehat{T}+1}^*; \underline{\bar{u}}_{(\widehat{T}+1)-}^*\right)\delta_1^{\widehat{T}+1}\right]\\
&= \sum_{\widehat{T}=\tau}^{T} \frac{\theta_{\widehat{T}}}{\sum_{\varsigma=\tau}^{T}\theta_\zeta}\left[\sum_{k=\tau}^{\widehat{T}} \beta_k^i\left(x_k^*; \underline{\bar{u}}_{k-}^*\right)\delta_1^k\right.\\
&\qquad\left. + q_{\widehat{T}+1}^i\left(x_{\widehat{T}+1}^*; \underline{\bar{u}}_{(\widehat{T}+1)}^*\right)\delta_1^{\widehat{T}+1}\right],
\end{aligned} \tag{6.5}$$

for $i \in N$ and $\tau \in \{1, 2, \ldots, T\}$.

Theorem VI. *A payment equaling*

$$\begin{aligned}
\beta_\tau^i(x_\tau^*; \underline{\bar{u}}_{\tau-}^*) = (\delta_1^\tau)^{-1}&\left[\xi^i\left(\tau, x_\tau^*; \underline{\bar{u}}_{\tau-}^*\right)\right.\\
&- \frac{\theta_\tau}{\sum_{\zeta=\tau}^{T}\theta_\zeta} q_{\tau+1}^i\left(x_{\tau+1}^*; \underline{\bar{u}}_{(\tau+1)-}^*\right)\delta_1^{\tau+1}\\
&\left.- \frac{\sum_{\zeta=\tau+1}^{T}\theta_\zeta}{\sum_{\zeta=\tau}^{T}\theta_\zeta}\xi^i\left(\tau+1, x_{\tau+1}^*; \underline{\bar{u}}_{(\tau+1)-}^*\right)\right]
\end{aligned} \tag{6.6}$$

given to player $i \in N$ at stage $\tau \in \{1, 2, \ldots, T\}$ would lead to the realization of the imputation vector $\xi(\tau, x_\tau^*; \underline{\bar{u}}_{\tau-}^*)$ in (6.4).

Proof. Using (6.5) we obtain

$$\begin{aligned}
\xi^i\left(\tau, x_\tau^*; \underline{u}_{\tau-}^*\right) &= \beta_\tau^i\left(x_\tau^*; \underline{u}_{\tau-}^*\right)\delta_1^\tau + \frac{\theta_\tau}{\sum_{\zeta=\tau}^{T}\theta_\zeta} q_{\tau+1}^i\left(x_{\tau+1}^*; \underline{u}_{(\tau+1)-}^*\right)\delta_1^{\tau+1}\\
&\quad + \frac{\sum_{\zeta=\tau+1}^{T}\theta_{\widehat{T}}}{\sum_{\zeta=\tau}^{T}\theta_\zeta}\left\{\beta_k^i\left(x_k^*; \underline{u}_{k-}^*\right)\delta_i^k\right.\\
&\quad \left. + q_{\widehat{T}+1}^i\left(x_{\widehat{T}+1}^*; \underline{u}_{(\widehat{T}+1)-}^*\right)\delta_1^{\widehat{T}+1}\right\}.
\end{aligned} \tag{6.7}$$

The term $\xi^i(\tau, x_\tau^*; \underline{u}_{\tau-}^*)$ in (6.7) can be expressed as

$$
\begin{aligned}
&\xi^i\left(\tau, x_\tau^*; \underline{u}_{\tau-}^*\right) \\
&= \beta_\tau^i\left(x_\tau^*; \underline{u}_{\tau-}^*\right)\delta_1^\tau + \frac{\theta_\tau}{\sum_{\zeta=\tau}^T \theta_\zeta} q_{\tau+1}^i\left(x_{\tau+1}^*; \underline{u}_{(\tau+1)-}^*\right)\delta_1^{\tau+1} \\
&+ \frac{\sum_{\zeta=\tau+1}^T \theta_\zeta}{\sum_{\zeta=\tau}^T \theta_\zeta} \frac{\sum_{\zeta=\tau+1}^T \theta_{\widehat{T}}}{\sum_{\zeta=\tau+1}^T \theta_\zeta} \left\{ \sum_{k=\tau+1}^{\widehat{T}} \beta_k^i\left(x_k^*; \underline{u}_{k-}^*\right)\delta_i^k \right. \\
&\left. + q_{\widehat{T}+1}^i\left(x_{\widehat{T}+1}^*; \underline{u}_{(\widehat{T}+1)-}^*\right)\delta_1^{\widehat{T}+1} \right\}. \qquad (6.8)
\end{aligned}
$$

Invoking (5.9), we can express $\xi^i(\tau+1, x_{\tau+1}^*; \underline{u}_{(\tau+1)-}^*)$ as follows:

$$
\begin{aligned}
&\xi^i(\tau+1, x_{\tau+1}^*; \underline{u}_{(\tau+1)-}^*) \\
&= E\left[\sum_{k=\tau+1}^{\widehat{T}} \beta_k^i\left(x_k^*; \underline{u}_{k-}^*\right)\delta_i^k + q_{\widehat{T}+1}^i\left(x_{\widehat{T}+1}^*; \underline{u}_{(\widehat{T}+1)-}^*\right)\delta_1^{\widehat{T}+1}\right] \\
&= \frac{\sum_{\widehat{T}=\tau+1}^T \theta_{\widehat{T}}}{\sum_{\zeta=\tau}^T \theta_\zeta}\left[\sum_{k=\tau+1}^{\widehat{T}} \beta_k^i\left(x_k^*; \underline{u}_{k-}^*\right)\delta_1^k \right. \\
&\left. \qquad + q_{\widehat{T}+1}^i\left(x_{\widehat{T}+1}^*; \underline{u}_{(\widehat{T}+1)}^*\right)\delta_1^{\widehat{T}+1}\right]. \qquad (6.9)
\end{aligned}
$$

Therefore, $\xi^i(\tau, x_\tau^*; \underline{u}_{\tau-}^*)$ in (6.8) can be expressed as follows:

$$
\begin{aligned}
\xi^i\left(\tau, x_\tau^*; \underline{u}_{\tau-}^*\right) &= \beta_\tau^i\left(x_\tau^*; \underline{u}_{\tau-}^*\right)\delta_1^\tau + \frac{\theta_\tau}{\sum_{\zeta=\tau}^T \theta_\zeta} q_{\tau+1}^i\left(x_{\tau+1}^*; \underline{u}_{(\tau+1)-}^*\right)\delta_1^{\tau+1} \\
&+ \frac{\sum_{\zeta=\tau+1}^T \theta_\zeta}{\sum_{\zeta=\tau}^T \theta_\zeta}\xi^i\left(\tau+1, x_{\tau+1}^*; \underline{u}_{(\tau+1)-}^*\right). \qquad (6.10)
\end{aligned}
$$

The term $\beta^i_\tau(x^*_\tau; \underline{u}^*_{\tau-})$ can be obtained as

$$\beta^i_\tau(x^*_\tau; \underline{u}^*_{\tau-}) = (\delta^\tau_1)^{-1} \left[\xi^i \left(\tau, x^*_\tau; \underline{u}^*_{\tau-}\right) \right.$$
$$- \frac{\theta_\tau}{\sum_{\zeta=\tau}^{T} \theta_\zeta} q^i_{\tau+1} \left(x^*_{\tau+1}; \underline{u}^*_{(\tau+1)-}\right) \delta^{\tau+1}_1$$
$$\left. - \frac{\sum_{\zeta=\tau+1}^{T} \theta_\zeta}{\sum_{\zeta=\tau}^{T} \theta_\zeta} \xi^i \left(\tau+1, x^*_{\tau+1}; \underline{u}^*_{(\tau+1)-}\right) \right]. \quad (6.11)$$

Hence, Theorem VI follows. □

Theorem VI was established by Yeung and Petrosyan (2021).

References

Petrosyan, L. A. & Yeung, D. W. K. (2020). Cooperative dynamic games with durable controls: Theory and application. *Dynamic Games and Applications*, 10, 872–896. DOI: 10.1007/s13235-019-00336-w.

Yeung, D. W. K. & Petrosyan, L. A. (2019). Cooperative dynamic games with control lags. *Dynamic Games and Applications*, 9(2), 550–567, https://doi.org/10.1007/s13235-018-0266-6.

Yeung, D. W. K. & Petrosyan, L. A. (2021). Generalized dynamic games with durable strategies under uncertain planning horizon. *Journal of Computational and Applied Mathematics*, 395, 113595. DOI: 10.1016/j.cam.2021.113595.

Yeung, D. W. K. & Petrosyan, L. A. (2022). *Durable-Strategies Dynamic Games: Theory, Solution Techniques and Applications.* Springer, Switzerland, 303 p.

Appendices

Mathematical Appendices for Computational Illustrations

Appendix A: Proof of Proposition 5.1 in Chapter 3

To prove Proposition 5.1 in Chapter 3, we first invoke Theorem II in the Solution Mechanisms Chapter to characterize the solution for game (5.1)–(5.3) in the following Corollary.

Corollary A.1. *A set of target controls* $\{q_k^{i**}, \bar{q}_k^{i**}, u_k^{i**}, \bar{w}_k^{i**};$ *for* $k \in \{1, 2, \ldots, T\}$ *and* $i \in N\}$ *provides a feedback Nash equilibrium solution to the game* (5.1)–(5.3), *if there exist functions* $V^i(k, x, \vartheta; \underline{q}_{k-}, \underline{\bar{w}}_{k-})$ *such that the following recursive relations are satisfied:*

$$\begin{aligned} &V^i(T+1, x, \vartheta; \bar{q}_{(T+1)-}, \bar{w}_{(T+1)-}) \\ &\quad = (A_{T+1}^i x_{T+1} + B_{T+1}^i \vartheta_{T+1} + C_{T+1}^i)\delta^T \\ &\quad = (-h_{T+1}^i x_{T+1} + y_{T+1}^i \vartheta_{T+1} + Q_{T+1}^i)\delta^T, \end{aligned} \tag{A.1}$$

$$\begin{aligned} &V^i(k, x, \vartheta; \underline{q}_{k-}, \underline{\bar{w}}_{k-}) \\ &= \max_{\substack{q_k^i, \bar{q}_k^i, u_k^i \\ \bar{w}_k^i}} \Big\{ \Big[\alpha_k^i \bar{q}_k^i - \beta_k^i (\bar{q}_k^i)^2 - c_k^i \bar{q}_k^i + 2\varpi_k (\vartheta_k q_k^i)^{\frac{1}{2}} - \varepsilon_k^i q_k^i \\ &\qquad + \chi_k \bar{w}_k^i - (\bar{w}_k^i)^2 - c_k^{(a)i} (u_k^i)^2 - h_k^i x_k \Big] \delta^{k-1} \\ &\qquad + V^i\left(k+1, x_{k+1}^{\neq i*}, \vartheta_{k+1}^{\neq i**}; \bar{q}_{k-}^i, \underline{q}_{k-}^{\neq i**}, \bar{w}_{k-}^i, \underline{\bar{w}}_{k-}^{\neq i**}\right)\Big\}, \end{aligned}$$

for $k \in \{1, 2, \ldots, T\}$ *and* $i \in N$; (A.2)

where

$$x_{k+1}^{\neq i**} = x_k + a_k^i \bar{q}_k^i + \sum_{\substack{j=1 \\ i \neq j}}^{n} a_k^j \bar{q}_k^{j**} + \sum_{j=1}^{n} a_k^{(1)j} \bar{q}_{k-1}^{j**}$$

$$+ \sum_{j=1}^{n} a_k^{(2)j} \bar{q}_{k-2}^{j**} - b_k^i u_k^i (x_k)^{\frac{1}{2}} - \sum_{\substack{j=1 \\ j \neq i}}^{n} b_j u_k^{j**} (x_k)^{\frac{1}{2}} - \gamma_k x_k,$$

$$\vartheta_{k+1}^{\neq i**} = \vartheta_k + z_k^i \bar{w}_k^i + \sum_{\substack{j=1 \\ j \neq i}}^{n} z_k^j \bar{w}_k^{j**} + \sum_{j=1}^{n} z_k^{j(1)} \bar{w}_{k-1}^{j**} - \lambda_k \vartheta_k.$$

Using (A.1), we obtain

$$A_{T+1}^i = -h_{T+1}^i, \quad B_{T+1}^i = y_{T+1}^i, \quad C_{T+1}^i = Q_{T+1}^i. \tag{A.3}$$

Using the technique of backward induction, we consider first the last period of operation, that is, period T. Using (A.2) and (A.3), we can obtain the game equilibrium target controls in period T as follows:

$$\bar{q}_T^i = \frac{\alpha_T^i - c_T^i + A_{T+1}^i \delta a_T^i}{2\beta_T^i}, \quad q_T^i = \frac{(\varpi_T^i)^2 \vartheta_T}{(\varepsilon_T^i)^2},$$

$$u_T^i = -\frac{b_T^i}{2c_T^{(a)i}} A_{T+1}^i \delta (x_T)^{1/2}, \quad \bar{w}_T^i = \frac{\chi_T^i + B_{T+1}^i \delta z_T^i}{2}, \quad \text{for } i \in N. \tag{A.4}$$

Substituting the game equilibrium target controls in (A.4) into the period T equation of (A.2), we obtain

$$V^i(T, x, \vartheta; \underline{\bar{q}}_{T-}, \underline{\bar{w}}_{T-})$$

$$= \left[\alpha_T^i \frac{\alpha_T^i - c_T^i + A_{T+1}^i \delta a_T^i}{2\beta_T^i} - \beta_T^i \left(\frac{\alpha_T^i - c_T^i + A_{T+1}^i \delta a_T^i}{2\beta_T^i} \right)^2 \right.$$

$$- c_T^i \frac{\alpha_T^i - c_T^i + A_{T+1}^i \delta a_T^i}{2\beta_T^i} + \frac{(\varpi_T^i)^2 \vartheta_T}{\varepsilon_T^i} + \chi_T \frac{\chi_T^i + B_{T+1}^i \delta z_T^i}{2}$$

$$- \left(\frac{\chi_T^i + B_{T+1}^i \delta z_T^i}{2} \right)^2$$

$$- c_T^{(a)i}\left(\frac{b_T^i}{2c_T^{(a)i}}A_{T+1}^i\delta(x_T)^{\frac{1}{2}}\right)^2 - h_T^i x_T\Bigg]\delta^{T-1}$$

$$+\Bigg[A_{T+1}^i\Bigg(x_T+\sum_{j=1}^n a_T^j\frac{\alpha_T^j - c_T^j + A_{T+1}^j\delta a_T^j}{2\beta_T^i}+\sum_{j=1}^n a_{T-1}^{(1)j}\bar{q}_{T-1}^{j**}$$

$$+\sum_{j=1}^n a_T^{(2)j}\bar{q}_{T-2}^{j**}+\sum_{j=1}^n\frac{(b_T^j)^2}{2c_T^{(a)j}}A_{T+1}^j\delta x_T-\gamma_T x_T\Bigg)$$

$$+B_{T+1}^i\Bigg(\vartheta_T+\sum_{j=1}^n z_T^j\frac{\chi_T^j+B_{T+1}^j\delta z_T^j}{2}$$

$$+\sum_{j=1}^n z_T^{j(1)}\bar{\varpi}_{T-1}^{j**}-\lambda_T\vartheta_T\Bigg)+C_{T+1}^i\Bigg]\delta^T,$$

for $i \in N$. (A.5)

The RHS of equation system (A.5) is a linear function of x and ϑ which can be expressed as $(A_T^i x_T + B_T^i\vartheta_T + C_T^i)\delta^{T-1}$, where

$$A_T^i x_T = -c_T^{(a)i}\left(\frac{b_T^i}{2c_T^{(a)i}}A_{T+1}^i\delta\right)^2 x_T - h_T^i x_T$$

$$+A_{T+1}^i\left[(1-\gamma_T)x_T+\sum_{j=1}^n\frac{(b_T^j)^2}{2c_T^{(a)j}}A_{T+1}^j\delta x_T\right]\delta,$$

$$B_T^i\vartheta_T = \frac{(\varpi_T^i)^2\vartheta_T}{\varepsilon_T^i}+B_{T+1}^i(1-\lambda_T)\vartheta_T\delta,$$

$$C_T^i = \alpha_T^i\frac{\alpha_T^i - c_T^i + A_{T+1}^i\delta a_T^i}{2\beta_T^i}-\beta_T^i\left(\frac{\alpha_T^i - c_T^i + A_{T+1}^i\delta a_T^i}{2\beta_T^i}\right)^2$$

$$- c_T^i\frac{\alpha_T^i - c_T^i + A_{T+1}^i\delta a_T^i}{2\beta_T^i}+\chi_T\frac{\chi_T^i+B_{T+1}^i\delta z_T^i}{2}$$

$$-\left(\frac{\chi_T^i+B_{T+1}^i\delta z_T^i}{2}\right)^2$$

$$+ A^i_{T+1}\left[\sum_{j=1}^{n} a^j_T \frac{\alpha^j_T - c^j_T + A^j_{T+1}\delta a^j_T}{2\beta^i_T}\right.$$

$$\left. + \sum_{j=1}^{n} a^{(1)j}_{T-1}\bar{q}^{j*}_{T-1} + \sum_{j=1}^{n} a^{(2)j}_{T}\bar{q}^{j*}_{T-2}\right]\delta$$

$$+ B^i_{T+1}\left[\sum_{j=1}^{n} z^j_T \frac{\chi^j_T + B^j_{T+1}z^j_T}{2} + \sum_{j=1}^{n} z^{j(1)}_T \bar{w}^{j*}_{T-1}\right]\delta;$$

for $i \in N$. (A.6)

In particular, C^i_T is a function involving target controls executed before period T, namely, $\bar{q}^j_{T-1}$, $\bar{q}^j_{T-2}$, and $\bar{w}^j_{T-1}$, for $j \in N$.

We consider the process for period $T-1$. Following the analysis in (A.1)–(A.6), we can obtain the game equilibrium target controls in period $T-1$ as follows:

$$\bar{q}^i_{T-1} = \frac{\alpha^i_{T-1} - c^i_{T-1} + A^i_T\delta a^i_{T-1} + A^i_{T+1}\delta^2 a^{(1)i}_{T-1}}{2\beta^i_{T-1}},$$

$$q^i_{T-1} = \frac{(\varpi^i_{T-1})^2\vartheta_{T-1}}{(\varepsilon^i_{T-1})^2},$$

$$u^i_{T-1} = -\frac{b^i_{T-1}}{2c^{(a)i}_{T-1}}A^i_T\delta(x_{T-1})^{1/2},$$

$$\bar{w}^i_{T-1} = \frac{\chi^i_{T-1} + B^i_T\delta z^i_{T-1} + B^i_{T+1}\delta^2 z^{(1)i}_{T-1}}{2}, \quad \text{for } i \in N. \quad (A.7)$$

Substituting the game equilibrium target controls in (A.7) into the period $T-1$ equation of (A.2) we obtain an equation system which RHS is a linear function of x and ϑ which can be expressed as $(A^i_{T-1}x_{T-1} + B^i_{T-1}\vartheta_{T-1} + C^i_{T-1})\delta^{T-2}$, where

$$A^i_{T-1} = -c^{(a)i}_{T-1}\left(\frac{b^i_{T-1}}{2c^{(a)i}_{T-1}}A^i_T\delta\right)^2 - h^i_{T-1}$$

$$+ A_T^i \left[(1-\gamma_{T-1}) + \sum_{j=1}^{n} \frac{(b_{T-1}^j)^2}{2c_{T-1}^{(a)j}} A_T^j \delta \right] \delta,$$

$$B_{T-1}^i = \frac{(\varpi_{T-1}^i)^2}{\varepsilon_{T-1}^i} + B_T^i(1-\lambda_{T-1})\delta, \tag{A.8}$$

and C_{T-1}^i is a function involving target controls executed before period $T-1$, namely, $\bar{q}_{T-2}^j$, $\bar{q}_{T-3}^j$, and $\bar{w}_{T-2}^j$, for $j \in N$.

We consider the processes for periods $T-2, T-3, \ldots, 1$. Following the analysis in (A.1)–(A.8), we can obtain the game equilibrium target controls in period $k \in \{T-2, T-3, \ldots, 1\}$ as follows:

$$\bar{q}_k^i = \frac{\alpha_k^i - c_k^i + A_{k+1}^i \delta a_k^i + A_{k+2}^i \delta^2 a_k^{(1)i} + A_{k+3}^i \delta^3 a_k^{(2)i}}{2\beta_k^i},$$

$$q_k^i = \frac{(\varpi_k^i)^2 \vartheta_k}{(\varepsilon_k^i)^2},$$

$$u_k^i = -\frac{b_k^i}{2c_k^{(a)i}} A_{k+1}^i \delta (x_k)^{1/2}, \quad \bar{w}_k^i = \frac{\chi_k^i + B_k^i \delta z_k^i + B_{k+2}^i \delta^2 z_k^{(1)i}}{2},$$

for $i \in N$. (A.9)

Substituting the game equilibrium target controls in (A.9) into the period k equation of (A.2) we obtain an equation system in which RHS is a linear function of x and ϑ which can be expressed as $(A_k^i x_k + B_k^i \vartheta_k + C_k^i)\delta^{k-1}$, where

$$A_k^i = -c_k^{(a)i} \left(\frac{b_k^i}{2c_k^{(a)i}} A_{k+1}^i \delta \right)^2 - h_k^i$$

$$+ A_{k+1}^i \left[(1-\gamma_k) \sum_{j=1}^{n} \frac{(b_k^j)^2}{2c_k^{(a)j}} A_{k+1}^j \delta \right] \delta,$$

$$B_k^i = \frac{(\varpi_k^i)^2}{\varepsilon_k^i} + B_{k+1}^i (1-\lambda_k)\delta, \tag{A.10}$$

and C_k^i is a function involving target controls executed before period k, namely, $\bar{q}_{k-1}^j$, $\bar{q}_{k-2}^j$, and $\bar{w}_{k-1}^j$, for $j \in N$.

Repeating the above analysis for $k \in \{T-2, T-3, \ldots, 1\}$, we can obtain the game equilibrium target controls as follows:

$$\bar{q}_T^{i**} = \frac{\alpha_T^i - c_T^i + A_{T+1}^i \delta a_T^i}{2\beta_T^i}, \quad q_T^{i**} = \frac{(\varpi_T^i)^2 \vartheta_T}{(\varepsilon_T^i)^2},$$

$$u_T^{i**} = -\frac{b_T^i}{2c_T^{(a)i}} A_{T+1}^i \delta (x_T)^{1/2}, \quad \bar{w}_T^{i**} = \frac{\chi_T^i + B_{T+1}^i \delta z_T^i}{2}, \quad \text{for } i \in N;$$

$$\bar{q}_{T-1}^{i**} = \frac{\alpha_{T-1}^i - c_{T-1}^i + A_T^i \delta a_{T-1}^i + A_{T+1}^i \delta^2 a_{T-1}^{(1)i}}{2\beta_{T-1}^i},$$

$$q_{T-1}^{i**} = \frac{(\varpi_{T-1}^i)^2 \vartheta_{T-1}}{(\varepsilon_{T-1}^i)^2},$$

$$u_{T-1}^{i**} = -\frac{b_{T-1}^i}{2c_{T-1}^{(a)i}} A_T^i \delta (x_{T-1})^{1/2},$$

$$\bar{w}_{T-1}^{i**} = \frac{\chi_{T-1}^i + B_T^i \delta z_{T-1}^i + B_{T+1}^i \delta^2 z_{T-1}^{(1)i}}{2}, \quad \text{for } i \in N;$$

$$\bar{q}_k^{i**} = \frac{\alpha_k^i - c_k^i + A_{k+1}^i \delta a_k^i + A_{k+2}^i \delta^2 a_k^{(1)i} + A_{k+3}^i \delta^3 a_k^{(2)i}}{2\beta_k^i},$$

$$q_k^{i**} = \frac{(\varpi_k^i)^2 \vartheta_k}{(\varepsilon_k^i)^2},$$

$$u_k^{i**} = -\frac{b_k^i}{2c_k^{(a)i}} A_{k+1}^i \delta (x_k)^{1/2}, \quad \bar{w}_k^{i**} = \frac{\chi_k^i + B_k^i \delta z_k^i + B_{k+2}^i \delta^2 z_k^{(1)i}}{2},$$

$$\text{for } i \in N, \text{ and } k \in \{T-2, T-3, \ldots, 1\}. \tag{A.11}$$

■

Appendix B: Proof of Proposition 8.1 in Chapter 4

To prove Proposition 8.1 in Chapter 4, we first invoke Theorem I in the Solution Mechanisms chapter to characterize the solution for collaboration game (8.1)–(8.3) in the following corollary:

Corollary B.1. *A set of target controls* $\{q_k^{i*}, \bar{q}_k^{i*}, u_k^{i*}, \bar{w}_k^{i*};$ *for* $k \in \{1, 2, \ldots, T\}$ *and* $i \in N\}$ *provides an optimal collaborative solution* *to the game* (8.1)–(8.3), *if there exist functions* $W(k,x,\vartheta;\underline{\bar{q}}_{k-},\underline{\bar{\varpi}}_{k-})$ *such that the following recursive relations are satisfied:*

$$W(T+1,x,\vartheta;\underline{\bar{q}}_{(T+1)-},\underline{\bar{\varpi}}_{(T+1)-})$$

$$=\sum_{i=1}^{n}(A_{T+1}x_{T+1}+B_{T+1}\vartheta_{T+1}+C_{T+1})\delta^{T}$$

$$=\sum_{i=1}^{n}(-h_{T+1}^{i}x_{T+1}+y_{T+1}^{i}\vartheta_{T+1}+Q_{T+1}^{i})\delta^{T}, \tag{B.1}$$

$$W(k,x,\vartheta;\underline{\bar{q}}_{k-},\underline{\bar{\varpi}}_{k-})$$

$$=\max_{\substack{q_k^i,\bar{q}_k^i,u_k^t\\ \bar{\varpi}_k^i;i\in N}}\left\{\sum_{i=1}^{n}\left[\alpha_k^i\bar{q}_k^i-\beta_k^i(\bar{q}_k^i)^2-c_k^i\bar{q}_k^i+2\varpi_k(\vartheta_k q_k^i)^{\frac{1}{2}}-\varepsilon_k^i q_k^i\right.\right.$$

$$\left.+\chi_k\bar{\varpi}_k^i-(\bar{\varpi}_k^i)^2-c_k^{(a)i}(u_k^i)^2-\alpha_k^i\bar{q}_k^i\right]\delta^{k-1}$$

$$\left.+V^i(k+1,x_{k+1},\vartheta_{k+1};\underline{\bar{q}}_{k-},\underline{\bar{\varpi}}_{k-})\right\},$$

for $k\in\{1,2,\ldots,T\}$ and $i\in N$, (B.2)

where

$$x_{k+1}=x_k+\sum_{j=1}^{n}a_k^j\bar{q}_k^j+\sum_{j=1}^{n}a_k^{(1)j}\bar{q}_{k-1}^j+\sum_{j=1}^{n}a_k^{(2)j}\bar{q}_{k-2}^j$$

$$-\sum_{j=1}^{n}b_j u_k^j(x_k)^{1/2}-\gamma_k x_k,$$

$$\vartheta_{k+1}=\vartheta_k+\sum_{j=1}^{n}z_k^j\bar{\varpi}_k^j+\sum_{j=1}^{n}z_k^{j(1)}\bar{\varpi}_{k-1}^j-\lambda_k\vartheta_k.$$

Using (B.1), we obtain

$$A_{T+1}=\sum_{i=1}^{n}-h_{T+1}^i,\quad B_{T+1}=\sum_{i=1}^{n}y_{T+1}^i,\quad C_{T+1}=\sum_{i=1}^{n}Q_{T+1}^i. \tag{B.3}$$

Using the technique of backward induction, we consider first the last period of operation, that is, period T. Using (B.2) and (B.3), we can obtain the optimal collaborative controls in period T as follows:

$$\bar{q}_T^{i*} = \frac{\alpha_T^i - c_T^i + A_{T+1}\delta a_T^i}{2\beta_T^i}, \quad q_T^{i*} = \frac{(\varpi_T^i)^2\vartheta_T}{(\varepsilon_T^i)^2},$$

$$u_T^{i*} = -\frac{b_T^i}{2c_T^{(a)i}}A_{T+1}\delta(x_T)^{1/2}, \quad \bar{w}_T^{i*} = \frac{\chi_T^i + B_{T+1}\delta z_T^i}{2}, \quad \text{for } i \in N. \tag{B.4}$$

Substituting the optimal target controls in (B.4) into the period T equation of (B.2), we obtain

$$\begin{aligned}
&W(T, x, \vartheta; \underline{\bar{q}}_{T-}, \underline{\bar{w}}_{T-}) \\
&= \sum_{i=1}^{n} \Bigg[\alpha_T^i \frac{\alpha_T^i - c_T^i + A_{T+1}^i\delta a_T^i}{2\beta_T^i} - \beta_T^i \left(\frac{\alpha_T^i - c_T^i + A_{T+1}^i\delta a_T^i}{2\beta_T^i} \right)^2 \\
&\qquad - c_T^i \frac{\alpha_T^i - c_T^i + A_{T+1}^i\delta a_T^i}{2\beta_T^i} + \frac{(\varpi_T^i)^2\vartheta_T}{\varepsilon_T^i} + \chi_T \frac{\chi_T^i + B_{T+1}^i\delta z_T^i}{2} \\
&\qquad - \left(\frac{\chi_T^i + B_{T+1}^i\delta z_T^i}{2} \right)^2 \\
&\qquad - c_T^{(a)i} \left(\frac{b_T^i}{2c_T^{(a)i}} A_{T+1}^i\delta(x_T)^{\frac{1}{2}} \right)^2 - h_T^i x_T \Bigg] \delta^{T-1} \\
&\quad + \Bigg[A_{T+1} \Bigg(x_T + \sum_{j=1}^{n} a_T^j \frac{\alpha_T^j - c_T^j + A_{T+1}^j\delta a_T^j}{2\beta_T^i} + \sum_{j=1}^{n} a_{T-1}^{(1)j}\bar{q}_{T-1}^{j**} \\
&\qquad + \sum_{j=1}^{n} a_T^{(2)j}\bar{q}_{T-2}^{j**} + \sum_{j=1}^{n} \frac{(b_T^j)^2}{2c_T^{(a)j}} A_{T+1}\delta x_T - \gamma_T x_T \Bigg) \\
&\quad + B_{T+1} \Bigg(\vartheta_T + \sum_{j=1}^{n} z_T^j \frac{\chi_T^j + B_{T+1}^j\delta z_T^j}{2} \\
&\qquad + \sum_{j=1}^{n} z_T^{j(1)}\bar{w}_{T-1}^{j**} - \lambda_T\vartheta_T \Bigg) + C_{T+1}^i \Bigg] \delta^T.
\end{aligned} \tag{B.5}$$

The RHS of equation system (B.5) is a linear function of x and ϑ which can be expressed as $(A_T x_T + B_T \vartheta_T + C_T)\delta^{T-1}$, where

$$A_T x_T = \sum_{i=1}^{n}\left[-c_T^{(a)i}\left(\frac{b_T^i}{2c_T^{(a)i}}A_{T+1}^i\delta\right)^2 x_T - h_T^i x_T\right]$$

$$+ A_{T+1}\left[(1-\gamma_T)x_T + \sum_{j=1}^{n}\frac{(b_T^j)^2}{2c_T^{(a)j}}A_{T+1}\delta x_T\right]\delta,$$

$$B_T\vartheta_T = \sum_{i=1}^{n}\frac{(\varpi_T^i)^2\vartheta_T}{\varepsilon_T^i} + B_{T+1}(1-\lambda_T)\vartheta_T\delta,$$

$$C_T = \sum_{i=1}^{n}\left[\alpha_T^i\frac{\alpha_T^i - c_T^i + A_{T+1}\delta a_T^i}{2\beta_T^i} - \beta_T\left(\frac{\alpha_T^i - c_T^i + A_{T+1}^i\delta a_T^i}{2\beta_T^i}\right)^2\right.$$

$$-c_T^i\frac{\alpha_T^i - c_T^i + A_{T+1}\delta a_T^i}{2\beta_T^i} + \chi_T\frac{\chi_T^i + B_{T+1}\delta z_T^i}{2}$$

$$\left.-\left(\frac{\chi_T^i + B_{T+1}\delta z_T^i}{2}\right)^2\right]$$

$$+ A_{T+1}\left(\sum_{j=1}^{n}\alpha_T^j\frac{\alpha_T^j - c_T^j + A_{T+1}\delta a_T^j}{2\beta_T^i} + \sum_{j=1}^{n}a_{T-1}^{(1)j}\bar{q}_{T-1}^j\right.$$

$$\left.+\sum_{j=1}^{n}a_T^{(2)j}\bar{q}_{T-2}^j)\right)\delta$$

$$+ B_{T+1}\left(\sum_{j=1}^{n}z_T^j\frac{\chi_T^j + B_{T+1}z_T^j}{2} + \sum_{j=1}^{n}z_T^{j(1)}\bar{\varpi}_{T-1}^j\right)\delta. \tag{B.6}$$

In particular, C_T is a function involving strategies executed before period T, namely, $\bar{q}_{T-1}^j$, $\bar{q}_{T-2}^j$, and $\bar{\varpi}_{T-1}^j$, for $j \in N$.

We consider the process for period $T-1$. Following the analysis in (B.1)–(B.6), we can obtain the optimal target controls in period

$T-1$ as follows:

$$\bar{q}^{i*}_{T-1} = \frac{\alpha^i_{T-1} - c^i_{T-1} + A_T\delta a^i_{T-1} + A_{T+1}\delta^2 a^{(1)i}_{T-1}}{2\beta^i_{T-1}},$$

$$q^{i*}_{T-1} = \frac{(\varpi^i_{T-1})^2\vartheta_{T-1}}{(\varepsilon^i_{T-1})^2},$$

$$u^{i*}_{T-1} = -\frac{b^i_{T-1}}{2c^{(a)i}_{T-1}} A_T\delta(x_{T-1})^{1/2},$$

$$\bar{w}^{i*}_{T-1} = \frac{\chi^i_{T-1} + B_T\delta z^i_{T-1} + B_{T+1}\delta^2 z^{(1)i}_{T-1}}{2}, \quad \text{for } i \in N. \quad \text{(B.7)}$$

Substituting the optimal target controls in (B.7) into the period $T-1$ equation of (B.2), we obtain an equation system which RHS is a linear function of x and ϑ which can be expressed as $(A_{T-1}x_{T-1} + B_{T-1}\vartheta_{T-1} + C_{T-1})\delta^{T-2}$, where

$$A_{T-1} = \sum_{i=1}^{n}\left[-c^{(a)i}_{T-1}\left(\frac{b^i_{T-1}}{2c^{(a)i}_{T-1}} A^i_{T+1}\delta\right)^2 - h^i_{T-1}\right] + A_T\left[(1-\gamma_{T-1}) + \sum_{j=1}^{n}\frac{(b^j_{T-1})^2}{2c^{(a)j}_{T-1}} A_T\delta\right]\delta,$$

$$B_{T-1} = \sum_{i=1}^{n}\frac{(\varpi^i_{T-1})^2}{\varepsilon^i_{T-1}} + B_T(1-\lambda_{T-1})\delta, \quad \text{(B.8)}$$

and C_{T-1} is a function involving strategies executed before period $T-1$, namely, $\bar{q}^j_{T-2}$, $\bar{q}^j_{T-3}$, and $\bar{w}^j_{T-2}$, for $j \in N$.

We consider the processes for periods $T-2, T-3, \ldots, 1$. Following the analysis in (B.1)–(B.8), we can obtain the optimal target controls in period $k \in \{T-2, T-3, \ldots, 1\}$ as follows:

$$\bar{q}^{i*}_k = \frac{\alpha^i_k - c^i_k + A_{k+1}\delta a^i_k + A_{k+2}\delta^2 a^{(1)i}_k + A_{k+3}\delta^3 a^{(2)i}_k}{2\beta^i_k},$$

$$q^{i*}_k = \frac{(\varpi^i_k)^2\vartheta_k}{(\varepsilon^i_k)^2},$$

$$u_k^{i*} = -\frac{b_k^i}{2c_k^{(a)i}} A_{k+1}\delta(x_k)^{1/2}, \quad \bar{w}_k^{i*} = \frac{\chi_k^i + B_k\delta z_k^i + B_{k+2}\delta^2 z_k^{(1)i}}{2},$$

for $i \in N$. (B.9)

Substituting the optimal target controls in (B.9) into the period k equation of (B.2), we obtain an equation system which RHS is a linear function of x and ϑ which can be expressed as $(A_k x_k + B_k \vartheta_k + C_k)\delta^{k-1}$, where

$$A_k = \sum_{i=1}^{n}\left[-c_k^{(a)i}\left(\frac{b_k^i}{2c_k^{(a)i}}A_{k+1}\delta\right)^2 - h_k^i\right] + A_{k+1}\left[(1-\gamma_k) + \sum_{j=1}^{n}\frac{(b_k^j)^2}{2c_k^{(a)j}}A_{k+1}\delta\right]\delta,$$

$$B_k = \sum_{i=1}^{n}\frac{(\varpi_k^i)^2}{\varepsilon_k^i} + B_{k+1}(1-\lambda_k)\delta, \tag{B.10}$$

and C_k is a function involving strategies executed before period k, namely, $\bar{q}_{k-1}^j$, $\bar{q}_{k-2}^j$, and $\bar{w}_{k-1}^j$, for $j \in N$.

Repeating the above analysis for $k \in \{T-2, T-3, \ldots, 1\}$, we obtain the cooperative optimal control targets as follows:

$$\bar{q}_T^{i*} = \frac{\alpha_T^i - c_T^i + A_{T+1}\delta a_T^i}{2\beta_T^i}, \quad q_T^{i*} = \frac{(\varpi_T^i)^2\vartheta_T}{(\varepsilon_T^i)^2},$$

$$u_T^{i*} = -\frac{b_T^i}{2c_T^{(a)i}}A_{T+1}\delta(x_T)^{1/2}, \quad \bar{w}_T^{i*} = \frac{\chi_T^i + B_{T+1}\delta z_T^i}{2}, \quad \text{for } i \in N,$$

$$\bar{q}_{T-1}^{i*} = \frac{\alpha_{T-1}^i - c_{T-1}^i + A_T\delta a_{T-1}^i + A_{T+1}\delta^2 a_{T-1}^{(1)i}}{2\beta_{T-1}^i},$$

$$q_{T-1}^{i*} = \frac{(\varpi_{T-1}^i)^2\vartheta_{T-1}}{(\varepsilon_{T-1}^i)^2},$$

$$u_{T-1}^{i*} = -\frac{b_{T-1}^i}{2c_{T-1}^{(a)i}}A_T\delta(x_{T-1})^{1/2},$$

$$\bar{w}_{T-1}^{i*} = \frac{\chi_{T-1}^i + B_T\delta z_{T-1}^i + B_{T+1}\delta^2 z_{T-1}^{(1)i}}{2}, \quad \text{for } i \in N,$$

$$\bar{q}_k^{i*} = \frac{\alpha_k^i - c_k^i + A_{k+1}\delta a_k^i + A_{k+2}\delta^2 a_k^{(1)i} + A_{k+3}\delta^3 a_k^{(2)i}}{2\beta_k^i},$$

$$q_k^{i*} = \frac{(\varpi_k^i)^2 \vartheta_k}{(\varepsilon_k^i)^2},$$

$$u_k^{i*} = -\frac{b_k^i}{2c_k^{(a)i}} A_{k+1}\delta(x_k)^{1/2}, \quad \bar{w}_k^{i*} = \frac{\chi_k^i + B_k\delta z_k^i + B_{k+2}\delta^2 z_k^{(1)i}}{2},$$

for $i \in N$ and $k \in \{T-2, T-3, \ldots, 1\}$. (B.11)

■

Appendix C: Proof of Propositions 6.1 and 6.2 in Chapter 5

Invoking Theorem 2.1 in Chapter 5, the equilibrium payoffs of the two coalitions and the equilibrium payoffs of unallied nations plus the equilibrium target controls of all nations can be characterized in the following corollary.

Corollary C.1. *Let $\underline{u}_k^{**} = (\underline{u}_k^{(1)**}, \underline{u}_k^{(2)**}, \underline{u}_k^{(3)**})$ and $\underline{\bar{u}}_k^{**} = (\underline{\bar{u}}_k^{(1)**}, \underline{\bar{u}}_k^{(2)**})$, where $\underline{u}_k^{(z)**} = (u_k^{(z)1**}, u_k^{(z)2**}, \ldots, u_k^{(z)n_z**})$, for $z \in \{1,2\}$, $\underline{u}_k^{(3)**} = (u_k^{(3)1**}, u_k^{(3)2**}, \ldots, u_k^{(3)n_3**})$, $\underline{\bar{u}}_k^{(z)**} = (\bar{u}_k^{(z)1**}, \bar{u}_k^{(z)2**}, \ldots, \bar{u}_k^{(z)n_z**})$, for $z \in \{1,2\}$, be the set of coalition equilibrium target controls and $W^z(k, x, y; \underline{\bar{u}}_{k-}^{**})$ be the equilibrium payoff of coalition $z \in \{1,2\}$, and $W^{3(\varsigma)}(k, x, y; \underline{\bar{u}}_{k-}^{**})$, for $\varsigma \in S^3$, be the equilibrium payoffs of the n_3 unallied nations at period k of the coalition game (3.6), then the functions $W^z(k, x, y; \underline{\bar{u}}_{k-}^{**})$ for $z \in \{1,2\}$, and $W^{3(\varsigma)}(k, x, y; \underline{\bar{u}}_{k-}^{**})$, for $\varsigma \in S^3$ satisfies the following recursive equations:*

$$W^1(T+1, x, y; \underline{\bar{u}}_{(T+1)-}^{**}) = \sum_{i \in S^1} (h_{T+1}^{(1)i} x_{T+1} - \hat{h}_{T+1}^{(1)i} y_{T+1} + Q_{T+1}^{(1)i})\delta^T,$$

$$W^2(T+1, x, y; \underline{\bar{u}}_{(T+1)-}^{**}) = \sum_{i \in S^2} (h_{T+1}^{(2)i} y_{T+1} - \hat{h}_{T+1}^{(2)i} x_{T+1} + Q_{T+1}^{(2)j})\delta^T,$$

$$W^{3(\varsigma)}(T+1, x, y; \underline{\bar{u}}_{(T+1)-}^{**})$$
$$= (Q_{T+1}^{(3)\zeta} - h_{T+1}^{(3)\zeta} x_{T+1} - \hat{h}_{T+1}^{(3)\zeta} y_{T+1})\delta^T, \quad \text{for } \varsigma \in S^3, \qquad \text{(C.1)}$$

$$\begin{aligned}
&W^1(k,x,y;\underline{\bar{u}}_{k-}^{**})\\
&= \max_{u_k^{(1)},\bar{u}_k^{(1)}} \left\{ \sum_{i\in S^1} \left[\left(-b_k^{(1)i} u_k^{(1)i} + \sum_{\substack{\ell\in S^1\\ \ell\neq i}} \varpi_k^{(1)i(\ell)} u_k^{(1)\ell} \right) u_k^{(1)i} \right.\right.\\
&\quad \left. - c_k^{(1)i}(u_k^{(1)i})^2 - w_k^{(1)i}(\bar{u}_k^{(1)i})^2 + h_k^{(1)i} x_k - \hat{h}_k^{(1)i} y_k \right] \delta^{k-1}\\
&\quad \left. + W^1\left[k+1, x_{k+1}^{(\neq i)**}, y_{k+1}^{**}; u_k^{(1)}, \bar{u}_k^{(1)}, \underline{\bar{u}}_k^{**(\neq 1)}, \underline{\bar{u}}_{(k+1)-}^{**} \cap \underline{\bar{u}}_{k-}^{**}\right] \right\},
\end{aligned} \tag{C.2}$$

where

$$x_{k+1}^{(\neq 1)**} = x_k + \sum_{\ell=1}^{n_1} \gamma_k^{(1)\ell} \bar{u}_k^{(1)\ell} + \sum_{\ell=1}^{n_1} \sigma_k^{(1)\ell} \bar{u}_{k-1}^{(1)\ell **} - \lambda_k^{(1)} x_k,$$

$$y_{k+1}^{**} = y_k + \sum_{\ell=1}^{n_1} \gamma_k^{(2)\ell} \bar{u}_k^{(2)\ell **} + \sum_{\ell=1}^{n_2} \sigma_k^{(2)\ell} \bar{u}_{k-1}^{(2)\ell **} - \lambda_k^{(2)} x_k,$$

$$\begin{aligned}
&W^2(k,x,y;\underline{\bar{u}}_{k-}^{**})\\
&= \max_{u_k^{(2)},\bar{u}_k^{(2)}} \left\{ \sum_{j\in S^2} \left[\left(p_k^{(2)i} - b_k^{(2)j(j)} u_k^{(2)j} + \sum_{\substack{\ell\in S^2\\ \ell\neq j}} \varpi_k^{(2)j(\ell)} u_k^{(2)\ell} \right) u_k^{(2)j} \right.\right.\\
&\quad \left. - c_k^{(2)j}(u_k^{(2)j})^2 - w_k^{(2)j}(\bar{u}_k^{(2)j})^2 + h_k^{(2)j} y_k - \hat{h}_k^{(2)j} x_k \right] \delta^{k-1}\\
&\quad \left. + W^2\left[k+1, x_{k+1}^{**}, y_{k+1}^{(\neq 2)**}; u_k^{(2)}, \bar{u}_k^{(2)}, \underline{\bar{u}}_k^{**(\neq 2)}, \underline{\bar{u}}_{(k+1)-}^{**} \cap \underline{\bar{u}}_{k-}^{**}\right] \right\},
\end{aligned} \tag{C.3}$$

where

$$x_{k+1}^{**} = x_k + \sum_{\ell=1}^{n_1} \gamma_k^{(1)\ell} \bar{u}_k^{(1)\ell**} + \sum_{\ell=1}^{n_1} \sigma_k^{(1)\ell} \bar{u}_{k-1}^{(1)\ell**} - \lambda_k^{(1)} x_k,$$

$$y_{k+1}^{(\neq 2)**} = y_k + \sum_{\ell=1}^{n_1} \gamma_k^{(2)\ell} \bar{u}_k^{(2)\ell} + \sum_{\ell=1}^{n_2} \sigma_k^{(2)\ell} \bar{u}_{k-1}^{(2)\ell**} - \lambda_k^{(2)} x_k,$$

$$\begin{aligned} &W^{(3)\varsigma}(k, x, y; \underline{\bar{u}}_{k-}^{**}) \\ &= \max_{u_k^{(3)\varsigma}} \Bigg\{ \Bigg[\Bigg(p_k^{(3)\varsigma} - b_k^{(3)\varsigma(\varsigma)} u_k^{(3)\varsigma} + \sum_{\substack{\ell \in S^3 \\ \ell \neq \varsigma}} \varpi_k^{(3)\varsigma(\ell)} u_k^{(3)\ell**} \Bigg) u_k^{(3)\varsigma} \\ &\qquad - c_k^{(3)\varsigma} (u_k^{(3)\varsigma})^2 - h_k^{(3)\varsigma} x_k - \hat{h}_k^{(3)\varsigma} y_k \Bigg] \delta^{k-1} \\ &\qquad + W^{(3)\varsigma} \left[k+1, x_{k+1}^{**}, y_{k+1}^{**}; u_k^{(3)\varsigma}, \underline{u}_k^{**}, \bar{u}_{(k+1)-}^{**} \cap \underline{\bar{u}}_{k-}^{**} \right] \Bigg\}, \end{aligned} \tag{C.4}$$

for $\varsigma \in \{1, 2, \ldots, n_3\}$,
where

$$x_{k+1}^{**} = x_k + \sum_{\ell=1}^{n_1} \gamma_k^{(1)\ell} \bar{u}_k^{(1)\ell**} + \sum_{\ell=1}^{n_1} \sigma_k^{(1)\ell} \bar{u}_{k-1}^{(1)\ell**} - \lambda_k^{(1)} x_k,$$

$$y_{k+1}^{**} = y_k + \sum_{\ell=1}^{n_1} \gamma_k^{(2)\ell} \bar{u}_k^{(2)\ell**} + \sum_{\ell=1}^{n_2} \sigma_k^{(2)\ell} \bar{u}_{k-1}^{(2)\ell**} - \lambda_k^{(2)} x_k.$$

Performing the indicated maximization operator in (C.2) with respect to $u_k^{(1)i}$ yields the first-order conditions

$$\begin{aligned} &p_k^{(1)i} - b_k^{(1)i(i)} 2u_k^{(1)i} + \sum_{\substack{\ell \in S^1 \\ \ell \neq i}} \varpi_k^{(1)i(\ell)} u_k^{(1)\ell} + \sum_{\substack{\ell \in S^1 \\ \ell \neq i}} \varpi_k^{(1)\ell(i)} u_k^{(1)\ell} \\ &\quad -2c_k^{(1)i} u_k^{(1)i} = 0, \\ &\quad \text{for } i \in \{1, 2, \ldots, n_1\}. \end{aligned} \tag{C.5}$$

The first-order conditions in (C.5) form a system of n_1 linear equations with n_1 variables $(u_k^{(1)1}, u_k^{(1)2}, \ldots, u_k^{(1)n_1})$, which can be written as follows:

$$\frac{p_k^{(1)i}}{2b_k^{(1)i(i)} + 2c_k^{(1)i}} = u_k^{(1)i} - \sum_{\substack{\ell \in S^1 \\ \ell \neq i}} \frac{\varpi_k^{(1)i(\ell)} + \varpi_k^{(1)\ell(i)}}{2b_k^{(1)i(i)} + 2c_k^{(1)i}} u_k^{(1)\ell}, \tag{C.6}$$

for $i \in \{1, 2, \ldots, n_1\}$.

The linear equation system in (C.6) can be obtained with standard techniques and we use $(u_k^{(1)1**}, u_k^{(1)2**}, \ldots, u_k^{(1)n_1**})$ to denote the solution of (C.6).

Similarly, performing the indicated maximization operator in (C.3) with respect to $u_k^{(2)j}$ yields the first-order conditions as a system of linear equations:

$$\frac{p_k^{(2)j}}{2b_k^{(2)j(j)} + 2c_k^{(2)j}} = u_k^{(2)j} - \sum_{\substack{\ell \in S^2 \\ \ell \neq j}} \frac{\varpi_k^{(2)j(\ell)} + \varpi_k^{(2)\ell(j)}}{2b_k^{(2)j(j)} + 2c_k^{(2)j}} u_k^{(2)\ell},$$

$$\text{for } j \in \{1, 2, \ldots, n_2\}. \tag{C.7}$$

The solution to (C.7) is denoted by $(u_k^{(2)1**}, u_k^{(2)2**}, \ldots, u_k^{(2)n_2**})$.

Performing the indicated maximization operator in (C.4) with respect to $u_k^{(3)\varsigma}$ yields the first-order conditions as a system of linear equations:

$$\frac{p_k^{(3)\varsigma}}{2b_k^{(3)\varsigma(\varsigma)} + 2c_k^{(3)\varsigma}} = u_k^{(3)\varsigma} - \sum_{\substack{\ell \in S^3 \\ \ell \neq \varsigma}} \frac{\varpi_k^{(3)\varsigma(\ell)}}{2b_k^{(3)\varsigma(\varsigma)} + 2c_k^{(3)\varsigma}} u_k^{(3)\ell},$$

$$\text{for } \varsigma \in \{1, 2, \ldots, n_3\}. \tag{C.8}$$

The solution to (C.8) is denoted by $(u_k^{(3)1**}, u_k^{(3)2**}, \ldots, u_k^{(3)n_2**})$.

Performing the indicated maximization operator in (C.2) with respect to $\bar{u}_k^{(1)i}$ yields the first-order conditions:

$$\frac{\partial W_{k+1}^1}{\partial x_{k+1}} \gamma_k^{(1)i} + \frac{\partial W_{k+1}^1}{\partial \bar{u}_k^{(1)i}} = 2w_k^{(1)i} \bar{u}_k^{(1)i} \delta^{k-1},$$

$$\text{for } i \in \{1, 2, \ldots, n_1\}, \tag{C.9}$$

where W^1_{k+1} stands for $W^1[k+1, x^{(\neq i)**}_{k+1} y^{**}_{k+1}); u^{(1)}_k, \bar{u}^{(1)}_k, \underline{\bar{u}}^{**(\neq 1)}_k$, $\underline{\bar{u}}^{**}_{(k+1)-} \cap \underline{\bar{u}}^{**}_{k-}]$.

Performing the indicated maximization operator in (C.3) with respect to $\bar{u}^{(2)j}_k$ yields the first-order conditions:

$$\frac{\partial W^2_{k+1}}{\partial x_{k+1}} \gamma^{(1)i}_k + \frac{\partial W^2_{k+1}}{\partial \bar{u}^{(2)j}_k} = 2w^{(2)j}_k \bar{u}^{(2)j}_k \delta^{k-1},$$
$$\text{for } j \in \{1, 2, \ldots, n_2\}, \tag{C.10}$$

where W^2_{k+1} stands for $W^2[k+1, x^{**}_{k+1} y^{(\neq j)}_{k+1}); u^{(2)}_k, \bar{u}^{(2)}_k, \underline{\bar{u}}^{(**\neq j)}_k$, $\underline{\bar{u}}^{**}_{(k+1)-} \cap \underline{\bar{u}}^{**}_{k-}]$.

Invoking the technique of backward induction, we consider first the last period of operation, that is, period T. The terminal payoffs in (C.1) can be expressed as follows:

$$\begin{aligned}
W^1(T+1, x, y; \underline{\bar{u}}^{**}_{(T+1)-}) &= \sum_{i \in S^1} (h^{(1)i}_{T+1} x_{T+1} - \hat{h}^{(1)i}_{T+1} y_{T+1} + Q^{(1)i}_{T+1}) \delta^T \\
&= (A^{(1)}_{T+1} x_{T+1} + B^{(1)}_{T+1} y_{T+1} + C^{(1)}_{T+1}) \delta^T, \\
W^2(T+1, x, y; \underline{\bar{u}}^{**}_{(T+1)-}) &= \sum_{j \in S^2} (h^{(2)j}_{T+1} y_{T+1} - \hat{h}^{(2)j}_{T+1} x_{T+1} + Q^{(2)j}_{T+1}) \delta^T \\
&= (A^{(2)}_{T+1} x_{T+1} + B^{(2)}_{T+1} y_{T+1} + C^{(2)}_{T+1}) \delta^T, \\
W^{3(\varsigma)}(T+1, x, y; \underline{\bar{u}}^{**}_{(T+1)-}) &= (Q^{(3)\varsigma}_{T+1} - h^{(3)\varsigma}_{T+1} x_{T+1} - \hat{h}^{(3)\varsigma}_{T+1} y_{T+1}) \delta^T \\
&= (A^{(3)\varsigma}_{T+1} x_{T+1} + B^{(3)\varsigma}_{T+1} y_{T+1} + C^{(3)\varsigma}_{T+1}) \delta^T, \\
&\quad \text{for } \varsigma \in S^3.
\end{aligned} \tag{C.11}$$

Consider the operations in Coalition 1. Using (C.9) and (C.11), we obtain

$$\begin{aligned}
\frac{\partial W^1_{T+1}}{\partial x_{T+1}} \gamma^{(1)i}_T + \frac{\partial W^1_{T+1}}{\partial \bar{u}^{(1)i}_{T+1}} &= 2w^{(1)i}_T \bar{u}^{(1)i}_T \delta^{T-1} \Rightarrow A^{(1)}_{T+1} \gamma^{(1)i}_T \delta^T \\
&= 2w^{(1)i}_T \bar{u}^{(1)i}_T \delta^{T-1}.
\end{aligned} \tag{C.12}$$

Hence, the game equilibrium durable target control of nation i in Coalition 1 can be obtained as follows:

$$\bar{u}_T^{(1)i**} = \frac{A_{T+1}^{(1)}\gamma_T^{(1)i}\delta}{2w_T^{(1)i}}, \quad \text{for } i \in \{1, 2, \ldots, n_1\}. \tag{C.13}$$

For operations in Coalition 2, we use (C.10) and (C.11) to obtain

$$\begin{aligned}\frac{\partial W_{T+1}^2}{\partial x_{T+1}}\gamma_T^{(1)i} + \frac{\partial W_{T+1}^2}{\partial \bar{u}_T^{(2)j}} &= 2w_T^{(2)j}\bar{u}_T^{(2)j}\delta^{T-1} \Rightarrow B_{T+1}^{(2)}\gamma_T^{(2)j}\delta^T \\ &= 2w_T^{(2)j}\bar{u}_T^{(2)j}\delta^{T-1}. \end{aligned} \tag{C.14}$$

Hence, the game equilibrium durable target control of nation j in Coalition 2 can be obtained as follows:

$$\bar{u}_T^{(2)j**} = \frac{B_{T+1}^{(2)}\gamma_T^{(2)j}\delta}{2w_T^{(2)j}}, \quad \text{for } j \in \{1, 2, \ldots, n_2\}. \tag{C.15}$$

Substituting the game equilibrium period T target controls from (C.6)–(C.8) and (C.13) and (C.15) into the period T equations of (C.2)–(C.4), we obtain

$$\begin{aligned} &W^1\left(T, x, y; \underline{\bar{u}}_{T-}^{**}\right) \\ &= \sum_{i\in S^1}\left[\left(p_T^{(1)i} - b_T^{(1)i(i)}u_T^{(1)i**} + \sum_{\substack{\ell\in S^1\\ \ell\neq i}}\varpi_T^{(1)i(\ell)}u_T^{(1)\ell**}\right)\right. \\ &\quad - c_T^{(1)i}(u_T^{(1)i**})^2 - w_T^{(1)i}(\bar{u}_T^{(1)i**})^2 + h_T^{(1)i}x_T - \hat{h}_T^{(1)i}y_T]\delta^{T-1} \\ &\quad \left. + \sum_{i\in S^1}(h_{T+1}^{(1)i}x_{T+1}^{**} - \hat{h}_{T+1}^{(1)i}y_{T+1}^{**} + Q_{T+1}^{(1)i}\right]\delta^T; \end{aligned} \tag{C.16}$$

$$W^2\left(T, x, y; \underline{\bar{u}}_{T-}^{**}\right)$$

$$= \sum_{j \in S^2} \left[\left(p_T^{(2)j} - b_T^{(2)j(j)} u_T^{(2)i**} + \sum_{\substack{\ell \in S^2 \\ \ell \neq j}} \varpi_T^{(2)j(\ell)} u_T^{(2)\ell**} \right) u_T^{(2)j**} \right.$$

$$\left. - c_T^{(2)j} (u_T^{(2)j**})^2 - w_T^{(2)j} (\bar{u}_T^{(2)j**})^2 + h_T^{(2)j} y_T - \hat{h}_T^{(2)j} x_T \right] \delta^{T-1}$$

$$+ \sum_{j \in S^2} (h_{T+1}^{(2)j} y_{T+1}^{**} - \hat{h}_{T+1}^{(2)j} x_{T+1}^{**} + Q_{T+1}^{(2)j}) \delta^T; \tag{C.17}$$

$$W^{(3)\varsigma}\left(T, x, y; \underline{\bar{u}}_{T-}^{**}\right)$$

$$= \left[\left(p_T^{(3)\varsigma} - b_T^{(3)\varsigma(\varsigma)} u_T^{(3)\varsigma**} + \sum_{\substack{\ell \in S^3 \\ \ell \neq \varsigma}} \varpi_T^{(3)\varsigma(l)} u_T^{(3)\ell**} \right) u_T^{(3)\varsigma**} \right.$$

$$\left. - c_T^{(3)\varsigma} (u_T^{(3)\varsigma**})^2 - h_T^{(3)\varsigma} x_T^{**} - \hat{h}_T^{(3)\varsigma} y_T^{**} \right] \delta^{T-1}$$

$$+ \left[Q_{T+1}^{(3)\zeta} - h_{T+1}^{(3)\zeta} x_{T+1}^{**} - \hat{h}_{T+1}^{(3)\zeta} y_{T+1}^{**} \right] \delta^T, \quad \text{for } \varsigma \in \{1, 2, \ldots, n_3\}, \tag{C.18}$$

where

$$x_{T+1}^{**} = x_T + \sum_{\ell=1}^{n_1} \gamma_T^{(1)\ell} \bar{u}_T^{(1)\ell**} + \sum_{\ell=1}^{n_1} \sigma_T^{(1)\ell} \bar{u}_{T-1}^{(1)\ell**} - \lambda_T^{(1)} x_T,$$

$$y_{T+1}^{**} = y_T + \sum_{\ell=1}^{n_1} \gamma_T^{(2)\ell} \bar{u}_T^{(2)\ell**} + \sum_{\ell=1}^{n_2} \sigma_T^{(2)\ell} \bar{u}_{T-1}^{(2)\ell**} - \lambda_T^{(2)} x_T.$$

The right-hand sides (C.16)–(C.18) are linear functions in (x, y), hence we can express (C.16)–(C.18) as follows:

$$W^1(T, x, y; \underline{\bar{u}}_{T-}^{**}) = (A_T^{(1)} x_T + B_T^{(1)} y_T + C_T^{(1)}) \delta^{T-1},$$

$$W^2(T, x, y; \underline{\bar{u}}_{T-}^{**}) = (A_T^{(2)} x_T + B_T^{(2)} y_T + C_T^{(2)}) \delta^{T-1},$$

$$W^{(3)\varsigma}(T,x,y;\underline{\bar{u}}^{**}_{T-}) = (A_T^{(3)\varsigma}x_T + B_T^{(3)\varsigma}y_T + C_T^{(3)\varsigma})\delta^{T-1},$$
$$\varsigma \in \{1,2,\ldots,n_3\}, \tag{C.19}$$

where

$$A_T^{(1)} = \sum_{i\in S^1} h_T^{(1)i} + A_{T+1}^{(1)}(1-\lambda_T^{(1)})\delta,$$

$$B_T^{(1)} = -\sum_{i\in S^1} \hat{h}_T^{(1)i} + B_{T+1}^{(1)}(1-\lambda_T^{(2)})\delta, \text{ and}$$

$C_T^{(1)}$ is the rest of the terms on the right-hand side of (C.16) which do not involve (x,y) but have previously executed controls $(\bar{u}_{T-}^{(1)}, \bar{u}_{T-}^{(2)})$;

$$A_T^{(2)} = -\sum_{j\in S^2} \hat{h}_T^{(2)j} + A_{T+1}^{(2)}(1-\lambda_T^{(1)})\delta,$$

$$B_T^{(2)} = \sum_{j\in S^2} h_T^{(2)j} + B_{T+1}^{(2)}(1-\lambda_T^{(2)})\delta, \text{ and}$$

$C_T^{(2)}$ is the rest of the terms on the right-hand side of (C.17) which do not involve (x,y) but have previously executed controls $(\bar{u}_{T-}^{(1)}, \bar{u}_{T-}^{(2)})$;

$$A_T^{(3)\varsigma} = -h_T^{(3)\varsigma} + A_{T+1}^{(3)\varsigma}(1-\lambda_T^{(1)})\delta,$$
$$B_T^{(3)\varsigma} = -\hat{h}_T^{(3)\varsigma} + B_{T+1}^{(3)\varsigma}(1-\lambda_T^{(2)})\delta, \text{ and}$$

$C_T^{(3)\varsigma}$ is the rest of the terms on the right-hand side of (C.18) which do not involve (x,y) but have previously executed controls $(\bar{u}_{T-}^{(1)}, \bar{u}_{T-}^{(2)})$. (C.20)

Now, we consider the operation in the second last period, that is, period $T-1$. Invoking the $T-1$ equation in (C.2), we have

$$W^1(T-1,x,y;\underline{\bar{u}}^{**}_{(T-1)-})$$
$$= \max_{u_{T-1}^{(1)},\bar{u}_{T-1}^{(1)}} \Bigg\{ \sum_{i\in S^1} \Bigg[\Bigg(p_{T-1}^{(1)i} - b_{T-1}^{(1)i(i)} u_{T-1}^{(1)i} + \sum_{\substack{\ell\in S^1\\ \ell\neq i}} \varpi_{T-1}^{(1)i(\ell)} u_{T-1}^{(1)\ell} \Bigg) u_{T-1}^{(1)\ell}$$
$$- c_{T-1}^{(1)i}\left(u_{T-1}^{(1)i}\right)^2 - w_{T-1}^{(1)i}\left(\bar{u}_{T-1}^{(1)i}\right)^2$$

$$\left. + h_{T-1}^{(1)i} x_{T-1} - \hat{h}_{T-1}^{(1)i} y_{T-1} \right] \delta^{T-2}$$

$$\left. + (A_T^{(1)} x_T - B_T^{(1)} y_T + C_T^{(1)}) \delta^{T-1} \right\} \tag{C.21}$$

where the term $(A_T^{(1)} x_T - B_T^{(1)} y_T + C_T^{(1)}) \delta^{T-1}$ can be expressed as

$$\begin{aligned}
&(A_T^{(1)} x_T - B_T^{(1)} y_T + C_T^{(1)}) \delta^{T-1} \\
&= \sum_{i \in S^1} \left[\left(p_T^{(1)i} - b_T^{(1)i(i)} u_T^{(1)i^{**}} + \sum_{\substack{l \in S^1 \\ l \neq i}} \varpi_T^{(1)i(l)} u_T^{(1)l^{**}} \right) u_T^{(1)i^{**}} \right. \\
&\qquad \left. - c_T^{(1)i} (u_T^{(1)i^{**}})^2 - w_T^{(1)i} (\bar{u}_T^{(1)i**})^2 + h_T^{(1)i} x_T - \hat{h}_T^{(1)i} y_T \right] \delta^{T-1} \\
&\quad + (A_{T+1}^{(1)} x_{T+1}^{**} - B_{T+1}^{(1)} y_{T+1}^{**} + C_{T+1}^{(1)}) \delta^T \\
&\equiv F^{i(1)^{**}} (\underline{u}_T^{(1)i**}, \underline{\bar{u}}_T^{(1)i**}) \delta^{T-1} \\
&\quad + \left[A_{T+1}^{(1)} \left(x_T + \sum_{\ell=1}^{n_1} \gamma_T^{(1)\ell} \bar{u}_T^{(1)\ell^{**}} + \sum_{\ell=1}^{n_1} \sigma_T^{(1)\ell} \bar{u}_{T-1}^{(1)\ell^{**}} - \lambda_T^{(1)} x_T \right) \right. \\
&\qquad + B_{T+1}^{(1)} \left(y_T + \sum_{\ell=1}^{n_1} \gamma_T^{(2)\ell} \bar{u}_T^{(2)\ell^{**}} + \sum_{\ell=1}^{n_2} \sigma_T^{(2)\ell} \bar{u}_{T-1}^{(2)\ell^{**}} - \lambda_T^{(2)} x_T \right) \\
&\qquad \left. + C_{T+1}^{(1)} \right] \delta^T.
\end{aligned} \tag{C.22}$$

Performing the maximization operator in (C.20) with respect to $\bar{u}_{T-1}^{(1)i}$ yields

$$A_T^{(1)} \gamma_{T-1}^{(1)i} \delta + A_{T+1}^{(1)} \sigma_T^{(1)i} \delta^2 = 2 w_{T-1}^{(1)i} \bar{u}_{T-1}^{(1)i},$$

$$\text{which implies } \bar{u}_{T-1}^{(1)i**} = \frac{A_T^{(1)}\gamma_{T-1}^{(1)i}\delta + A_{T+1}^{(1)}\sigma_T^{(1)i}\delta^2}{2w_{T-1}^{(1)i}}, \quad \text{(C.23)}$$

for $i \in \{1, 2, \ldots, n_1\}$.

Following the above analysis in performing the maximization operator in (C.3) with respect to $\bar{u}_{T-1}^{(2)j}$ yields

$$\bar{u}_{T-1}^{(2)j**} = \frac{B_T^{(2)}\gamma_{T-1}^{(2)j}\delta + B_{T+1}^{(2)}\sigma_T^{(2)j}\delta^2}{2w_{T-1}^{(2)j}},$$

$$\text{for } j \in \{1, 2, \ldots, n_2\}. \quad \text{(C.24)}$$

The game equilibrium target controls: $(u_{T-1}^{(1)1**}, u_{T-1}^{(1)2**}, \ldots, u_{T-1}^{(1)n_1**})$, $(u_{T-1}^{(2)1**}, u_{T-1}^{(2)2**}, \ldots, u_{T-1}^{(2)n_2**})$, and $(u_{T-1}^{(3)1**}, u_{T-1}^{(3)2**}, \ldots, u_{T-1}^{(3)n_3**})$ can be obtained from (C.6)–(C.8) with $k = T-1$. The game equilibrium payoffs of the two coalitions and n_3 unallied nations at period $T-1$ can be obtained as follows:

$$W^1(T-1, x, y; \underline{\bar{u}}_{(T-1)T-}^{**}) = (A_{T-1}^{(1)}x_{T-1} + B_{T-1}^{(1)}y_{T-1} + C_{T-1}^{(1)})\delta^{T-2},$$

$$W^2(T-1, x, y; \underline{\bar{u}}_{(T-1)-}^{**}) = (A_{T-1}^{(2)}x_{T-1} + B_{T-1}^{(2)}y_{T-1} + C_{T-1}^{(2)})\delta^{T-2},$$

$$W^{(3)\varsigma}(T-1, x, y; \underline{\bar{u}}_{(T-1)-}^{**}) = (A_{T-1}^{(3)\varsigma}x_{T-1} + B_{T-1}^{(3)\varsigma}y_{T-1} + C_{T-1}^{(3)\varsigma})\delta^{T-2},$$

$$\varsigma \in \{1, 2, \ldots, n_3\}. \quad \text{(C.25)}$$

The values of $A_{T-1}^{(1)}, B_{T-1}^{(1)}y_{T-1}, C_{T-1}^{(1)}, A_{T-1}^{(2)}, B_{T-1}^{(2)}, C_{T-1}^{(2)}, A_{T-1}^{(3)\varsigma}, B_{T-1}^{(3)\varsigma}, C_{T-1}^{(3)\varsigma}$ are similar to those given in (C.20) with the period T parameters replaced by the period $T-1$ parameters.

Repeating the analysis for periods $k \in \{T-2, T-3, \ldots, 1\}$, we can obtain the game equilibrium payoffs of the two coalitions and n_3 unallied nations at period $k \in \{1, 2, \ldots, T\}$ as follows:

$$W^1(k, x, y; \underline{\bar{u}}_{k-}^{**}) = (A_k^{(1)}x_k + B_k^{(1)}y_k + C_k^{(1)})\delta^{k-1},$$

$$W^2(k, x, y; \underline{\bar{u}}_{k-}^{**}) = (A_k^{(2)}x_k + B_k^{(2)}y_k + C_k^{(2)})\delta^{k-1},$$

$$W^{(3)\varsigma}(k, x, y; \underline{\bar{u}}_{k-}^{**}) = (A_k^{(3)\varsigma}x_k - B_k^{(3)\varsigma}y_k + C_k^{(3)\varsigma})\delta^{k-1},$$

$$\varsigma \in \{1, 2, \ldots, n_3\}, \quad \text{(C.26)}$$

where

$$A_k^{(1)} = \sum_{i \in S^1} h_k^{(1)i} + A_{k+1}^{(1)}(1 - \lambda_k^{(1)})\delta,$$

$$B_k^{(1)} = -\sum_{i \in S^1} \hat{h}_k^{(1)i} + B_{k+1}^{(1)}(1 - \lambda_k^{(2)})\delta, \text{ and}$$

$C_k^{(1)}$ is a term which does not involve (x, y) but has previously executed controls $(\bar{u}_{k-}^{(1)}, \bar{u}_{k-}^{(2)})$;

$$A_k^{(2)} = -\sum_{j \in S^2} \hat{h}_k^{(2)j} + A_{k+1}^{(2)}(1 - \lambda_k^{(1)})\delta,$$

$$B_k^{(2)} = \sum_{j \in S^2} h_k^{(2)j} + B_{k+1}^{(2)}(1 - \lambda_k^{(2)})\delta, \text{ and}$$

$C_k^{(2)}$ is a term which does not involve (x, y) but has previously executed controls $(\bar{u}_{k-}^{(1)}, \bar{u}_{k-}^{(2)})$;

$$A_k^{(3)\varsigma} = -h_k^{(3)\varsigma} + A_{k+1}^{(3)\varsigma}(1 - \lambda_k^{(1)})\delta,$$

$$B_k^{(3)\varsigma} = -\hat{h}_k^{(3)\varsigma} + B_{k+1}^{(3)\varsigma}(1 - \lambda_k^{(2)})\delta, \text{ and}$$

$C_k^{(3)\varsigma}$ is a term which does not involve (x, y) but has previously executed controls $(\bar{u}_{k-}^{(1)}, \bar{u}_{k-}^{(2)})$.

(C.27)

Finally, the game equilibrium target controls can be obtained as follows.

The game equilibrium target controls of Coalition 1 are $(u_k^{(1)1**}, u_k^{(1)2**}, \ldots, u_k^{(1)n_1**})$ which is the solution of the linear equation system

$$\frac{p_k^{(1)i}}{2b_k^{(1)i(i)} + 2c_k^{(1)i}} = u_k^{(1)i} - \sum_{\substack{\ell \in S^1 \\ \ell \neq i}} \frac{\varpi_k^{(1)i(\ell)} + \varpi_k^{(1)\ell(i)}}{2b_k^{(1)i(i)} + 2c_k^{(1)i}} u_k^{(1)\ell},$$

for $i \in \{1, 2, \ldots, n_1\}$, and

$$\bar{u}_T^{(1)i**} = \frac{A_{T+1}^{(1)} \gamma_T^{(1)i} \delta}{2w_T^{(1)i}},$$

$$\bar{u}_k^{(1)i**} = \frac{A_{k+1}^{(1)}\gamma_k^{(1)i}\delta + A_{k+2}^{(1)}\sigma_k^{(1)i}\delta^2}{2w_k^{(1)i}},$$

for $k \in \{1, 2, \ldots, T-1\}$ and $i \in \{1, 2, \ldots, n_1\}$. (C.28)

The game equilibrium target controls of Coalition 2 are $(u_k^{(2)1**}, u_k^{(2)2**}, \ldots, u_k^{(2)n_2**})$ which is the solution of the linear equation system

$$\frac{p_k^{(2)j}}{2b_k^{(2)j(j)} + 2c_k^{(2)j}} = u_k^{(2)j} - \sum_{\substack{\ell \in S^2 \\ \ell \neq j}} \frac{\varpi_k^{(2)j(\ell)} + \varpi_k^{(2)\ell(j)}}{2b_k^{(2)j(j)} + 2c_k^{(2)j}} u_k^{(2)\ell},$$

for $j \in \{1, 2, \ldots, n_2\}$, and

$$\bar{u}_T^{(2)j**} = \frac{B_{T+1}^{(2)}\gamma_T^{(2)j}\delta}{2w_T^{(2)k}},$$

$$\bar{u}_k^{(2)j**} = \frac{B_{k+1}^{(2)}\gamma_k^{(2)j}\delta + B_{k+2}^{(2)}\sigma_k^{(2)j}\delta^2}{2w_k^{(2)j}},$$

for $k \in \{1, 2, \ldots, T-1\}$ and $j \in \{1, 2, \ldots, n_2\}$. (C.29)

The game equilibrium target controls of unallied ς nations are $(u_k^{(3)1**}, u_k^{(3)2**}, \ldots, u_k^{(3)n_2**})$ which is the solution of the linear equation system

$$\frac{p_k^{(3)\varsigma}}{2b_k^{(3)\varsigma(\varsigma)} + 2c_k^{(3)\varsigma}} = u_k^{(3)\varsigma} - \sum_{\substack{\ell \in S^3 \\ \ell \neq \varsigma}} \frac{\varpi_k^{(3)\varsigma(\ell)}}{2b_k^{(3)\varsigma(\varsigma)} + 2c_k^{(3)\varsigma}} u_k^{(3)\ell},$$

for $\varsigma \in \{1, 2, \ldots, n_3\}$. (C.30)

■

Appendix D: Proof of Propositions 6.5 and 6.6 in Chapter 5

Invoking Theorem 3.1 of Chapter 5, the solution of the grand coalition game (6.22) can be characterized as follows. Let $\underline{u}_k^* = (\underline{u}_k^{(1)*}, \underline{u}_k^{(2)*}, \underline{u}_k^{(3)*})$ and $\underline{\bar{u}}_k^* = (\underline{\bar{u}}_k^{(1)*}, \underline{\bar{u}}_k^{(2)*})$ be the set of Pareto optimal target controls under the grand coalition in period k, and $W(k,x,y;\underline{\bar{u}}^*_{k-})$ be the maximized joint payoff of all nations at period k. The function $W(k,x,y;\underline{\bar{u}}^*_{k-})$ satisfies the following recursive equations:

$$
\begin{aligned}
&W(T+1,x,y;\underline{\bar{u}}^*_{(T+1)-})\\
&\quad=\left[\sum_{i\in S^1}(h^{(1)i}_{T+1}x_{T+1}-\hat{h}^{(1)i}_{T+1}y_{T+1}+Q^{(1)i}_{T+1})\right.\\
&\qquad+\sum_{j\in S^2}(h^{(2)j}_{T+1}y_{T+1}-\hat{h}^{(2)j}_{T+1}x_{T+1}+Q^{(2)j}_{T+1})\\
&\qquad\left.+\sum_{\varsigma\in S^3}(Q^{(3)\varsigma}_{T+1}-h^{(3)\varsigma}_{T+1}x_{T+1}-\hat{h}^{(3)\varsigma}_{T+1}y_{T+1})\right]\delta^T
\end{aligned}
\tag{D.1}
$$

$$
\begin{aligned}
&W(k,x,y;\underline{\bar{u}}^*_{k-})\\
&=\max_{\substack{u^{(1)i}_k,\bar{u}^{(1)i}_k,i\in S^1\\ u^{(2)j}_k,\bar{u}^{(2)j}_k,j\in S^2\\ u^{(3)\varsigma}_k,\varsigma\in S^3}}\left\{\sum_{i\in S^1}\left[\left(p^{(1)i}_k-b^{(1)i(i)}_k u^{(1)i}_k+\sum_{\substack{\ell\in S^1\\ \ell\neq i}}\varpi^{(1)i(\ell)}_k u^{(1)\ell}_k\right)u^{(1)i}_k\right.\right.\\
&\left.-c^{(1)i}_k(u^{(1)i}_k)^2-w^{(1)i}_k(\bar{u}^{(1)i}_k)^2+h^{(1)i}_k x_k-\hat{h}^{(1)i}_k y_k\right]\delta^{k-1}\\
&+\sum_{j\in S^2}\left[\left(p^{(2)j}_k-b^{(2)j(j)}_k u^{(2)j}_k+\sum_{\substack{\ell\in S^2\\ \ell\neq j}}\varpi^{(2)j(\ell)}_k u^{(2)\ell}_k\right)u^{(2)j}_k\right.\\
&\left.-c^{(2)j}_k(u^{(2)j}_k)^2-w^{(2)j}_k(\bar{u}^{(2)j}_k)^2+h^{(2)j}_k y_k-\hat{h}^{(2)j}_k x_k\right]\delta^{k-1}\\
&+\sum_{\varsigma\in S^3}\left[\left(-b^{(3)\varsigma(\varsigma)}_k u^{(3)\varsigma}_k+\sum_{\substack{\ell\in S^3\\ \ell\neq\varsigma}}\varpi^{(3)\varsigma(\ell)}_k u^{(3)\varsigma}_k\right)u^{(3)\varsigma}_k-c^{(3)\varsigma}_k(u^{(3)\varsigma}_k)^2\right.
\end{aligned}
$$

$$\left. - h_k^{(3)\varsigma} x_k - \hat{h}_k^{(3)\varsigma} y_k \right] \delta^{k-1} + W(k+1, x_{k+1}, y_{k+1}; \underline{\bar{u}}^*_{(k+1)-}) \Bigg\}, \tag{D.2}$$

where

$$x_{t+1} = x_t + \sum_{\ell=1}^{n_1} \gamma_t^{(1)\ell} \bar{u}_t^{(1)\ell} + \sum_{\ell=1}^{n_1} \sigma_t^{(1)\ell} \bar{u}_{t-1}^{(1)\ell*} - \lambda_t^{(1)} x_t,$$

$$y_{t+1} = y_t + \sum_{\ell=1}^{n_1} \gamma_t^{(2)\ell} \bar{u}_t^{(2)\ell} + \sum_{\ell=1}^{n_2} \sigma_t^{(2)\ell} \bar{u}_{t-1}^{(2)\ell*} - \lambda_k^{(2)} x_k.$$

Performing the indicated maximization operators in (D.1)–(D.2) sequentially backward from period T to period 1 yields the grand coalition cooperative payoff at period $k \in \{1, 2, \ldots, T\}$:

$$W(k, x, y; \underline{\bar{u}}^*_{k-}) = (A_k x_k + B_k y_k + C_k)\delta^{k-1}, \tag{D.3}$$

where

$$A_k = \sum_{i \in S^1} h_k^{(1)i} - \sum_{j \in S^2} \hat{h}_k^{(2)j} - \sum_{\varsigma \in S^3} h_k^{(3)\varsigma} + A_{k+1}(1 - \lambda_k^{(1)})\delta,$$

$$B_k = \sum_{j \in S^2} h_k^{(2)j} - \sum_{j \in S^1} \hat{h}_k^{(1)i} - \sum_{\varsigma \in S^3} \hat{h}_k^{(3)\varsigma} + B_{k+1}(1 - \lambda_k^{(2)})\delta, \text{ and}$$

$C_k^{(1)}$ is a term which does not involve (x, y) but has previously executed controls $(\bar{u}_{k-}^{(1)}, \bar{u}_{k-}^{(2)})$,
with

$$A_{T+1} = \sum_{i \in S^1} h_{T+1}^{(1)i} - \sum_{j \in S^2} h_{T+1}^{(2)j} - \sum_{\varsigma \in S^3} h_{T+1}^{(3)\varsigma},$$

$$B_{T+1} = \sum_{j \in S^2} h_{T+1}^{(2)j} - \sum_{i \in S^1} \hat{h}_{T+1}^{(1)i} - \sum_{\varsigma \in S^3} \hat{h}_{T+1}^{(3)\varsigma},$$

$$C_{T+1} = \sum_{i \in S^1} Q_{T+1}^{(1)i} + \sum_{j \in S^2} Q_{T+1}^{(2)j} + \sum_{\varsigma \in S^3} Q_{T+1}^{(3)\varsigma}.$$

The set of Pareto optimal target controls under the grand coalition can be obtained as follows:

$(u_k^{(1)1*}, u_k^{(1)2*}, \ldots, u_k^{(1)n_1*})$ which is the solution of the linear equation system

$$\frac{p_k^{(1)i}}{2b_k^{(1)i(i)} + 2c_k^{(1)i}} = u_k^{(1)i} - \sum_{\substack{\ell \in S^1 \\ \ell \neq i}} \frac{\varpi_k^{(1)i(\ell)} + \varpi_k^{(1)\ell(i)}}{2b_k^{(1)i(i)} + 2c_k^{(1)i}} u_k^{(1)\ell},$$

for $i \in \{1, 2, \ldots, n_1\}$, and

$$\bar{u}_T^{(1)i*} = \frac{A_{T+1}\gamma_T^{(1)i}\delta}{2w_T^{(1)i}},$$

$$\bar{u}_k^{(1)i*} = \frac{A_{k+1}\gamma_k^{(1)i}\delta + A_{k+2}\sigma_k^{(1)i}\delta^2}{2w_k^{(1)i}},$$

$$\text{for } k \in \{1, 2, \ldots, T-1\} \text{ and } i \in \{1, 2, \ldots, n_1\}; \quad \text{(D.4)}$$

$(u_k^{(2)1*}, u_k^{(2)2*}, \ldots, u_k^{(2)n_2*})$ which is the solution of the linear equation system

$$\frac{p_k^{(2)j}}{2b_k^{(2)j(j)} + 2c_k^{(2)j}} = u_k^{(2)j} - \sum_{\substack{\ell \in S^2 \\ \ell \neq j}} \frac{\varpi_k^{(2)j(\ell)} + \varpi_k^{(2)\ell(j)}}{2b_k^{(2)j(j)} + 2c_k^{(2)j}} u_k^{(2)\ell},$$

for $j \subset \{1, 2, \ldots, n_2\}$, and

$$\bar{u}_T^{(2)j*} = \frac{B_{T+1}\gamma_T^{(2)j}\delta}{2w_T^{(2)k}},$$

$$\bar{u}_k^{(2)j*} = \frac{B_{k+1}\gamma_k^{(2)j}\delta + B_{k+2}\sigma_k^{(2)j}\delta^2}{2w_k^{(2)j}},$$

$$\text{for } k \in \{1, 2, \ldots, T-1\} \text{ and } j \in \{1, 2, \ldots, n_2\}, \quad \text{(D.5)}$$

$(u_k^{(3)1*}, u_k^{(3)2*}, \ldots, u_k^{(3)n_2*})$ which is the solution of the linear equation system

$$\frac{p_k^{(3)\varsigma}}{2b_k^{(3)\varsigma(\varsigma)} + 2c_k^{(3)\varsigma}} = u_k^{(3)\varsigma} - \sum_{\substack{\ell \in S^3 \\ \ell \neq \varsigma}} \frac{\varpi_k^{(3)\varsigma(\ell)}}{2b_k^{(3)\varsigma(\varsigma)} + 2c_k^{(3)\varsigma}} u_k^{(3)\ell},$$

$$\text{for } \varsigma \in \{1, 2, \ldots, n_3\}. \quad \text{(D.6)}$$

Hence, Propositions 6.5 and 6.6 follow. ■

Appendix E: Proof of Proposition 5.1 in Chapter 6

Using (5.4), we have

$$A_{11}^{(i)i} = Q_{11}^{(i)i}, \quad A_{11}^{(i)j} = 0,$$
$$C_{11}^{i} = \sum_{t=11-3}^{10} \bar{p}_{11}^{(t)i} \bar{u}_t^{(4)i**} - \sum_{t=11-2}^{10} \bar{w}_{11}^{(t)i} \bar{u}_t^{(3)i**} + \varpi_{11}^{i}, \tag{E.1}$$

for $i, j \in \{1, 2\}$ and $i \neq j$.

Invoking the technique of backward induction, we consider first the last operational stage 10. Using Proposition 5.1 and (5.4), we can express the game equilibrium strategies in stage 10 as follows:

$$u_{10}^{Ii**} = \frac{A_{11}^{(i)i} \varepsilon_{10}^{i} \delta}{2c_{10}^{Ii}}, \quad u_{10}^{IIi**} = \frac{(R_{10}^{i})^2}{4(c_{10}^{IIi})^2} x^i,$$
$$\bar{u}_{10}^{(3)i**} = \frac{(\chi_{10}^{i})^2}{4\left(w_{10}^{|10|i} + \delta \bar{w}_{11}^{|10|i}\right)^2},$$
$$\bar{u}_{10}^{(4)i**} = \frac{p_{10}^{|10|i} + \delta \bar{q}_{11}^{|10|i}}{2\phi_{10}^{i}}, \quad \text{for } i \in \{1, 2\}. \tag{E.2}$$

Substituting the optimal strategies in (E.2) into the stage 10 equation of (4.5), we obtain

$$V^i(10, x; \underline{\bar{u}}_{10-}^{(3)**}, \underline{\bar{u}}_{10-}^{(4)**})$$
$$= (A_{10}^{(i)i} x^i + A_{10}^{(i)j} x^j + C_{10}^{i}) \delta^9$$
$$= \left[\frac{(R_{10}^{i})^2}{4c_{10}^{IIi}} x^i - \frac{(A_{11}^{(i)i} \varepsilon_{10}^{i} \delta)^2}{4c_{10}^{Ii}} + p_{10}^{|10|i} \frac{p_{10}^{|10|i} + \delta \bar{p}_{11}^{|10|i}}{2\phi_{10}^{i}} \right.$$
$$+ \sum_{t=10-3}^{9} p_{10}^{|t|i} \bar{u}_t^{(4)i**} - \frac{\left(p_{10}^{|10|i} + \delta \bar{p}_{11}^{|10|i}\right)^2}{4\varphi_{10}^{i}} + \frac{(\chi_{10}^{i})^2}{2\left(w_{10}^{|10|i} + \delta \bar{w}_{11}^{|10|i}\right)}$$
$$\left. - w_{10}^{|10|i} \frac{(\chi_{10}^{i})^2}{4\left(w_{10}^{|10|i} + \delta \bar{w}_{11}^{|10|i}\right)^2} - \sum_{t=10-2}^{9} w_{10}^{|t|i} \bar{u}_t^{(3)i**} \right] \delta^9$$

$$+\left[A_{11}^{(i)i}\left(x^i+\varepsilon_{10}^i\frac{A_{11}^{(i)i}\varepsilon_{10}^i\delta}{2c_{10}^{Ii}}-\lambda_{10}^i x^i+\gamma_{10}^{(i)j}x^j\right)\right.$$

$$+p_{11}^{|10|i}\frac{p_{10}^{|10|i}+\delta\bar{p}_{11}^{|10|i}}{2\phi_{10}^i}+\sum_{t=11-3}^{9}\bar{p}_{11}^{|t|i}\bar{u}_t^{(4)i**}$$

$$\left.-\bar{w}_{11}^{|10|i}\frac{(\chi_{10}^i)^2}{4\left(w_{10}^{|10|i}+\delta\bar{w}_{11}^{|10|i}\right)^2}-\sum_{t=11-2}^{9}\bar{w}_{11}^{|t|i}\bar{u}_t^{(3)i**}+\varpi_{11}^i\right]\delta^{10},$$

$$i,j\in\{1,2\}\text{ and }i\neq j. \tag{E.3}$$

System (E.3) yields a system of equations in which the right-hand side and left-hand side are linear functions of (x^1, x^2). For (E.3) to hold, it is required that

$$A_{10}^{(i)i}x^i=\frac{(R_{10}^i)^2x^i}{4c_{10}^{IIi}}+\delta A_{11}^{(i)i}(1-\lambda_{10}^i)x^i,$$

$$A_{10}^{(i)j}x^j=\delta A_{11}^{(i)i}\gamma_{10}^{(i)j}x^j. \tag{E.4}$$

From (E.4), we obtain

$$A_{10}^{(i)i}=\frac{(R_{10}^i)^2}{4c_{10}^{IIi}}+\delta A_{11}^{(i)i}(1-\lambda_{10}^i),$$

$$A_{10}^{(i)j}=\delta A_{11}^{(i)i}\gamma_{10}^{(i)j}. \tag{E.5}$$

In addition,

$$C_{10}^i=\left[p_{10}^{|10|i}\frac{p_{10}^{|10|i}+\delta\bar{p}_{11}^{|10|i}}{2\varphi_{10}^i}-\frac{(A_{11}^{(i)i}\varepsilon_{10}^i\delta)^2}{4c_{10}^{Ii}}+\sum_{t=10-3}^{9}p_{10}^{|t|i}\bar{u}_t^{(4)i**}\right.$$

$$-\frac{\left(p_{10}^{|10|i}+\delta\bar{p}_{11}^{|10|i}\right)^2}{4\varphi_{10}^i}+\frac{(\chi_{10}^i)^2}{2\left(w_{10}^{|10|i}+\delta\bar{w}_{11}^{|10|i}\right)}$$

$$\left.-w_{10}^{|10|i}\frac{(\chi_{10}^i)^2}{4\left(w_{10}^{|10|i}+\delta\bar{w}_{11}^{|10|i}\right)^2}-\sum_{t=10-2}^{9}w_{11}^{|t|i}\bar{u}_t^{(3)i**}\right]$$

$$+\left[A_{11}^{(i)i}\left[\varepsilon_{10}^i\frac{A_{11}^{(i)i}\varepsilon_{10}^i\delta}{2c_{10}^{Ii}}\right]+p_{11}^{|10|i}\frac{p_{10}^{|10|i}+\delta\bar{p}_{11}^{|10|i}}{2\varphi_{10}^i}\right.$$

$$+ \sum_{t=11-3}^{9} \bar{p}_{11}^{|t|i} \bar{u}_t^{(4)i**} - \bar{w}_{11}^{|10|i} \frac{(\chi_{10}^i)^2}{4\left(w_{10}^{|10|i} + \delta \bar{w}_{11}^{|10|i}\right)^2}$$

$$\left. - \sum_{t=11-2}^{9} \bar{w}_{11}^{|t|i} \bar{u}_t^{(3)i**} + \varpi_{11}^i \right] \delta, \tag{E.6}$$

which is an expression containing previously executed controls $(\underline{\bar{u}}_{10-}^{(3)**}, \underline{\bar{u}}_{10-}^{(4)**})$.

Then we move to stage 9.

Using $V^i(10, x; \underline{\bar{u}}_{10-}^{(3)**}, \underline{\bar{u}}_{10-}^{(4)**}) = (A_{10}^{(i)i} x^i + A_{10}^{(i)j} x^j + C_{10}^i)\delta^9$ derived in (E.3)–(E.6) and the stage 9 equation in (5.6), the game equilibrium strategies in stage 9 can be expressed as follows:

$$u_9^{Ii**} = \frac{A_{10}^{(i)i} \varepsilon_9^i \delta}{2c_9^{Ii}}, \quad u_9^{IIi**} = \frac{(R_9^i)^2}{4(c_9^{IIi})^2} x^i,$$

$$\bar{u}_9^{(3)i**} = (\chi_9^i)^2 \div 4\left[w_9^{|9|i} + \frac{\theta_9}{\sum_{\zeta=9}^{10} \theta_\zeta} \bar{w}_{10}^{|9|i} \delta + \frac{\sum_{\zeta=10}^{10} \theta_\zeta}{\sum_{\zeta=9}^{10} \theta_\zeta} \left(w_{10}^{|9|i} \delta + \bar{w}_{11}^{|9|i} \delta^2\right)\right]^2,$$

$$\bar{u}_9^{(4)i**} = \frac{1}{2\varphi_9^i} \left[p_9^{|9|i} + \frac{\theta_9}{\sum_{\zeta=9}^{10} \theta_\zeta} \bar{p}_{10}^{|9|i} \delta + \frac{\sum_{\zeta=10}^{10} \theta_\zeta}{\sum_{\zeta=9}^{10} \theta_\zeta} \left(p_{10}^{|9|i} \delta + \bar{p}_{11}^{|9|i} \delta^2\right)\right],$$

for $i \in \{1, 2\}$. (E.7)

Substituting the optimal strategies in (E.7) into the stage 9 equation of (5.6), we obtain

$$V^i(9, x; \underline{\bar{u}}_{9-}^{(3)**}, \underline{\bar{u}}_{9-}^{(4)**})$$

$$= (A_9^{(i)i} x^i + A_9^{(i)j} x^j + C_9^i)\delta^8$$

$$= \left\{ \frac{(R_9^i)^2}{4c_9^{IIi}} x^i - \frac{(A_{10}^{(i)i} \varepsilon_9^i \delta)^2}{4c_9^{Ii}} + p_9^{|9|i} \frac{1}{2\varphi_9^i} \right.$$

$$\times \left[p_9^{|9|i} + \frac{\theta_9}{\sum_{\zeta=9}^{10} \theta_\zeta} \bar{p}_{10}^{|9|i} \delta + \frac{\sum_{\zeta=10}^{10} \theta_\zeta}{\sum_{\zeta=9}^{10} \theta_\zeta} \left(p_{10}^{|9|i} \delta + \bar{p}_{11}^{|9|i} \delta^2\right)\right]$$

$$+ \sum_{t=9-3}^{8} p_9^{|t|i} \bar{u}_t^{(4)i**} - \frac{1}{4\varphi_9^i}$$

$$\times \left[p_9^{|9|i} + \frac{\theta_9}{\sum_{\zeta=9}^{10} \theta_\zeta} \bar{p}_{10}^{|9|i} \delta + \frac{\sum_{\zeta=10}^{10} \theta_\zeta}{\sum_{\zeta=9}^{10} \theta_\zeta} \left(p_{10}^{|9|i} \delta + \bar{p}_{11}^{|9|i} \delta^2 \right) \right]^2 + (\chi_9^i)^2$$

$$\div 2 \left[w_9^{|9|i} \frac{\theta_9}{\sum_{\zeta=9}^{10} \theta_\zeta} \bar{w}_{10}^{|9|i} \delta + \frac{\sum_{\zeta=10}^{10} \theta_\zeta}{\sum_{\zeta=9}^{10} \theta_\zeta} \left(w_{10}^{|9|i} \delta + \bar{w}_{11}^{|9|i} \delta^2 \right) \right]$$

$$-w_9^{|9|i} (\chi_9^i)^2 \div 4 \left[w_9^{|9|i} + \frac{\theta_9}{\sum_{\zeta=9}^{10} \theta_\zeta} \bar{w}_{10}^{|9|i} \delta \right.$$

$$\left. + \frac{\sum_{\zeta=10}^{10} \theta_\zeta}{\sum_{\zeta=9}^{10} \theta_\zeta} \left(w_{10}^{|9|i} \delta + \bar{w}_{11}^{|9|i} \delta^2 \right) \right]^2$$

$$\left. - \sum_{t=9-2}^{8} w_9^{|t|i} \bar{u}_t^{(3)i**} \right\} \delta^8 + \frac{\theta_9}{\sum_{\zeta=9}^{10} \theta_\zeta}$$

$$\times \left\{ Q_{10}^{(i)i} \left(x^i + \varepsilon_9^i \frac{A_{10}^{(i)i} \varepsilon_9^i \delta}{2c_9^{Ii}} - \lambda_9^i x^i + \gamma_9^{(i)j} x^j \right) \right.$$

$$+ p_{10}^{|9|i} \frac{1}{2\varphi_9^i} \left[p_{10}^{|9|i} + \frac{\theta_9}{\sum_{\zeta=9}^{10} \theta_\zeta} \bar{p}_{10}^{|9|i} \delta + \frac{\sum_{\zeta=10}^{10} \theta_\zeta}{\sum_{\zeta=9}^{10} \theta_\zeta} \left(p_{10}^{|9|i} + \bar{p}_{11}^{|9|i} \delta^2 \right) \right]$$

$$+ \sum_{t=10-3}^{8} \bar{p}_{10}^{|t|i} \bar{u}_t^{(4)i**} - \bar{w}_{10}^{|9|i} (\chi_9^i)^2$$

$$\div 4 \left[w_9^{|9|i} + \frac{\theta_9}{\sum_{\zeta=9}^{10} \theta_\zeta} \bar{w}_{10}^{|9|i} \delta + \frac{\sum_{\zeta=10}^{10} \theta_\zeta}{\sum_{\zeta=9}^{10} \theta_\zeta} \left(w_{10}^{|9|i} \delta + \bar{w}_{11}^{|9|i} \delta^2 \right) \right]^2$$

$$\left. - \sum_{t=10-2}^{8} \bar{w}_{10}^{|t|i} \bar{u}_t^{(3)i**} + \varpi_{10}^i \right\} \delta^9 + \frac{\sum_{\zeta=10}^{10} \theta_\zeta}{\sum_{\zeta=9}^{10} \theta_\zeta}$$

$$\times \left\{ A_{10}^{(i)i} \left[x^i + \varepsilon_9^i \frac{A_{10}^{(i)i} \varepsilon_9^i \delta}{2c_9^{Ii}} - \lambda_9^i x^i + \gamma_9^{(i)j} x^j \right] \right.$$

$$\left. + A_{10}^{(i)j} \left[x^j + \frac{A_{10}^{(j)j} \varepsilon_9^j \delta}{2c_9^{Ij}} - \lambda_9^j x^j + \gamma_9^{(j)i} x^i \right] + C_{10}^{i***} \right\} \delta^9, \tag{E.8}$$

where C_{10}^{i***} is C_{10}^{i} in (E.6) with $\bar{u}_9^{(3)i**}$, $\bar{u}_9^{(3)j**}$, $\bar{u}_9^{(4)i**}$, and $u_9^{(4)j**}$ being given in (E.7).

System (E.8) yields a system of equations in which the right-hand side and left-hand side are linear functions of (x^1, x^2). For (E.8) to hold, it is required that

$$\begin{aligned}
A_9^{(i)i} x^i &= \frac{(R_9^i)^2}{4c_9^{IIi}} x^i + \frac{\theta_9}{\sum_{\zeta=9}^{10} \theta_\zeta} \delta Q_{10}^{(i)i} (1 - \lambda_9^i) x^i \\
&\quad + \frac{\sum_{\zeta=10}^{10} \theta_\zeta}{\sum_{\zeta=9}^{10} \theta_\zeta} (A_{10}^{(i)i} (1 - \lambda_9^i) x^i + A_{10}^{(i)j} \gamma_9^{(j)i} x^i) \delta, \\
A_9^{(i)j} x^j &= \frac{\theta_9}{\sum_{\zeta=9}^{10} \theta_\zeta} (\delta Q_{10}^{(i)i} \gamma_9^{(i)j} x^j) \\
&\quad + \frac{\sum_{\zeta=10}^{10} \theta_\zeta}{\sum_{\zeta=9}^{10} \theta_\zeta} (A_{10}^{(i)i} \gamma_9^{(i)j} x^j + A_{10}^{(i)j} (1 - \lambda_9^j) x^j) \delta, \\
&\quad i, j \in \{1, 2\} \text{ and } i \neq j.
\end{aligned} \tag{E.9}$$

From (E.9), we obtain

$$\begin{aligned}
A_9^{(i)i} &= \frac{(R_9^i)^2}{4c_9^{IIi}} + \frac{\theta_9}{\sum_{\zeta=9}^{10} \theta_\zeta} \delta Q_{10}^{(i)i} (1 - \lambda_9^i) \\
&\quad + \frac{\sum_{\zeta=10}^{10} \theta_\zeta}{\sum_{\zeta=9}^{10} \theta_\zeta} [A_{10}^{(i)i} (1 - \lambda_9^i) + A_{10}^{(i)j} \gamma_9^{(j)i}] \delta, \\
A_9^{(i)j} &= \frac{\theta_9}{\sum_{\zeta=9}^{10} \theta_\zeta} (\delta Q_{10}^{(i)i} \gamma_9^{(i)j}) + \frac{\sum_{\zeta=10}^{10} \theta_\zeta}{\sum_{\zeta=9}^{10} \theta_\zeta} [A_{10}^{(i)i} \gamma_9^{(i)j} + A_{10}^{(i)j} (1 - \lambda_9^i)] \delta, \\
&\quad i, j \in \{1, 2\} \text{ and } i \neq j.
\end{aligned} \tag{E.10}$$

In addition, C_9^i is an expression including all the terms on the right-hand side of (E.8) which do not involve x^1 or x^2. It is a function of previously executed controls $(\underline{u}_{9-}^{(3)**}, \underline{u}_{9-}^{(4)**})$.

Then, we move to stage 8.

Using $V^i(9, x; \underline{\bar{u}}_{9-}^{(3)**}, \underline{\bar{u}}_{9-}^{(4)**}) = (A_9^{(i)i} x^i + A_9^{(i)j} x^j + C_9^i) \delta^9$ derived in (E.8)–(E.10) and the stage 8 equation in (5.6), the game equilibrium strategies in stage 8 can be expressed as follows:

$$
\begin{aligned}
u_8^{Ii**} &= \frac{A_9^{(i)i}\varepsilon_8^i\delta}{2c_8^{Ii}}, \quad u_8^{IIi**} = \frac{(R_8^i)^2}{4(c_8^{IIi})^2}x^i, \\
\bar{u}_8^{(3)i**} &= (\chi_8^i)^2 \div 4\left[w_8^{|8|i} + \frac{\theta_8}{\sum_{\zeta=8}^{10}\theta_\zeta}\bar{w}_9^{|8|i}\delta + \frac{\sum_{\zeta=9}^{10}\theta_\zeta}{\sum_{\zeta=8}^{9}\theta_\zeta}\left(w_9^{|8|i}\delta + \bar{w}_{10}^{|8|i}\delta^2\right)\right]^2, \\
\bar{u}_8^{(4)i**} &= \frac{1}{2\varphi_8^i}\left[p_8^{|8|i} + \frac{\theta_8}{\sum_{\zeta=8}^{10}\theta_\zeta}\bar{p}_9^{|8|i}\delta + \frac{\sum_{\zeta=9}^{10}\theta_\zeta}{\sum_{\zeta=8}^{10}\theta_\zeta}\left(p_9^{|8|i}\delta + p_{10}^{|8|i}\delta^2 + \bar{p}_{11}^{|8|i}\delta^2\right)\right],
\end{aligned}
\tag{E.11}
$$

for $i \in \{1, 2\}$.

Substituting the optimal strategies in (E.11) into the stage 8 equation of (5.6), we obtain

$$
\begin{aligned}
&V^i(8, x; \underline{\bar{u}}_{8-}^{(3)**}, \underline{\bar{u}}_{8-}^{(4)**}) \\
&= (A_8^{(i)i}x^i + A_8^{(i)j}x^j + C_8^i)\delta^7 \\
&= \left\{\frac{(R_8^i)^2}{4c_8^{IIi}}x^i - \frac{(A_9^{(i)i}\varepsilon_8^i\delta)^2}{4c_8^{Ii}} + p_8^{|8|i}\frac{1}{2\varphi_8^i}\right. \\
&\quad \times \left[p_8^{|8|i} + \frac{\theta_8}{\sum_{\zeta=8}^{10}\theta_\zeta}\bar{p}_9^{|8|i}\delta + \frac{\sum_{\zeta=9}^{10}\theta_\zeta}{\sum_{\zeta=8}^{10}\theta_\zeta}\left(p_9^{|8|i}\delta + p_{10}^{|8|i}\delta^2 + \bar{p}_{11}^{|8|i}\delta^2\right)\right] \\
&\quad + \sum_{t=8-3}^{7} p_8^{|t|i}\bar{u}_t^{(4)i**} \\
&\quad - \frac{1}{4\varphi_8^i}\left[p_8^{|8|i} + \frac{\theta_8}{\sum_{\zeta=8}^{10}\theta_\zeta}\bar{p}_9^{|8|i}\delta + \frac{\sum_{\zeta=9}^{10}\theta_\zeta}{\sum_{\zeta=8}^{10}\theta_\zeta}\left(p_9^{|8|i}\delta + p_{10}^{|8|i}\delta^2 + \bar{p}_{11}^{|8|i}\delta^2\right)\right]^2 \\
&\quad + (\chi_8^i)^2 \div 2\left[w_8^{|8|i} + \frac{\theta_8}{\sum_{\zeta=8}^{10}\theta_\zeta}\bar{w}_9^{|8|i}\delta + \frac{\sum_{\zeta=9}^{10}\theta_\zeta}{\sum_{\zeta=8}^{10}\theta_\zeta}\left(w_9^{|8|i}\delta + \bar{w}_{10}^{|8|i}\delta^2\right)\right] \\
&\quad - w_8^{|8|i}(\chi_8^i)^2 \div 4\left[w_8^{|8|i} + \frac{\theta_8}{\sum_{\zeta=8}^{10}\theta_\zeta}\bar{w}_9^{|8|i}\delta + \frac{\sum_{\zeta=9}^{10}\theta_\zeta}{\sum_{\zeta=8}^{10}\theta_\zeta}\left(w_9^{|8|i}\delta + \bar{w}_{10}^{|8|i}\delta^2\right)\right]^2 \\
&\quad \left. - \sum_{t=8-2}^{7} w_8^{|t|i}\bar{u}_t^{(3)i**}\right\}\delta^7 + \frac{\theta_8}{\sum_{\zeta=8}^{10}\theta_\zeta}
\end{aligned}
$$

$$\times \left\{ Q_9^{(i)i} \left[x^i + \varepsilon_8^i \frac{A_9^{(i)i} \varepsilon_8^i \delta}{2c_8^{Ii}} - \lambda_8^i x^i + \gamma_8^{(i)j} x^j \right] + \bar{p}_9^{|8|i} \frac{1}{2\varphi_8^i} \right.$$

$$\times \left[p_8^{|8|i} + \frac{\theta_8}{\sum_{\zeta=8}^{10} \theta_\zeta} \bar{p}_9^{|8|i} \delta + \frac{\sum_{\zeta=9}^{10} \theta_\zeta}{\sum_{\zeta=8}^{10} \theta_\zeta} \left(p_9^{|8|i} \delta + p_{10}^{|8|i} \delta^2 + \bar{p}_{11}^{|8|i} \delta^2 \right) \right]$$

$$+ \sum_{t=10-3}^{8} \bar{p}_{10}^{|t|i} \bar{u}_t^{(4)i**} - \bar{w}_9^{|8|i} (\chi_8^i)^2$$

$$\div 4 \left[w_8^{|8|i} + \frac{\theta_8}{\sum_{\zeta=8}^{10} \theta_\zeta} \bar{w}_9^{|8|i} \delta + \frac{\sum_{\zeta=9}^{10} \theta_\zeta}{\sum_{\zeta=8}^{10} \theta_\zeta} \left(w_9^{|8|i} \delta + \bar{w}_{10}^{|8|i} \delta^2 \right) \right]^2$$

$$\left. - \sum_{t=10-2}^{8} \bar{w}_{10}^{|t|i} \bar{u}_t^{(3)i**} + \varpi_{10}^i \right\} \delta^9 + \frac{\sum_{\zeta=9}^{10} \theta_\zeta}{\sum_{\zeta=8}^{10} \theta_\zeta}$$

$$\times \left\{ A_9^{(i)i} \left[x^i + \varepsilon_8^i \frac{A_9^{(i)i} \varepsilon_8^i \delta}{2c_8^{Ii}} - \lambda_8^i x^i + \gamma_8^{(i)j} x^j \right] \right.$$

$$\left. + A_9^{(i)j} \left[x^j + \varepsilon_8^j \frac{A_9^{(j)j} \varepsilon_8^j \delta}{2c_8^{Ij}} - \lambda_8^j x^j + \gamma_8^{(j)i} x^i \right] + C_9^{i***} \right\} \delta^8,$$

$i, j \in \{1, 2\}$ and $i \neq j$, (E.12)

where $\bar{u}_8^{(3)i**}$, $\bar{u}_8^{(3)j**}$, $\bar{u}_8^{(4)i**}$, and $\bar{u}_8^{(4)j**}$ in C_9^{i***} are given in (E.11).

System (E.12) yields a system of equations in which the right-hand side and left-hand side are linear functions of (x^1, x^2). For (E.12) to hold, it is required that

$$A_8^{(i)i} x^i = \frac{(R_8^i)^2}{4c_8^{IIi}} x^i + \frac{\theta_8}{\sum_{\zeta=8}^{10} \theta_\zeta} \delta Q_9^{(i)i} (1 - \lambda_8^i) x^i$$

$$+ \frac{\sum_{\zeta=9}^{10} \theta_\zeta}{\sum_{\zeta=8}^{10} \theta_\zeta} [A_9^{(i)i} (1 - \lambda_8^i) x^i + A_9^{(i)j} \gamma_8^{(j)i} x^i] \delta,$$

$$A_8^{(i)j} x^j = \frac{\theta_8}{\sum_{\zeta=8}^{10} \theta_\zeta} (\delta Q_9^{(i)i} \gamma_8^{(i)j} x^j)$$

$$+ \frac{\sum_{\zeta=9}^{10} \theta_\zeta}{\sum_{\zeta=8}^{10} \theta_\zeta} [A_9^{(i)i} \gamma_8^{(i)j} x^j + A_9^{(i)j} (1 - \lambda_8^j) x^j] \delta,$$

$i, j \in \{1, 2\}$ and $i \neq j$. (E.13)

From (E.13), we obtain

$$A_8^{(i)i} = \frac{(R_8^i)^2}{4c_8^{IIi}} + \frac{\theta_8}{\sum_{\zeta=8}^{10}\theta_\zeta}\delta Q_9^{(i)i}(1-\lambda_8^i)$$
$$+ \frac{\sum_{\zeta=9}^{10}\theta_\zeta}{\sum_{\zeta=8}^{10}\theta_\zeta}[A_9^{(i)i}(1-\lambda_8^j) + A_9^{(i)j}\gamma_8^{(j)i}]\delta,$$

$$A_8^{(i)j} = \frac{\theta_8}{\sum_{\zeta=8}^{10}\theta_\zeta}(\delta Q_9^{(i)i}\gamma_8^{(i)j}) + \frac{\sum_{\zeta=9}^{10}\theta_\zeta}{\sum_{\zeta=8}^{10}\theta_\zeta}[A_9^{(i)i}\gamma_8^{(i)j} + A_9^{(i)j}(1-\lambda_8^j)]\delta,$$
$$i, j \in \{1, 2\} \text{ and } i \neq j. \tag{E.14}$$

In addition, C_8^i is an expression including all the terms on the right-hand side of (E.12) which do not involve x^1 or x^2. It is an expression containing previously executed controls $(\underline{\bar{u}}_{8-}^{(3)**}, \underline{\bar{u}}_{8-}^{(4)**})$.

Repeating the process for stages $\tau \in \{7, 6, \ldots, 1\}$, we obtain

$$u_\tau^{Ii**} = \frac{A_{\tau+1}^{(i)i}\varepsilon_\tau^i\delta}{2c_\tau^{Ii}}, \quad u_\tau^{IIi**} = \frac{(R_\tau^i)^2}{4(c_\tau^{IIi})^2}x^i,$$

$$\bar{u}_\tau^{(3)i**} = (\chi_\tau^i)^2 \div 4\left[w_\tau^{|\tau|i} + \frac{\theta_\tau}{\sum_{\zeta=\tau}^{10}\theta_\zeta}\bar{w}_9^{|8|i}\delta\right.$$
$$\left. + \frac{\sum_{\zeta=\tau+1}^{10}\theta_\zeta}{\sum_{\zeta=\tau}^{10}\theta_\zeta}\left(w_{\tau+1}^{|\tau|i}\delta + \bar{w}_{\tau+2}^{|\tau|i}\delta^2\right)\right]^2,$$

$$\bar{u}_\tau^{(4)i**} = \frac{1}{2\varphi_\tau^i}\left[p_\tau^{|\tau|i} + \frac{\theta_\tau}{\sum_{\zeta=\tau}^{10}\theta_\zeta}\bar{p}_{\tau+1}^{|\tau|i}\delta\right.$$
$$\left. + \frac{\sum_{\zeta=\tau+1}^{10}\theta_\zeta}{\sum_{\zeta=\tau}^{10}\theta_\zeta}\left(p_{\tau+1}^{|\tau|i}\delta + p_{\tau+2}^{|\tau|i}\delta^2 + \bar{p}_{\tau+3}^{|\tau|i}\delta^3\right)\right], \tag{E.15}$$

for $i \in \{1, 2\}$.

The game equilibrium payoff of agent i for stage $\tau \in \{7, 6, \ldots, 1\}$ can be obtained as

$$V^i(\tau, x; \underline{\bar{u}}_{\tau-}^{(3)**}, \underline{\bar{u}}_{\tau-}^{(4)**}) = (A_\tau^{(i)i}x^i + A_\tau^{(i)j}x^j + C_\tau^i)\delta^{\tau-1},$$

where

$$A_\tau^{(i)i} = \frac{(R_\tau^i)^2}{4c_\tau^{IIi}} + \frac{\theta_\tau}{\sum_{\zeta=\tau}^{10}\theta_\zeta}\delta Q_{\tau+1}^{(i)i}(1-\lambda_\tau^i)$$
$$+ \frac{\sum_{\zeta=\tau+1}^{10}\theta_\zeta}{\sum_{\zeta=\tau}^{10}\theta_\zeta}[A_{\tau+1}^{(i)i}(1-\lambda_\tau^j) + A_{\tau+1}^{(i)j}\gamma_\tau^{(j)i}]\delta,$$
$$A_\tau^{(i)j} = \frac{\theta_\tau}{\sum_{\zeta=\tau}^{10}\theta_\zeta}(\delta Q_{\tau+1}^{(i)i}\gamma_\tau^{(i)j})$$
$$+ \frac{\sum_{\zeta=\tau+1}^{10}\theta_\zeta}{\sum_{\zeta=\tau}^{10}\theta_\zeta}[A_{\tau+1}^{(i)i}\gamma_\tau^{(i)j} + A_{\tau+1}^{(i)j}(1-\lambda_\tau^j)]\delta,$$
$$i, j \in \{1,2\} \text{ and } i \neq j. \tag{E.16}$$

In addition, C_τ^i is an expression containing previously executed controls $(\underline{\bar{u}}_{\tau-}^{(3)**}, \underline{\bar{u}}_{\tau-}^{(4)**})$. ■

Appendix F: Proof of Proposition 6.1 in Chapter 6

Using (6.3), we have

$$A_{11}^1 = Q_{T+1}^{(1)1}, \quad A_{11}^2 = Q_{T+1}^{(2)2},$$
$$C_{11} = \sum_{i=1}^{2}\left(\sum_{t=11-2}^{10}\bar{v}_{11}^{(t)i}u_t^{(3)i} + \sum_{t=11-3}^{10}\bar{p}_{11}^{(t)i}u_t^{(4)i} + \varpi_{11}^i\right). \tag{F.1}$$

Invoking the technique of backward induction, we consider first the last operational stage 10. Using Proposition 6.1 and (6.4), we can express the optimal cooperative strategies in stage 10 as follows:

$$u_{10}^{(1)i} = \frac{(R_{10}^i)^2}{4(c_{10}^i)^2}x^i, \quad u_{10}^{(3)i} = \frac{A_{11}^i\varepsilon_{10}^{(10)i} + \bar{v}_{11}^{(10)i}}{2w_{10}^i}\delta,$$
$$u_{10}^{(4)i} = \frac{p_{10}^{(10)i} + \delta\bar{p}_{11}^{(10)i}}{2\varphi_{10}^i}, \quad \text{for } i \in \{1,2\}. \tag{F.2}$$

Substituting the optimal strategies in (B.2) into the stage 10 equation of (6.4), we obtain

$$
\begin{aligned}
&W(10, x; \underline{u}^{(3)}_{10-}, \underline{u}^{(4)}_{10-}) \\
&\quad = (A^1_{10}x^1 + A^2_{10}x^2 + C_{10})\delta^9 \\
&\quad = \sum_{i=1}^{2}\Bigg\{\frac{(R^i_{10})^2}{4(c^i_{10})}x^i - \frac{(A^i_{11}\varepsilon^{(10)i}_{10} + \bar{v}^{(10)i}_{11})^2\delta^2}{4w^i_{10}} - \frac{(p^{(10)i}_{10} + \delta\bar{p}^{(10)i}_{11})^2}{4\varphi^i_{10}} \\
&\qquad + \frac{p^{(10)i}_{10}(p^{(10)i}_{10} + \delta\bar{p}^{(10)i}_{11})}{2\varphi^i_{10}} + \sum_{t=10-3}^{9} p^{(t)i}_{10}u^{(4)i}_{t}\Bigg\}\delta^9 \\
&\qquad + \Bigg\{A^1_{11}\Bigg[x^1 + \varepsilon^{(10)1}_{10}\frac{(A^1_{11}\varepsilon^{(10)1}_{10} + \bar{v}^{(10)1}_{11})\delta}{2w^1_{10}} \\
&\qquad\quad + \sum_{t=10-2}^{9}\varepsilon^{(t)1}_{10}u^{(3)1}_{t} - \lambda^1_{10}x^1 + \gamma^{(1)2}_{10}x^2\Bigg] \\
&\qquad + A^2_{11}\Bigg[x^2 + \varepsilon^{(10)2}_{10}\frac{(A^2_{11}\varepsilon^{(10)2}_{10} + \bar{v}^{(10)2}_{11})\delta}{2w^2_{10}} \\
&\qquad\quad + \sum_{t=10-2}^{9}\varepsilon^{(t)2}_{10}u^{(3)2}_{t} - \lambda^2_{10}x^2 + \gamma^{(2)1}_{10}x^1\Bigg] \\
&\qquad + \sum_{i=1}^{2}\Bigg[\bar{v}^{(10)i}_{11}\frac{(A^i_{11}\varepsilon^{(10)i}_{10} + \bar{v}^{(10)i}_{11})\delta}{2w^i_{10}} + \sum_{t=10-2}^{9}\bar{v}^{(t)i}_{11}u^{(3)i}_{t} \\
&\qquad\quad + \frac{\bar{p}^{(10)i}_{11}(p^{(10)i}_{10} + \delta\bar{p}^{(10)i}_{11})}{2\varphi^i_{10}} + \sum_{t=11-3}^{9}\bar{p}^{(t)i}_{11}u^{(4)i}_{t} + \varpi^i_{11}\Bigg]\Bigg\}\delta^{10}.
\end{aligned}
\tag{F.3}
$$

(F.3) is an equation in which the right-hand side and left-hand side are linear functions of (x^1, x^2). For (F.3) to hold, it is required that

$$
\begin{aligned}
A^1_{10}x^1 &= \frac{(R^1_{10})^2x^1}{4c^1_{10}} + \delta A^1_{11}(1 - \lambda^1_{10})x^1 + \delta A^2_{11}\gamma^{(2)1}_{10}x^1, \\
A^2_{10}x^2 &= \frac{(R^2_{10})^2x^2}{4c^2_{10}} + \delta A^2_{11}(1 - \lambda^2_{10})x^2 + \delta A^1_{11}\gamma^{(1)2}_{10}x^2.
\end{aligned}
\tag{F.4}
$$

From (F.4), we obtain

$$A_{10}^1 = \sum_{i=1}^{2} \frac{(R_{10}^i)^2}{4c_{10}^i} + \delta A_{11}^1(1-\lambda_{10}^1) + \delta A_{11}^2 \gamma_{10}^{(2)1},$$

$$A_{10}^2 = \sum_{i=1}^{2} \frac{(R_{10}^i)^2}{4c_{10}^i} + \delta A_{11}^2(1-\lambda_{10}^2) + \delta A_{11}^1 \gamma_{10}^{(1)2}. \tag{F.5}$$

In addition,

$$\begin{aligned} C_{10} = \sum_{i=1}^{2} \Bigg(& \frac{p_{10}^{(10)i}(p_{10}^{(10)i} + \delta \bar{p}_{11}^{(10)i})}{2\varphi_{10}^i} + \sum_{t=10-3}^{9} p_{10}^{(t)i} u_t^{(4)i} \\ & - \frac{(A_{11}^{(i)i}\varepsilon_{10}^{(10)i} + \bar{v}_{11}^{(10)i})^2 \delta^2}{4w_{10}^i} - \frac{(p_{10}^{(10)i} + \delta \bar{p}_{11}^{(10)i})^2}{4\varphi_{10}^i} \Bigg) \\ & + \Bigg[A_{11}^1 \left[\varepsilon_{10}^{(10)1} \frac{(A_{11}^1 \varepsilon_{10}^{(10)1} + \bar{v}_{11}^{(10)1})\delta}{2w_{10}^1} + \sum_{t=10-2}^{9} \varepsilon_{10}^{(t)1} u_t^{(3)1} \right] \\ & + A_{11}^2 \left[\varepsilon_{10}^{(10)1} \frac{(A_{11}^2 \varepsilon_{10}^{(10)2} + \bar{v}_{11}^{(10)2})\delta}{2w_{10}^2} + \sum_{t=10-2}^{9} \varepsilon_{10}^{(t)2} u_t^{(3)2} \right] \\ & + \sum_{i=1}^{2} \Bigg(\bar{v}_{11}^{(10)i} \frac{(A_{11}^{(i)i}\varepsilon_{10}^{(10)i} + \bar{v}_{11}^{(10)i})\delta}{2w_{10}^i} + \sum_{t=10-2}^{9} \bar{v}_{11}^{(t)i} u_t^{(3)i**} \\ & + \frac{\bar{p}_{11}^{(10)i}(p_{10}^{(10)i} + \delta \bar{p}_{11}^{(10)i})}{2\varphi_{10}^i} + \sum_{t=11-3}^{9} \bar{p}_{11}^{(t)i} u_t^{(4)i**} + \varpi_{11}^i \Bigg) \Bigg] \delta, \end{aligned} \tag{F.6}$$

which is an expression containing previously executed controls $(\underline{u}_{10-}^{(3)}, \underline{u}_{10-}^{(4)})$.

Then we move to stage 9.

Using $W(10, x; \underline{u}_{10-}^{(3)}, \underline{u}_{10-}^{(4)}) = (A_{10}^1 x^1 + A_{10}^2 x^2 + C_{10})\delta^9$ derived in (F.3)–(F.6) and the stage 9 equation in (6.5), the optimal cooperative strategies in stage 9 can be expressed as follows:

$$u_9^{(1)i} = \frac{(R_9^i)^2}{4(c_9^i)} x^i,$$

$$
\begin{aligned}
u_9^{(3)i} &= \frac{1}{2w_9^i}\left[\frac{\theta_9}{\sum_{\zeta=9}^{10}\theta_\zeta}(Q_{10}^{(i)i}\varepsilon_9^{(9)i}\delta + \bar{v}_{10}^{(9)i}\delta)\right.\\
&\quad \left. + \frac{\sum_{\zeta=10}^{10}\theta_\zeta}{\sum_{\zeta=9}^{10}\theta_\zeta}(A_{10}^i\varepsilon_{10}^{(9)i}\delta + A_{11}^i\varepsilon_{10}^{(9)i}\delta^2 + \bar{v}_{11}^{(9)i}\delta^2)\right],\\
u_9^{(4)i} &= \frac{1}{2\varphi_9^i}\left[p_9^{(9)i} + \frac{\theta_9}{\sum_{\zeta=9}^{10}\theta_\zeta}\bar{p}_{10}^{(9)i}\delta + \frac{\sum_{\zeta=10}^{10}\theta_\zeta}{\sum_{\zeta=9}^{10}\theta_\zeta}(p_{10}^{(9)i}\delta + \bar{p}_{11}^{(9)i}\delta^2)\right],
\end{aligned}
\tag{F.7}
$$

for $i \in \{1, 2\}$.

Substituting the optimal strategies in (F.7) into the stage 9 equation of (6.5), we obtain

$$
\begin{aligned}
&W(9, x; \underline{u}_{9-}^{(3)}, \underline{u}_{9-}^{(4)})\\
&= (A_9^1 x^1 + A_9^2 x^2 + C_9)\delta^8\\
&= \sum_{i=1}^{2}\left(\frac{(R_9^i)^2}{4(c_9^i)}x^i - w_9^i(u_9^{(3)i})^2 - \varphi_9^i(u_9^{(4)i})^2\right.\\
&\quad \left. + p_9^{(9)i}u_9^{(4)i} + \sum_{t=9-3}^{8} p_9^{(t)i}u_t^{(4)i}\right)\delta^8 + \sum_{i=1}^{2}\frac{\theta_9}{\sum_{\zeta=9}^{10}\theta_\zeta}\\
&\times\left(Q_{10}^{(i)i}\left[x^i + \varepsilon_9^{(9)i}u_9^{(3)i} + \sum_{t=9-2}^{8}\varepsilon_9^{(t)i}u_t^{(3)i} - \lambda_9^i x^i + \gamma_9^{(i)j}x^j\right]\right.\\
&\quad + \bar{v}_{10}^{(9)i}u_9^{(3)i} + \sum_{t=9-2}^{8}\bar{v}_{10}^{(t)i}u_t^{(3)i} + \bar{p}_{10}^{(9)i}u_9^{(4)i}\\
&\quad \left. + \sum_{t=9+1-3}^{8}\bar{p}_{10}^{(t)i}u_t^{(4)i} + \varpi_{10}^i\right)\delta^9 + \frac{\sum_{\zeta=10}^{10}\theta_\zeta}{\sum_{\zeta=9}^{10}\theta_\zeta}\\
&\times\left(A_{10}^1\left[x^1 + \varepsilon_9^{(9)1}u_9^{(3)1} + \sum_{t=9-2}^{8}\varepsilon_9^{(t)1}u_t^{(3)1} - \lambda_9^1 x^1 + \gamma_9^{(1)2}x^2\right]\right.
\end{aligned}
$$

$$+ A_{10}^2 \left[x^2 + \varepsilon_9^{(9)2} u_9^{(3)2} + \sum_{t=9-2}^{8} \varepsilon_9^{(t)2} u_t^{(3)2} - \lambda_9^2 x^2 + \gamma_9^{(2)1} x^1 \right]$$

$$\left. + C_{10} \right) \delta^9, \tag{F.8}$$

where $u_9^{(3)1}$, $u_9^{(3)2}$, $u_9^{(4)1}$, and $u_9^{(4)2}$ are given in (F.7).

(F.8) is an equation in which the right-hand side and left-hand side are linear functions of (x^1, x^2). For (F.8) to hold, it is required that

$$\begin{aligned} A_9^1 x^1 &= \frac{(R_9^1)^2 x^1}{4c_9^1} + \frac{\theta_9}{\sum_{\zeta=9}^{10} \theta_\zeta} [Q_{10}^{(1)1}(1-\lambda_9^1)x^1 + Q_{10}^{(2)2}\gamma_9^{(2)1}x^1]\delta \\ &\quad + \frac{\sum_{\zeta=10}^{10} \theta_\zeta}{\sum_{\zeta=9}^{10} \theta_\zeta} [A_{10}^1(1-\lambda_9^1)x^1 + A_{10}^2\gamma_9^{(2)1}x^1]\delta, \\ A_9^2 x^2 &= \frac{(R_9^2)^2 x^2}{4c_9^2} + \frac{\theta_9}{\sum_{\zeta=9}^{10} \theta_\zeta} [Q_{10}^{(2)2}(1-\lambda_9^2)x^2 + Q_{10}^{(1)1}\gamma_9^{(1)2}x^2]\delta \\ &\quad + \frac{\sum_{\zeta=10}^{10} \theta_\zeta}{\sum_{\zeta=9}^{10} \theta_\zeta} [A_{10}^2(1-\lambda_9^2)x^2 + A_{10}^1\gamma_9^{(1)2}x^2]\delta. \end{aligned} \tag{F.9}$$

From (F.9), we obtain

$$\begin{aligned} A_9^1 &= \frac{(R_9^1)^2}{4c_9^1} + \frac{\theta_9}{\sum_{\zeta=9}^{10} \theta_\zeta} [Q_{10}^{(1)1}(1-\lambda_9^1) + Q_{10}^{(2)2}\gamma_9^{(2)1}]\delta \\ &\quad + \frac{\sum_{\zeta=10}^{10} \theta_\zeta}{\sum_{\zeta=9}^{10} \theta_\zeta} [A_{10}^1\left(1-\lambda_9^1\right) + A_{10}^2\gamma_9^{(2)1}]\delta, \\ A_9^2 &= \frac{(R_9^2)^2}{4c_9^2} + \frac{\theta_9}{\sum_{\zeta=9}^{10} \theta_\zeta} [Q_{10}^{(2)2}(1-\lambda_9^2) + Q_{10}^{(1)1}\gamma_9^{(1)2}]\delta \\ &\quad + \frac{\sum_{\zeta=10}^{10} \theta_\zeta}{\sum_{\zeta=9}^{10} \theta_\zeta} [A_{10}^2(1-\lambda_9^2) + A_{10}^1\gamma_9^{(1)2}]\delta. \end{aligned} \tag{F.10}$$

In addition, C_9 is an expression including all the terms on the right-hand side of (B.8) which do not involve x^1 or x^2. It is an expression containing previously executed controls $(\underline{u}_{9-}^{(3)}, \underline{u}_{9-}^{(4)})$.

Then, we move to stage 8.

Using $W(9, x; \underline{u}_{9-}^{(3)}, \underline{u}_{9-}^{(4)}) = (A_9^1 x^1 + A_9^2 x^2 + C_9)\delta^8$ derived in (F.8)–(F.10) and the stage 8 equation in (6.5), the optimal cooperative strategies in stage 8 can be expressed as follows:

$$
\begin{aligned}
u_8^{(1)i} &= \frac{(R_8^i)^2}{4(c_8^i)} x^i, \\
u_8^{(3)i} &= \frac{1}{2w_8^i}\left[\frac{\theta_8}{\sum_{\zeta=8}^{10}\theta_\zeta}(Q_9^{(i)i}\varepsilon_8^{(8)i}\delta + \bar{v}_9^{(8)i}\delta) + \frac{\sum_{\zeta=9}^{10}\theta_\zeta}{\sum_{\zeta=8}^{10}\theta_\zeta}\right. \\
&\qquad \left. \times\, (A_9^i\varepsilon_8^{(8)i}\delta + A_{10}^i\varepsilon_9^{(8)i}\delta^2 + A_{11}^i\varepsilon_{10}^{(8)i}\delta^3 + \bar{v}_{11}^{(8)i}\delta^3)\right], \\
u_8^{(4)i} &= \frac{1}{2\varphi_8^i}\left[p_8^{(8)i} + \frac{\theta_8}{\sum_{\zeta=8}^{10}\theta_\zeta}\bar{p}_9^{(8)i}\delta\right. \\
&\qquad \left. + \frac{\sum_{\zeta=9}^{10}\theta_\zeta}{\sum_{\zeta=8}^{10}\theta_\zeta}(p_9^{(8)i}\delta + p_{10}^{(8)i}\delta^2 + \bar{p}_{11}^{(8)i}\delta^3)\right],
\end{aligned}
\tag{F.11}
$$

for $i \in \{1, 2\}$.

Substituting the optimal strategies in (F.11) into the stage 8 equation of (6.5), we obtain

$$
\begin{aligned}
&W(8, x; \underline{u}_{8-}^{(3)}, \underline{u}_{8-}^{(4)}) \\
&\quad = (A_8^1 x^1 + A_8^2 x^2 + C_8)\delta^7 \\
&\quad = \sum_{i=1}^{2}\left[\frac{(R_8^i)^2}{4(c_8^i)}x^i - w_8^i(u_8^{(3)i})^2 - \varphi_8^i(u_8^{(4)i})^2 + p_8^{(8)i}u_8^{(4)i}\right. \\
&\qquad \left. + \sum_{t=8-3}^{7} p_8^{(t)i}u_t^{(4)i}\right]\delta^7 + \sum_{i=1}^{2}\frac{\theta_8}{\sum_{\zeta=8}^{10}\theta_\zeta} \\
&\qquad \times \left[Q_9^{(i)i}\left[x^i + \varepsilon_8^{(8)i}u_8^{(3)i} + \sum_{t=8-2}^{7}\varepsilon_8^{(t)i}u_t^{(3)i} - \lambda_8^i x^i + \gamma_8^{(i)j}x^j\right]\right.
\end{aligned}
$$

$$+\bar{v}_9^{(8)i}u_8^{(3)i}+\sum_{t=8-2}^{7}\bar{v}_8^{(t)i}u_t^{(3)i}+\bar{p}_9^{(8)i}u_8^{(4)i}$$

$$+\sum_{t=8+1-3}^{7}\bar{p}_9^{(t)i}u_t^{(4)i}+\varpi_9^i\Bigg]\delta^8+\frac{\sum_{\zeta=9}^{10}\theta_\zeta}{\sum_{\zeta=8}^{10}\theta_\zeta}$$

$$\times\left[A_9^1\left[x^1+\varepsilon_8^{(8)1}u_8^{(3)1}+\sum_{t=8-2}^{7}\varepsilon_8^{(t)1}u_t^{(3)1}-\lambda_8^1x^1+\gamma_8^{(1)2}x^2\right]\right.$$

$$+A_9^2\left[x^2+\varepsilon_8^{(8)2}u_8^{(3)2}+\sum_{t=8-2}^{7}\varepsilon_8^{(t)2}u_t^{(3)2}-\lambda_8^2x^2+\gamma_8^{(2)1}x^1\right]$$

$$\left.+C_9\right]\delta^8, \tag{F.12}$$

where $u_8^{(3)1}$, $u_8^{(3)2}$, $u_8^{(4)1}$, and $u_8^{(4)2}$ are given in (F.11).

System (F.12) yields a system of equations in which the right-hand side and left-hand side are linear functions of (x^1, x^2). For (F.12) to hold, it is required that

$$A_8^1=\frac{(R_8^1)^2}{4c_8^1}+\frac{\theta_8}{\sum_{\zeta=8}^{10}\theta_\zeta}[Q_9^{(1)1}(1-\lambda_8^1)+Q_9^{(2)2}\gamma_8^{(2)1}]\delta$$
$$+\frac{\sum_{\zeta=9}^{10}\theta_\zeta}{\sum_{\zeta=8}^{10}\theta_\zeta}[A_9^1(1-\lambda_8^1)+A_9^2\gamma_8^{(2)1}]\delta,$$
$$A_8^2=\frac{(R_8^2)^2}{4c_8^2}+\frac{\theta_8}{\sum_{\zeta=8}^{10}\theta_\zeta}[Q_9^{(2)2}(1-\lambda_8^2)+Q_9^{(1)1}\gamma_8^{(1)2}]\delta$$
$$+\frac{\sum_{\zeta=9}^{10}\theta_\zeta}{\sum_{\zeta=8}^{10}\theta_\zeta}[A_9^2(1-\lambda_8^2)+A_9^1\gamma_8^{(1)2}]\delta. \tag{F.13}$$

In addition, C_8 is an expression including all the terms on the right-hand side of (F.12) which do not involve x^1 or x^2. It is an expression containing previously executed controls $(\underline{u}_{8-}^{(3)}, \underline{u}_{8-}^{(4)})$.

Repeating the process for stages $\tau \in \{7, 6, \ldots, 1\}$, we obtain the optimal cooperative strategies in stage τ as follows:

$$u_\tau^{(1)i} = \frac{(R_\tau^i)^2}{4(c_\tau^i)} x^i,$$

$$u_\tau^{(3)i} = \frac{1}{2w_\tau^i} \left[\frac{\theta_\tau}{\sum_{\zeta=\tau}^{10} \theta_\zeta} (Q_{\tau+1}^{(i)i} \varepsilon_\tau^{(\tau)i} \delta + \bar{v}_{\tau+1}^{(\tau)i} \delta) + \frac{\sum_{\zeta=\tau+1}^{10} \theta_\zeta}{\sum_{\zeta=\tau}^{10} \theta_\zeta} \left(\sum_{t=\tau}^{\tau+2} A_{t+1}^i \varepsilon_t^{(\tau)i} \delta^{t+1-\tau} \right) \right],$$

$$u_\tau^{(4)i} = \frac{1}{2\varphi_\tau^i} \left[p_\tau^{(\tau)i} + \frac{\theta_\tau}{\sum_{\zeta=\tau}^{10} \theta_\zeta} \bar{p}_{\tau+1}^{(\tau)i} \delta + \frac{\sum_{\zeta=\tau+1}^{10} \theta_\zeta}{\sum_{\zeta=\tau}^{10} \theta_\zeta} \left(\sum_{t=\tau}^{\tau+2} p_{t+1}^{(\tau)i} \delta^{t+1-\tau} \right) \right], \quad \text{(F.14)}$$

for $i \in \{1, 2\}$.

The optimal expected cooperative joint payoff at stage $\tau \in \{7, 6, \ldots, 1\}$ can be obtained as

$$W(\tau, x; \underline{u}_{\tau-}^{(3)}, \underline{u}_{\tau-}^{(4)}) = (A_\tau^1 x^1 + A_\tau^2 x^2 + C_\tau)\delta^{\tau-1},$$

where

$$A_\tau^1 = \frac{(R_\tau^1)^2}{4c_\tau^1} + \frac{\theta_\tau}{\sum_{\zeta=\tau}^{10} \theta_\zeta} \left[Q_{\tau+1}^{(1)1}(1 - \lambda_\tau^1) + Q_{\tau+1}^{(2)2} \gamma_\tau^{(2)1} \right] \delta + \frac{\sum_{\zeta=\tau+1}^{10} \theta_\zeta}{\sum_{\zeta=\tau}^{10} \theta_\zeta} \left[A_{\tau+1}^1 (1 - \lambda_\tau^1) + A_{\tau+1}^2 \gamma_\tau^{(2)1} \right] \delta,$$

$$A_\tau^2 = \frac{(R_\tau^2)^2}{4c_\tau^2} + \frac{\theta_\tau}{\sum_{\zeta=\tau}^{10} \theta_\zeta} \left[Q_{\tau+1}^{(2)2}(1 - \lambda_\tau^2) + Q_\tau^{(1)1} \gamma_\tau^{(1)2} \right] \delta + \frac{\sum_{\zeta=\tau+1}^{10} \theta_\zeta}{\sum_{\zeta=\tau}^{10} \theta_\zeta} \left[A_{\tau+1}^2 (1 - \lambda_\tau^2) + A_{\tau+1}^1 \gamma_\tau^{(1)2} \right] \delta. \quad \text{(F.15)}$$

In addition, C_τ is an expression containing previously executed controls $(\underline{u}_{\tau-}^{(3)}, \underline{u}_{\tau-}^{(4)})$. ■

Index

Printed in the USA
CPSIA information can be obtained
at www.ICGtesting.com
CBHW061429241024
15752CB00011B/34

9 789811 294556